Lecture Notes in Control and Information Sciences

Edited by A. V. Balakrishnan and M. Thoma

61

Filtering and Control of Random Processes

Proceedings of the E.N.S.T.-C.N.E.T. Colloquium
Paris, France, February 23–24, 1983

Edited by
H. Korezlioglu, G. Mazziotto, and J. Szpirglas

Springer-Verlag
Berlin Heidelberg New York Tokyo 1984

Editors
Hayri Korezlioglu
E.N.S.T.
46, Rue Barrault
75634 Paris Cedex 13
France

Gérald Mazziotto
Jacques Szpirglas

C.N.E.T.-PAA/TIM/MTI
38–40, Rue du Général Leclerc
92131 Issy les Moulineaux
France

Library of Congress Cataloging in Publication Data
E.N.S.T.-C.N.E.T. Colloquium (1983 : Paris, France)
Filtering and control of random processes.
(Lecture notes in control and information sciences; 61)
1. Control theory -- Congresses. 2. Stochastic processes -- Congresses.
3. Filters (Mathematics) -- Congresses.
I. Korezlioglu, H. (Hayri).
II. Mazziotto, G. (Gerald).
III. Szpirglas, J. (Jacques)
IV. Ecole nationale supérieure des télécommunications (France)
V. Centre national d'éttudes des télécommunications (France)
VI. Title.
VII. Series.
QA402.3.E15 1983 519.2 84-1420

AMS Subject Classifications (1980): 60 G 35 – 60 G 40 – 93 E 11 – 93 E 20

ISBN 3-540-13270-8 Springer-Verlag Berlin Heidelberg New York Tokyo
ISBN 0-387-13270-8 Springer-Verlag New York Heidelberg Berlin Tokyo

Offsetprinting: Mercedes-Druck, Berlin
Binding: Lüderitz und Bauer, Berlin
2061/3020-543210

FOREWORD

The present volume englobes the papers presented at the ENST-CNET Colloquium on "Filtering and Control of Random Processes" held in Paris on 23-24 February 1983 and sponsored by the Centre National d'Etudes des Télécommunications (CNET) and the Ecole Nationale Supérieure des Télécommunications (ENST).

The papers cover the following areas: diffusion processes in bounded regions and their control; approximations for diffusion processes, for their filtering and control; study of the unnormalized filtering equation; control of partially observed diffusions; stochastic games; optimal stopping; and different topics related to the subject.

Many of the papers overlap several of these areas. Thinking that a classification by subject would seem artificial, we have chosen to present them in the alphabetic order of the authors' names.

We would like to express our acknowledgement to the CNET and the ENST. Particular thanks go to J. LE MEZEC, M. URIEN, B. AYRAULT and C. GUEGUEN for their encouragement and material support.

H. KOREZLIOGLU, G. MAZZIOTTO, J. SZPIRGLAS.

TABLE OF CONTENTS

J. AGUILAR-MARTIN : Projective Markov processes. 1

A. BENSOUSSAN : On the stochastic maximum priciple for infinite dimensional equations and applications to the control of Zakai equation. 13

R.K. BOEL : Some comments on control and estimation problems for diffusions in bounded regions. 24

N. CHRISTOPEIT and K. HELMES : The separation principle for partially observed linear control systems: a general framework. ... 36

G.B. DI MASI and W.J. RUNGGALDIER : Approximations for discrete time partially observable stochastic control problems. 61

M. HAZEWINKEL , S.I. MARCUS and H.J. SUSSMANN : Nonexistence of finite dimensional filters for conditional statistics of the cubic sensor problem. 76

D.P. KENNEDY : An extension of the prophet inequality. 104

H. KOREZLIOGLU and C. MARTIAS : Martingale representation and nonlinear filtering equation for distribution-valued processes. ..111

J.P. LEPELTIER and M.A. MAINGUENEAU : Jeu de Dynkin avec coût dépendant d'une stratégie continue. 138

P.L. LIONS : Optimal control of reflected diffusion processes. ...157

E. MAYER-WOLF and M. ZAKAI : On a formula relating the Shannon information to the Fisher information for the filtering problem. ..164

G. MAZZIOTTO : Optimal stopping of bi-Markov processes. 172

E. PARDOUX : Equations du lissage non linéaire. 206

J. PICARD : Approximation of nonlinear filtering problems and order of convergence. 219

G. PICCI and J.H. VAN SCHUPPEN : On the weak finite stochastic realization problem. 237

M. PONTIER and J. SZPIRGLAS : Controle linéaire sous contrainte avec observation partielle. 243

C. STRICKER : Quelques remarques sur les semimartingales gaussiennes et le problème de l'innovation. 260

J. SZPIRGLAS : Sur les propriétés markoviennes du processus de filtrage. 277

D. TALAY : Efficient numerical schemes for the approximation of expectations of functionals of the solution of a S.D.E., and applications. 294

A.S. USTUNEL : Distributions-valued semimartingales and applications to control and filtering. 314

PROJECTIVE MARKOV PROCESSES

J. AGUILAR-MARTIN
Laboratoire d'Automatique et d'Analyse des Systèmes
du C.N.R.S.
7, avenue du Colonel Roche
31400 TOULOUSE, France

0. GENERAL COMMENTS

We shall give here the fundamentals of what could be called "optimal polynomial regression", that is the orthogonal projection of a given random variable on the space of polynomial combinations of a group of possibility observable random variables. The vectorial case will be at once studied and therefore we need to use tensorial contracted notation (or Einstein's convention).

The optimal polynomial regression estimator or, shortly, polynomial estimator is a mere extension of the well known linear least squares estimator; and similar to Doob [1953] we shall define a Markov property based on the independence of the past conditionnally to the present, giving rise to wide sense N-polynomial projective Markov processes or Projective Markov Processes in the N-polynomial sense (PMPN).

The special case N=2 : Projective Markov Process in the Quadratic sense, (PMPQ), will be given special attention. It gives an useful dynamical model for diffusion processes encountered frequently when flows interact, as in thermic, biological or ecological processes.

I. POLYNOMIAL ESTIMATION

I.1 Basic definition and theorems

Let $\mathcal{E}$ be a space of square integrable random variables defined on (Ω, γ, P), and $\mathcal{E}^n$ be the space of n-dimensional random vectors, the components of which are in $\mathcal{E}$.

We shall distinguish between a collection of possibly observable random variables $\{X_i\}_{i=1}^{m}$, $X_i \in \mathcal{E}^{n_i}$ and the random variable upon which the estimation deals, $Y \in \mathcal{E}^n$. We shall recall here two well known fundamental results on probabilistic estimation.

THEOREM 1 : Optimality of conditional expectation

Let us consider the measurable functions $F\left(\left\{X_i\right\}_{i=1}^{m}\right)$ such that $F \in L^2\left[(\Omega, \gamma_x, P), \mathbb{R}^n\right]$ where γ_x is the σ-algebra generated by $\left\{X_i\right\}$. For any $\lambda \in \mathbb{R}^n$.

$$E\left[\left(\lambda^T (Y-F)\right)^2\right] \geqslant E\left[\left(\lambda^T (Y-\hat{Y})\right)^2\right]$$

where

$$\hat{Y} = E\left[Y \;/\; \left\{X\right\}_{i=1}^{N}\right]$$

COROLLARY 1 : Stochastic orthogonality of estimation error

On the same conditions as in the previous theorem

$$E\left[(Y-\hat{Y})\, F^T\right] = 0$$

Therefore $\hat{Y}$ is the orthogonal projection of Y on $L^2\left[(\Omega, \gamma_x, P), \mathbb{R}^n\right]$

Proofs of theorem 1 and its corollary can be found in all elementary books on probability.

DEFINITION 1

Let H_N be the Hilbert space of all the $\mathbb{R}^n$- valued polynomial functions of degree up to N of the components of $\left\{X_i\right\}_{i=1}^{m}$. (It is supposed that these polynomial functions belong to $\mathcal{E}^n$).

LEMMA 1

Let us denote by H the closure of $\bigcup_{N \geqslant 0} H_N$. We suppose that there is a positive number a such that for all $A \in \mathbb{R}^n$ of norm not greater than a

$$\int_\Omega e^{|A^T X|}\, dP < \infty \; . \; . \; . \; . \; . \; . \; . \; . \; . \; . \quad (G_\infty).$$

Then

$$H = L^2\left[(\Omega, \gamma_x, P), \mathbb{R}^n\right]$$

Proof : Condition (G_∞) implies that the characteristic function

$$\varphi(A) = \int_\Omega e^{i\, A^T X}\, dP, \qquad A \in \mathbb{R}^n$$

is such that for each A, the function $\varphi(tA)$ is an analytic function of $t \in \mathbb{R}$ in a certain neighborhood of zero. According to a lemma of GIHMAN - SKOROHOD [1974], the mentioned equality of spaces holds. □

We shall state later a fundamental limit theorem concerning the polynomial estimators.

We can write the random vector $Z_N \in H_N$ as a linear combination of tensor products; so its ℓth component will be

$$Z_N^{\ell} = \sum_{k=1}^{m} \sum_{r_1,\ldots,r_k=1}^{N} (\lambda^{\ell}_{j_1\ldots j_r})\, x_{r_1}^{j_1} \ldots x_{r_k}^{j_r} \in \mathcal{E}^n$$

The orthogonality strictly written involves scalar product. It is convenient to widen this concept to tensor product in order to be able to develop simpler equations.

I.2 Orthogonality and projections

THEOREM 2

Let $Z_N \in H_N$ and $W \in \mathcal{E}^n$

the following propositions are equivalent

A) W is orthogonal to H_N, that is, $\forall\, Z_N \in H_N$, $E\left[W^T Z_N\right] = 0$

(or $E\left[w_i\ z_N^i\right] = 0$)

B) W is such that $E\left[W \otimes Z_N\right] = 0$

(or $E\left[w^i\ z_N^j\right] = 0$)

Proof : B is equivalent to $E\left[W \otimes X_{r_1} \otimes \ldots \otimes X_{r_k}\right] = 0$

for all $r_1 \ldots r_k = 1 \longrightarrow m$ and $k = 1 \longrightarrow N$

Let e_i be the canonical basis of R^n, then,

$$E\left[W^T (x_{r_1}^{j_1} \ldots x_k^{j_k})\, e_i\right) = 0$$

for all $i = 1 \longrightarrow n$, $j_p = 1 \longrightarrow n_p$ $\quad p=1 \longrightarrow m$.

and as $(x_{r_1}^{j_1} \ldots x_{r_k}^{j_k})\ e_i$ is a basis of H_N.

Therefore W is orthogonal to H_N. □

This theorem leads us to a constructive theorem for the optimal polynomial estimator analogous to the Wiener Hopf equation in linear estimation.

THEOREM 3 : Generalized Wiener Hopf equation

The optimal N-polynomial estimation of Y based on the observation of $\{X_i\}$ is $\hat{Y}_N$, orthogonal projection of Y on H_N : for all $Z_N \in H_N$ it is such that.

$$E\left[(Y - \hat{Y}_N) \otimes Z_N\right] = 0 \quad \ldots\ldots\ldots\ldots \quad (*)$$

Therefore if it exists, it is solution of the following system of equations

$$E\left[Y \otimes X_{r_1} \ldots \otimes X_{r_k}\right] = E\left[\hat{Y} \otimes X_{r_1} \ldots \otimes X_{r_k}\right] \ldots \quad (**)$$

$r_1 \ldots r_k = 1, \ldots m \qquad k=1,\ldots, N$

Proof : A well known result that can be found in all elementary books on geometry, states that the orthogonal projection minimizes all quadratic errors, similarly to theorem 1. Equation (**) follows from Theorem 2 then we can write it for the tensor components defining

$$\hat{y}^{\ell} = \sum_{k=1}^{m} \sum_{r_1,\ldots,r_k=1}^{N} \alpha^{\ell r_1 \ldots r_k}_{j_1 \ldots j_{m_k}} x^{j_1}_{r_1} \ldots x^{j_{m_k}}_{r_k}$$

$$E\left[y^{\ell} x^{j_1}_{s_1} \ldots x^{j_{m_k}}_{s_k}\right] = \sum_{k=1}^{N} \alpha^{\ell r_1 \ldots r_k}_{j_1 \ldots j_{m_k}} E\left[x^{j_1}_{r_1} \ldots x^{j_{m_k}}_{r_k} \; x^{i_1}_{s_1} \ldots x^{i_{m_k}}_{s_k}\right] \qquad (***)$$

Equation (***) is a square system of equation for α and therefore its solution depends only on the moments up to order N+1 between Y and $\{X_i\}$ and up to 2N between $\{X_i\}$.

We shall use the notation $\hat{Y}_N$ = N-proj $[Y \mid X_1 \ldots X_m]$

I.3 Limit property

THEOREM 4

Let us consider the sequence of spaces $\{H_N\}$ as well as the sequence of projections $\{\hat{Y}_N\}$ and assume that condition G_∞ holds.

Then

$$\lim_{N \to \infty} \| \hat{Y} - \hat{Y}_N \|_2 = 0$$

Proof. Lemma 1 states that we may define the estimation error as

$$\delta = \| Y - \hat{Y} \|_2 = \inf_{Z \in H} \| Y - Z \|_2$$

Then

$\delta = \lim_{N \to \infty} \inf \| Y - \hat{Y}_N \|_2$ and as $(Y-\hat{Y})$ is orthogonal to $(\hat{Y}-\hat{Y}_N)$ by construction of $\hat{Y}_N$ therefore

$$\| Y - \hat{Y}_N \|_2^2 = \| Y - \hat{Y} \|_2^2 + \| \hat{Y} - \hat{Y}_N \|_2^2$$ and the result follows.

Remarks :

1) Linear projection is $\hat{Y}_1$ and gives the well known linear least squares estimator.
2) Wide sense Markov processes in Doob [1953] derive from a Markov property that will be written here as :

$$1\text{-proj } [X_{t+1} \mid X_t, X_{t-1} \ldots X_o] = 1\text{-proj } [X_{t+1} \mid X_t]$$

This property involves second order conditions on the sequence $\{X_i\}$.

II. PROJECTIVE MARKOV PROCESSES IN THE N-POLYNOMIAL SENSE (PMPN)

II.1 General definition

DEFINITION 2

Let $X(t) \in \mathcal{E}^n$ be a discrete time random process for t = 1,2,..., T, such that all moments up to order 2N are finite. $\{X(t)\}$ is a PMPN if an only if

$$\text{N-proj}\left[X(t) \underbrace{\otimes \ldots \otimes}_{m} X(t) \;\middle|\; \{X(r)\}_{r \leq s}\right]$$

$$= \text{N-proj}\left[X(t) \underbrace{\otimes \ldots \otimes}_{m} X(t) \;\middle|\; X(s)\right]$$

for m = 1,2,...,N

We shall notice here that the projective markov condition involves not only the process itself, m=1, but all its polynomial combinations up to degree N. The "state" of a PMPN is not X(t) but the augmented vector.

$$S^T(t) = \left[X(t), \ldots, X(t) \overbrace{\otimes \ldots \otimes}^{N} X(t)\right]$$

And any recursive model, or internal representation of it shall be a recursive vector equation on S(t).

II.2 Projective markov processes in the quadratic sense (PMPQ)

From now on we shall deal with N=2, that is PMPQ's and we shall use tensor component notation.

DEFINITION 3

n-Vector valued random process $\{x^{\ell}(t)\}_{t=0}^{\infty}$ is a PMPQ if an only if

$$\text{2-Proj}\left[x^{\ell}(t) \;\middle|\; \{x^j(r)\}_{\substack{r \leq s\\ j=1,2,\ldots,n}}\right]$$

$$= \text{2-Proj}\left[x^{\ell}(t) \;\middle|\; \{x^j(s)\}_{j=1,2,\ldots,n}\right]$$

and

$$\text{2-Proj}\left[x^{\ell}(t)\; x^c(t) \;\middle|\; \{x^j(r)\}_{\substack{r \leq s\\ j=1,2,\ldots,n}}\right]$$

$$= \text{2-Proj}\left[x^{\ell}(t)\; x^c(t) \;\middle|\; \{x^j(s)\}_{j=1,2,\ldots,n}\right]$$

Let us denote as follows the moments up to the 4th order.

$$\mu^{\ell}(s) = E\left[x^{\ell}(s)\right]$$
$$\mu^{\ell c}(s,t) = E\left[x^{\ell}(s)\,.\,x^{c}(t)\right]$$
$$\mu^{\ell ck}(s,t,u) = E\left[x^{\ell}(s)\,.\,x^{c}(t)\,.\,x^{k}(u)\right]$$
$$\mu^{\ell ckj}(s,t,u,v) = E\left[x^{\ell}(s)\,.\,x^{c}(t)\,.\,x^{k}(u)\,.\,x^{j}(v)\right]$$

and let us adopt the arithmetic ordering of components, that is

$$\overset{k}{\overline{ij}} \Longleftrightarrow j + (i-1).n = k$$

Using this convention we define the following reordered vectors and matrices by its components.

vectors	$\mu_i(s) = \mu^{i}(s)$	$: \underline{\mu}(s)$
	$\mu_k(uv) = \mu^{\overset{k}{\overline{ij}}}(u,v)$	$: \underline{\mu}(u,v)$
nxn matrix	$\mu_{ij}(u,s) = \mu^{ij}(u,s)$	$: \mu_*(u,s)$
n^2xn matrix	$\mu_{\ell k}(u,v,s) = \mu^{\overset{\ell}{\overline{ij}}k}(u,v,s)$	$: \mu_*(u,v,s)$
n^2xn^2 matrix	$\mu_{\ell k}(u,v,s,t) = \mu^{\overset{\ell}{\overline{ij}}\;\overset{k}{\overline{i'j'}}}(u,v,s,t)$	$: \mu_*(u,v,s,t)$

Then we can define the following matrix

$$M(s,u,v) = \begin{bmatrix} 1 & \underline{\mu}^T(s) & \underline{\mu}^T(s,s) \\ \underline{\mu}(u) & \mu_*(u,s) & \mu_*^T(u,s,s) \\ \underline{\mu}(u,v) & \mu_*(u,v,s) & \mu_*(u,v,s,s) \end{bmatrix}$$

PROPOSITION 1

Let be a measurable function $\mathbb{R}^n \longrightarrow \mathbb{R}$ such that

$$E\left[f(x(t))\;\; x^i(s)\, x^j(s)\right] < \infty$$

for all i, j and $S < t$.

Then $\hat{\hat{f}}(X(t))\big|_s =$

$$\text{2-proj}\left[f(X(t))\;\middle|\; X(s)\right] = d + a_i\, x^i(s) + b_{ij}\, x^i(s)\, x^j(s)$$

and the coefficients can be obtained from the following linear system of equations

$$\begin{bmatrix} E & [f(X(t)] \\ E & [f(X(t)) \, . \, x^i(s)] \\ E & [f(X(t)) \, . \, x^i(s) \, . \, x^j(s)] \end{bmatrix} = M\,(s,s,s) \begin{bmatrix} d \\ \hline a_i \\ \hline b_{\overline{ij}} \end{bmatrix}$$

the proof is a mere construction.

THEOREM 5

Characteristic property : $\{X(t)\}$ is a PMPQ if an only if for any u,v,s,t such that $u,v \leq s < t$. The following relations are satisfied by the moments :

$$\begin{bmatrix} \underline{\mu}^T(t) \\ \mu_*^T(u,t) \\ \mu_*^T(u,v,t) \end{bmatrix} = M(s,u,v)\; M^{-1}(s,s,s) \begin{bmatrix} \underline{\mu}^T(t) \\ \mu_*(s,t) \\ \mu_*(s,s,t) \end{bmatrix}$$

$$\begin{bmatrix} \underline{\mu}^T(t)t \\ \mu_*^T(u,t,t) \\ \mu_*^T(u,v,t,t) \end{bmatrix} = M(s,u,v)\; M^{-1}(s,s,s) \begin{bmatrix} \underline{\mu}^T(t,t) \\ \mu_*^T(s,t,t) \\ \mu_*^T(s,s,t,t) \end{bmatrix}$$

Proof : If α, β, γ are the components of $\hat{X}(t)^T_{|s} = [\alpha \mid \beta \mid \gamma]^T$

and $\quad \widehat{\widehat{X(t) \otimes X(t)}}_{|s} = [a \mid b \mid c]^T$

for a fixed component

α^ℓ and a^ℓ are scalars

β^ℓ and b^ℓ are n-vectors components $\beta^{\ell i}$

γ^ℓ and c^ℓ are n^2-vectors components $\gamma^{\ell \overline{ij}}$

As X(t) is a PMPQ for each $s < t$

$$X^\ell(t) - \widehat{\widehat{X^\ell(t)}}_{|s} \text{ and } X^\ell(t)\, X^c(t) - \widehat{\widehat{X^\ell(t)\, X^c(t)}}_{|s}$$

are centered random variables orthogonal to X(u) and $X_u \otimes X_v$ for any $u,v < s$

So we have :

$$\begin{bmatrix} \mu^\ell(t) \\ \mu^{\ell i}(u,t) \\ \mu^{\ell \overline{ij}}(u,v,t) \end{bmatrix} = \begin{bmatrix} 1 & \mu^i(s) & \mu^{\overline{ii'}}(s,s) \\ \mu^j(u) & \mu^{ji}(u,s) & \mu^{j\,\overline{ii'}}(u,s,s) \\ \mu^{\overline{jj'}}(u,v) & \mu^{\overline{jj'}\,i}(u,v,s) & \mu^{\overline{jj'}\,\overline{ii'}}(u,s,s,s) \end{bmatrix} \begin{bmatrix} \alpha^\ell \\ \beta_i^\ell \\ \gamma_{\overline{ij}}^\ell \end{bmatrix}$$

and similarly for

$$\mu^{\ell c}(t,t) \qquad \mu^{i \ell c}(u,t,t) \quad \text{and} \quad \mu^{\overline{ij}\;\ell c}(u,v,t,t)$$

with respect to a, b and c.

Then using Proposition 1 the theorem follows.

Remarks.

1) For wide sense markov and gauss-markov process this characteristic theorem gives

$$\mu^T(u,t) = \mu^T(s,u) \quad \mu^{-1}(s,s) \quad \mu^T(s,t)$$

or $$\Phi(t \leftarrow u) = \Phi(t \leftarrow s) \,.\, \Phi(s \leftarrow u)$$

using the transition operator

$$\Phi(t \leftarrow u) = \mu(t,u) \quad \mu^{-1}(u,u)$$

2) Unfortunately for PMPQ's it is more difficult to define transition operators.

II.3 Recurrent models

Let us consider the increments of a PMPQ with respect to a sequence of instants $\{t\}_{t=1}^{\infty}$

$$\varepsilon(t) = X(t) - \hat{\hat{X}}(t)|_s$$

$$\eta(t) = X(t) \otimes X(t) - \widehat{\widehat{X(t) \otimes X(t)}}|_s$$

Thus we define a sequence of random vectors and a sequence of random matrices, centered, and orthogonal, as they are, by construction orthogonal to any

$$X(u) \text{ and } X(u) \otimes X(v) \quad \text{for } u,v < t.$$

Moreover they are orthogonal to their quadratic values, for $u,v < t$

i.e. $$E\left[\varepsilon(t) \otimes \varepsilon(v) \otimes \varepsilon(u)\right] = 0$$

and $$E\left[\varepsilon(t) \otimes \varepsilon(t) \otimes \varepsilon(v) \otimes \varepsilon(u)\right] = 0$$

and similarly for $\eta(t)$.

Then we shall state the recurrent model for X(t). From now on we shall adopt the following notation.

$$\overset{+}{x} = x(t+1) \qquad x = x(t)$$

THEOREM 6

If (X(t)) is a PMPQ then it can be generated by the following equations

$$\overset{+}{x}{}^{\ell} = \beta^{\ell} + \varphi_i^{\ell} \, x^i + \psi_{ij}^{\ell} \, x^i x^j + \varepsilon^{\ell} \qquad \text{(I)}$$

$$\overset{+}{x}{}^{\ell} \overset{+}{x}{}^{k} = d^{\ell k} + a_i^{\ell k} \, x^i + b_{ij}^{\ell k} \, x^i x^j + \eta^{\ell k} \qquad \text{(II)}$$

Proof :

1) Tensor coefficients β, φ, ψ, a, b, d are deterministic function that derive from the solution of the system of equations of theorem 5.

2) $\{\varepsilon(t)\}$ and $\{\eta(t)\}$ are stochastic processes such that for any s, u $\neq$ t

$$E\left[\varepsilon^{\ell}(t)\right] = 0 \qquad E\left[\varepsilon^{\ell}(t) \quad \varepsilon^{k}(s)\right] = 0$$

$$E\left[\varepsilon^{\ell}(t) \quad \varepsilon^{k}(s) \quad \varepsilon^{m}(u)\right] = 0 \qquad E\left[\varepsilon^{\ell}(t) \ \varepsilon^{c}(t) \ \varepsilon^{k}(s) \ \varepsilon^{m}(u)\right] = 0$$

and

$$E\left[\eta^{\ell c}(t)\right] = 0 \qquad E\left[\eta^{\ell c}(t) \quad \eta^{km}(s)\right] = 0$$

$$E\left[\varepsilon^{\ell}(t) \quad \eta^{km}(s)\right] = 0 \qquad E\left[\varepsilon^{\ell}(t) \ \varepsilon^{c}(t)\, \eta^{km}(s)\right] = 0$$

We shall now define the following moments :

$$E\left[\varepsilon^{\ell}(t) \quad \varepsilon^{k}(t)\right] = \Gamma^{\ell k}$$

$$E\left[\eta^{\ell c}(t) \quad \eta^{km}(t)\right] = H^{\ell ckm}$$

$$E\left[\varepsilon^{\ell}(t) \quad \eta^{km}(t)\right] = \Sigma^{\ell km}$$

Then we derive some theorems concerning the evolution of the centered moments of x

$$\bar{x}^{\ell} = E\left[x^{\ell}\right] \qquad \tilde{x}^{\ell} = x - \bar{x}^{\ell}$$

$$P^{ij} = E\left[\tilde{x}^{i} \quad \tilde{x}^{j}\right]$$

$$S^{ijk} = E\left[\tilde{x}^{i}\ \tilde{x}^{j}\ \tilde{x}^{k}\right]$$

$$Q^{ijk\ell} = E\left[\tilde{x}^{i}\ \tilde{x}^{j}\ \tilde{x}^{k}\ \tilde{x}^{\ell}\right]$$

Thus, a model of the mean evolution and of the diffusion of x will be obtained.

<u>THEOREM 7</u>

The mean value of x satisfies the following recurrent quadratic equation in $\bar{x}$.

$$\overset{+}{x}{}^{\ell} = \varphi^{\ell}_{i}\ \bar{x}^{i} + \psi^{\ell}_{ij} \quad \bar{x}^{i}\ \bar{x}^{j} + \beta^{\ell} + \psi^{\ell}_{ij}\ P^{ij} \tag{m}$$

<u>Proof</u> : replace x by $\tilde{x} + \bar{x}$ in (I)

<u>THEOREM 8</u>

The covariance P satisfies the following linear equation in P.

$$\overset{+}{P}{}^{\ell k} = a_{i}^{\ell k}\ \bar{x}^{i} + b^{\ell k}_{ij} \quad (\bar{x}^{i}\ \bar{x}^{j} + P^{ij}) + d^{\ell k} - \overset{+}{x}{}^{\ell} \cdot \overset{+}{x}{}^{k} \tag{PI}$$

<u>Proof</u> : replace x by $\tilde{x} + \bar{x}$ in II

THEOREM 9

The third and fourth order moments S and Q satisfy the following coupled linear equations in S and Q.

$$\overset{+}{S}^{\ell km} = h_i^{\ell k} \lambda_{i'}^m P^{ii'} + (h_i^{\ell k} \psi_{i'j'}^m + f_{i'j'}^{\ell k} \lambda_i^m) S^{ii'j'}$$

$$+ \Sigma^{\ell km} + f_{ij}^{\ell k} \psi_{i'j'}^m (Q^{iji'j'} - P^{ij}P^{i'j'}) - (\overset{+}{x}^k \Gamma^{\ell m} + \overset{+}{x}^\ell \Gamma^{km}) \qquad (S)$$

$$\overset{+}{Q}^{\ell kmr} = f_{ij}^{\ell k} f_{i'j'}^{mr} (Q^{iji'j'} - P^{ij} P^{i'j'}) + h_i^{\ell k} h_j^{mr} P^{ij}$$

$$+ (h_i^{\ell k} f_{i'j'}^{mr} + f_{i'j'}^{\ell k} h_i^{mr}) S^{ii'j'} + \overset{+}{P}^{\ell k} \overset{+}{P}^{mr} \qquad (Q)$$

$$+ H^{\ell kmr} - (|\overset{+}{x}^k \Sigma^{mr\ell}| - |\overset{+}{x}^k \overset{+}{x}^m \Gamma^{\ell r}|)$$

Where $|z^{k\ell mr}|$ = sum of all possible permutations

and

$$\lambda_i^m = \varphi_i^m + 2\psi_{ji}^m \bar{x}^j$$

$$f_{ij}^{\ell k} = b_{ij}^{\ell k} = \overset{+}{x}^\ell \psi_{ij}^k - \overset{+}{x}^k \psi_{ij}^\ell$$

$$h_i^{\ell k} = a_i^{\ell k} + 2 b_{ij}^{\ell k} \bar{x}^j - \overset{+}{x}^\ell \lambda_i^k - \overset{+}{x}^k \lambda_i^\ell$$

$$\gamma_{ijj'}^{\ell k} = \varphi_i^\ell \psi_{jj'}^k + \varphi_i^k \psi_{jj'}^\ell$$

There is a second way to determine the evolution of P using equation I alone. We get then

THEOREM 10

The variance P satisfies the following quadratic recurrence in P.

$$\overset{+}{P}^{\ell k} = (\varphi_i^\ell + 2\psi_{ij}^\ell \bar{x}^j) P^{ik} + \beta_{ijj'}^{\ell k} S^{ijj'} \qquad (PII)$$

$$+ \psi_{ij}^\ell \psi_{i'j'}^k (Q^{iji'j'} - P^{ij} P^{i'j}) + \Gamma^{\ell k}$$

where $\beta_{ijj'}^{\ell k} = \lambda_i^\ell \psi_{jj'}^k + \lambda_i^k \psi_{jj'}^\ell$

Proof : replacing x by x + x in (I)

Comments :

1) Two systems of equations can be developped

(m) + (PII) + (S) + (Q) or (m) + (PI) + (S) + (Q).

Their equivalence gives the constraints to be satisfied by a, b, d, φ, ψ, β, Γ, Σ, H. It is a fastidious task to developp this point here.

2) Equation (m) + (PII) are sufficient to establish the evolution of the 2 first moments but (S) and (Q) must be satisfied in order to have compatibility.

At this point we are able to extend slightly the concept of PMPQ to processes that, by construction, satisfy equation I.

DEFINITION 4

Let a stochastic process z(t) be generated by equation (I) (parameters β, φ, ψ, Γ given).
We call $Z(t) = \{ z(o), \ldots, z(t) \}$ and by extension the σ-algebra generated by those random variables.
We say that z(t) behaves a posteriori as a PMPQ (projective post process quadratic or PPPQ) if and only if there exist 3 martingales a*(t), b*(t); d*(t) adapted to the flow $\{ Z(t) \}$ such that

$$\eta^{*\ell c} = \overset{+}{z}{}^{\ell} \, \overset{+}{z}{}^{c} - d^{\ell k} - \overset{*}{a}{}^{\ell c}_{i} \, z^{i} - \overset{*}{b}{}^{\ell k}_{ij} \, z^{i} \, z^{j} \qquad \text{II}^{*}$$

and for t > S

$$E\left[\eta^{*\ell c}(t) \, \eta^{*\, km}(s)\right] = 0 \qquad E\left[\eta^{*\ell c}\right] = 0$$

$$E\left[\varepsilon^{\ell}(t) \, \eta^{*\ell c}(s)\right] = 0 \qquad E\left[\varepsilon^{\ell}(t) \, \varepsilon^{c}(t) \, \eta^{*km}(s)\right] = 0$$

II.4 A practical modelisation tool

We suggest that PPPQ's can be used for modeling physical processes.

We assume that some mathematical dynamical model gives a recurrent equation of type (I) but where ε represents an error in equation corresponding to neglected terms or forces. From an engineering point of view one can assume that ε is a white noise. Either a simulation or a measurement during a period $[t_o \; t_f]$ gives a sequence $\{\bar{\bar{x}}\}_{t_o}^{t_f}$ of measured values. Estimators of different moments can be computed and a choice be made for $\bar{\bar{a}}$ $\bar{\bar{b}}$ $\bar{\bar{d}}$ in order to establish an equation of type (II) where $\bar{\bar{\eta}}$ will be the residual error. There is a priori no reason to get the same property for $\bar{\bar{\eta}}$ as needed for η^{*} in Definition 4, but as we dispose of a sequence $\{\bar{\bar{\eta}}\}_{t_o}^{t_f}$, we can test it, by estimating its moments and cross moments with ε similarly to experimental correlation functions.

REFERENCES

DOOB JL (1953), Stochastic processes. John Wiley, London

GIHMAN I and SKOROHOD A (1974), The theory of stochastic processes T1. Springer Verlag, Berlin.

NUALART-RODON D et AGUILAR-MARTIN J (1977), Estimation optimale en Puissances de degré N. CRAS Paris t. 284, 3 Janvier 1977.

Acknowledgements

The results shown here are part of the common work with D. NUALART-RODON and G. SALUT and more recently the application handled by G. STAVRAKAKIS. The description of an industrial application of § II.4 will appear in the thesis of the latter at the end of 1983.

We thank also the most valuable comments of Professor H. KOREZLIOGLU.

ON THE STOCHASTIC MAXIMUM PRINCIPLE FOR INFINITE DIMENSIONAL EQUATIONS AND APPLICATION TO THE CONTROL OF ZAKAI EQUATION

A. Bensoussan
University Paris Dauphine
and
INRIA
Domaine de Voluceau
BP 105 - Rocquencourt
78153 LE CHESNAY Cédex (France)

INTRODUCTION

The problem of optimal control of partially observed diffusions can be reformulated as a control problem with complete observation but for an infinite dimensional stochastic system. This system corresponds to the evolution of the unnormalized conditional probability and is governed by Zakaï equation.

H. KWAKERNAAK [1] was the first to derive a Pontryagin's maximum principle for a similar problem, the control of Kushner equation. This treatment was however formal. The control of Zakaï equation was treated rigorously, as far obtaining a stochastic maximum principle, in a previous paper of the author (cf. A. BENSOUSSAN [3]). The method was relying on the robust form approach. Later on we have derived a stochastic maximum principle for systems governed by evolution equations in Hilbert spaces (cf. A. BENSOUSSAN [4]). There is nothing equivalent to the robust form here and our method was completely different. We used a Galerkin finite dimensional approximation We show here that a slightly extented version of this approach can also lead to the stochastic maximum principle for Zakaï equation. This has many advantages. The use of the robust form is not natural and requires regularity properties which are unnecessary. Moreover one obtains here a slightly better result, namely an additional regularity of the adjoint processes.

1 - SETTING OF THE PROBLEM

1.1. Notations

Let V, H be two separable Hilbert spaces such that $V \subset H$, V dense in H with continuous injection.

Identifying H to its dual and considering V' to be the dual of V, we have the sequence

$$V \subset H \subset V'$$

each space being dense in the next one with continuous injection.

Let $A(t) \in L^{\infty}(0,T ; \mathscr{L}(V;V'))$ such that

$$(1.1) \qquad \langle A(t)z,z\rangle + \lambda|z|^2 \geq \alpha||z||^2, \ \forall z \in V, \ \ \alpha > 0 \qquad (^1)$$

Let next $(\Omega,\mathscr{A},P)$ be a probability space, equipped with a filtration F^t, $(\mathscr{A} = F^{\infty})$, such that there exists a F^t Wiener with values in E, (separable Hilbert spaces) with covariance operator Q.

We know that Q is a nuclear operator on E (cf. A. BENSOUSSAN [1], [2] for more details). We consider an orthonormal basis of E, made of eigenvectors of Q, namely e_n, and

$$(1.2) \qquad Qe_n = \lambda_n e_n, \ \lambda_n \geq 0, \ \Sigma\lambda_n < \infty$$

Let next

$$a(\xi,s) \ , \ b_n(\xi,s)$$

be stochastic processes depending on $\xi \in V$, such that

$$(1.3) \qquad \forall\xi \in V, \ a(\xi,s) \in L^2_F(0,T ; H), \ b_n(\xi,s) \in L^2_F(0,T ; H)$$

----

(1) $|\ |$ denotes the norm on H, $||\ ||$ the norm on V

$$(1.4)\quad \begin{cases} |a(\xi_1,s) - a(\xi_2,s)|_H \le C||\xi_1-\xi_2||_V \\ 2<A(s)(\xi_1-\xi_2),\ \xi_1-\xi_2> - \sum_n \lambda_n \mid b_n(\xi_1,s) - b_n(\xi_2,s)|_H^2 + \beta|\xi_1-\xi_2|_H^2 \\ \ge \gamma||\xi_1 - \xi_2||_V^2, \qquad \gamma > 0,\ \ \beta \ge 0,\ \ \forall \xi_1,\xi_2 \in V \end{cases}$$

where the constants C, β, γ are deterministic,

$$(1.5)\qquad |a(o,s)|_H \le C$$

$$(1.6)\quad \begin{cases} |b_n(o,s)|_H \le C \\ |b_n(\xi_1,s) - b_n(\xi_2,s)|_H \le C||\xi_1-\xi_2||_V \end{cases}$$

I.2. A non linear stochastic infinite dimensional problem

We have the following

Theorem 1.1 . Under the assumptions (1.3), (1.4), (1.5), (1.6) there exists one and only one process $\xi(t)$ satisfying

$$(1.7)\quad \begin{cases} \xi \in L^2_F(0,T\ ;\ V) \cap L^2(\Omega,\mathcal{A},P\ ;\ C(0,T\ ;\ H)) \\ d\xi + A(t)\xi(t)dt = a(\xi(t),t)dt + \sum_n b_n(\xi(t),t)d(w(t),e_n) \\ \xi(o) = \xi_0 \end{cases}$$

where $\xi_0 \in H$ is given

Proof

We first solve (1.7) in the case when the following additional assumption is made

$$(1.8)\qquad |b_n(\xi_1,s) - b_n(\xi_2,s)|_H \le C|\xi_1-\xi_2|_H$$

It is sufficient to solve the equation

$$(1.9)\quad \begin{cases} d\xi + (A(t)+k)\xi\ dt = a(\xi(t),t)dt + \sum_n b_n(\xi(t),t)\ d(w(t),e_n) \\ \xi(o) = \xi_0 \end{cases}$$

where k is chosen arbitrarily.

Let $\eta \in L^2_F(0,T ; F)$, we solve the equation

$$(1.10)\quad \begin{cases} d\xi + (A(t)+k)\xi dt + a(\eta(t),t)dt + \sum_n b_n(\eta(t),t)d(w(t),e_n) \\ \xi(o) = \xi_0 \\ \xi \in L^2_F(0,T ; V) \cap L^2(\Omega,\mathcal{A},P ; C(0,T ; H)) \end{cases}$$

This defines a map $\eta \to \xi$ from $L^2_F(0,T ; V)$ into a subspace of itself. Let us show that this map is a contraction. Let η_1, η_2 be given and ξ_1, ξ_2 be the corresponding solutions of (1.10). From the energy equality we have

$$(1.11)\quad \begin{cases} E|\xi_1(t) - \xi_2(t)|^2 + 2\,E \int_o^t <A(s)(\xi_1(s) - \xi_2(s)), \xi_1(s) - \xi_2(s)> ds \\ + 2k\, E \int_o^t |\xi_1(s) - \xi_2(s)|^2 ds = 2\,E \int_o^t <\xi_1(s) - \xi_2(s), a(\eta_1,s) - a(\eta_2,s)> ds \\ + E \sum_n \lambda_n \int_o^t |b_n(\eta_1(s),s) - b_n(\eta_2(s),s)|^2 ds \end{cases}$$

Choosing in V the norm $|||v||| = (||v||^2 + \rho|v|^2)^{1/2}$ which is equivalent to $||v||$, and picking a convenient choice of ρ,k we deduce the contraction property.

The general case (1.6) is obtained by Galerkin approximation method (cf. E. PARDOUX [1]). □

1.3. A control problem

Let U be a Hilbert space, called the space of controls

(1.12) $\quad U_{ad}$ closed convex, non empty subset of U

Let also

$$(1.13)\quad \begin{cases} g(\xi,v) : V \times U \to H \\ \sigma_n(\xi,v) : V \times U \to H \end{cases}$$

g, σ_n are Gateaux differentiable with continuous bounded derivations g_ξ, g_v, $\sigma_{n,\xi}$, $\sigma_{n,v}$

$$(1.14)\quad \begin{cases} 2<A(s)(\xi_1-\xi_2),\ \xi_1-\xi_2> - \sum_n \lambda_n\ |\sigma_{n,\xi}(\xi,v)(\xi_1-\xi_2)|_H^2 \\ + \beta|\xi_1-\xi_2|_H^2\ \geq \gamma||\xi_1-\xi_2||_V^2,\ \gamma > 0,\ \ \beta \geq 0 \\ \forall \xi_1,\xi_2,\ \xi \in V,\ v \in U \end{cases}$$

$$15)\qquad |g(o,v)|_H,\quad |\sigma_n(o,v)|_H \leq C$$

Le ılso

$$(1.16)\quad \begin{cases} \ell(\xi,v)\ :\ H \times U \to R \\ h(\xi)\ :\ H \to R \end{cases}$$

ℓ,h are Gateaux differentiable ; ℓ_ξ, ℓ_v, h_ξ are continuous

$$|h_\xi|\ ,\ |\ell_\xi|\ ,\ |\ell_v| \leq C(|\xi| + |v| + 1)$$

An admissible control is an adapted process with values in U_{ad}. To any such process one associates the solution of the equation

$$(1.17)\quad \begin{cases} dz + A(t)zdt = g(z(t),v(t))dt + \sum_n\ \sigma_n(z(t),v(t))d(w(t),e_n) \\ z(o)\ = y_0 \\ z \in L_F^2(0,T\ ;\ V) \cap L^2(\Omega,\mathcal{A},P\ ;\ C(0,T\ ;\ H)) \end{cases}$$

and we define the functional

$$(1.18)\qquad J(v(.)) = E[\int_o^T \ell(z(t),v(t))dt + h(z(T))]$$

2 - PRELIMINARIES

2.1. Gateaux derivative of the cost

We shall denote by u(.) an optimal control and by y(.) the corresponding state

$$(2.1)\quad \begin{cases} dy + A(t)y(t)dt = g(y(t),u(t))dt + \sum_n \sigma_n(y(t),u(t))d(w(t),e_n) \\ y(o) = y_0 \\ y \in L^2_F(0,T\ ;\ V) \cap L^2(\Omega,\mathcal{A},P\ ;\ C(0,T\ ;\ H)) \end{cases}$$

Lemma 2.1. The functional J(v(.)) is Gateaux differentiable and the following formula holds

$$(2.2)\quad \begin{cases} \frac{d}{d\theta} J(u(.)+\theta v(.))|_{\theta=0} = E\{ \int_o^T [(\ell_x(y(t),u(t)),z(t))dt \\ + (\ell_v(y(t),u(t)),v(t))]dt + (h_x(y(T)),\ z(T))\} \end{cases}$$

where z is the solution of the linear equation

$$(2.3)\quad \begin{cases} dz + A(t)zdt = [g_z(y(t),u(t))z(t) + g_v(y(t),u(t))v(t)]dt \\ + \sum_n (\sigma_{n,z}(y(t),u(t))z(t) + \sigma_{n,v}(y(t),u(t))v(t))d(w(t),e_n) \\ z(o) = z_0 \\ z \in L^2_F(0,T\ ;\ V) \cap L^2(\Omega,\mathcal{A},P\ ;\ C(0,T\ ;\ H)) \ \square \end{cases}$$

The proof is an easy adaptation of A. BENSOUSSAN [4].

2.2. Abstract definition of the adjoint process

Let $\phi \in L^2_F(0,T\ ;\ V')$, $\psi_n \in L^2_F(0,T\ ;\ H)$ with

$$(2.4)\quad \sum_n \lambda_n E \int_o^T |\psi_n(t)|^2 dt < \infty$$

We solve the equation

$$(2.5)\quad \begin{cases} d\rho + A(t)\rho\, dt = (g_z(y(t),u(t))\rho(t) + \phi)dt + \sum_n (\sigma_{n,z}(y(t),u(t))\ \rho(t) + \\ + \psi_n(t))d(w(t),e_n) \\ \rho(o) = 0 \\ \rho \in L^2_F(0,T\ ;\ V) \cap L^2(\Omega,\mathcal{A},P\ ;\ C(0,T\ ;\ H)) \end{cases}$$

This equation differs slightly from (1.7) since it requires $a(\xi,t)$ to belong to V' instead of H. However $a(\xi_1,t) - a(\xi_2,t) \in H$, which suffices to derive Theorem 1.1.

The map $\phi, \psi \to \rho$ is linear continuous. Therefore we can define in a unique way stochastic processes

$$(2.6)\quad \begin{cases} p \in L^2_F(0,T\ ;\ V),\ K_n \in L^2_F(0,T\ ;\ H) \\ \sum_n \lambda_n E \int_o^T |K_n(t)|^2 dt < \infty \end{cases}$$

such that the following relation holds

$$(2.7)\quad \begin{cases} E \int_o^T (\ell_z(y(t),u(t)),\rho(t))dt + E(h_z(y(T)),\rho(T)) \\ = E \int_o^T \langle p(t),\phi(t)\rangle dt + \sum_n \lambda_n E \int_o^T (K_n(t),\psi_n(t))dt \end{cases}$$

We immediately deduce from (2.7) and Lemma 2.2 that

Lemma 2.2. We have

$$(2.8)\quad \begin{cases} \frac{d}{d\theta} J(u(.) + \theta v(.))|_{\theta=0} = E \int_o^T [(\ell_v(y(t),u(t)) + g_v^*(y(t),u(t))p(t) \\ + \sum_n \lambda_n \sigma^*_{n,v}(y(t),u(t))K_n(t),\ v(t))]dt \end{cases}$$

3 - STOCHASTIC MAXIMUM PRINCIPLE

3.1. Statement of the result

Theorem 3.1. We assume (1.1), (1.12), (1.13), (1.14), (1.15), (1.16). If a(.) is an optimal control for (1.17), (1.18) and y(.) is the corresponding trajectory, then defining p, K_n in a unique way by (2.7), the following condition holds

$$(3.1)\quad \begin{cases} (\ell_v(y(t),u(t)) + g_v^*(y(t),u(t))p(t) + \sum_n \lambda_n \sigma_{n,v}^*(y(t),u(t))K_n(t), v-u(t)) \geq 0 \\ \forall v \in U_{ad}, \text{ a.e.t, a.s.} \end{cases}$$

Proof

It follows easily from formula (2.8) and a classical localization argument (cf. for instance A. BENSOUSSAN [2]). □

3.2. Equation for the adjoint processes

We shall assume here that

$$(3.2)\qquad F^t = \sigma(w(s), s \leq t)$$

We state the following

Theorem 3.2. We make the assumptions of Theorem 3.1. and (3.2). Then the processes p(t), $K_n(t)$ satisfy

$$(3.3)\quad \begin{cases} p \in L_F^2(0,T ; V) \cap L^2(\Omega,\mathcal{A},P ; C(0,T ; H)) \\ K_n \in L_F^2(0,T ; H), \sum_n \lambda_n E \int_0^T |K_n(t)|^2 dt < \infty \end{cases}$$

$$(3.4)\quad \begin{cases} p(T) = h_z(y(T)) \\ - dp + A^*(t)p(t)dt = [g_z^*(y(t),u(t))p(t) + \ell_z(y(t),u(t)) + \\ \quad + \sum_n \lambda_n \sigma_{n,z}^*(y(t),u(t))K_n(t)]dt - \sum_n K_n(t)\, d(w(t),e_n) \end{cases}$$

Moreover the processes p, K_n are uniquely characterized by the conditions (3.3),(3.4).

The proof can be found in A. BENSOUSSAN [4].

It is a technical one, relying on an approximation procedure in a finite dimensional space.

4 - APPLICATION TO THE PROBLEM OF STOCHASTIC CONTROL UNDER PARTIAL INFORMATIONS

4.1. Setting of the problem

The problem of stochastic control under partial informations can be reformulated as a stochastic control problem for an infinite dimensional system. This reformulation can be found in A. BENSOUSSAN [3], for instance. Here, we shall limit ourselves to stating the infinite dimensional stochastic control problem, which will appear as a particular case of (1.17), (1.18).

We consider

(4.1) $$H = L^2(R^n),\ V = H^1(R^n)$$

(4.2) $$\begin{cases} A(t) = -\sum_{i,j} \frac{\partial}{\partial x_i}\, a_{ij}(x,t) \frac{\partial}{\partial x_j} \\ \text{where} \\ a_{ij} = a_{ji} \in L^\infty(R^n \times (0,T)) \\ \Sigma a_{ij}\xi_j\xi_i \geq \alpha|\xi|^2,\ \alpha > 0,\ \forall \xi \in R^n \end{cases}$$

(4.3) $$\begin{cases} g(z,v) = \sum_{i=1}^{n} \frac{\partial}{\partial x_i} (a_i(x,v)z) \\ \forall z \in H^1(R^n),\ v \in U_{ad} \subset R^k \\ U_{ad} \text{ convex closed} \\ a_i(x,v),\ \frac{\partial a_i}{\partial x_i}(x,v),\ \frac{\partial a_i}{\partial v}(x,v),\ \frac{\partial^2 a_i}{\partial x \partial v} \quad \text{bounded} \end{cases}$$

(4.4) $$\begin{cases} \sigma_i(z,v) = z(x)h_i(x),\ i = 1, \dots, n \\ \forall z \in L^2(R^n),\ h_i \in L^\infty \end{cases}$$

(4.5) $$\begin{cases} \ell(z,v) = \int_{R^n} f(x,v)z(x)dx = (f_v,z)_H \text{ where } f_v \in H \qquad |f_v|_H \leq C,\ \forall v \in U_{ad} \\ h(z) = (m,z),\ m \in H. \end{cases}$$

4.2. Optimality conditions

We apply Theorem 3.1. and 3.2.. There exist uniquely defined processes

$$(4.\ )\qquad p \in L^2_F(0,T\ ;\ H^1(R^n)),\ K_i(t) \in L^2_F(0,T\ ;\ L^2(R^n)),\ i = 1, \ldots, n$$

such that (3.4) holds. We note that

$$(4.7)\qquad g_z(z,v)z_1 = \sum_i \frac{\partial}{\partial x_i}(a_i(x,v)z_1)$$

$$(4.8)\qquad g^*_z(z,v)q = -\sum_i a_i(x,v)\frac{\partial q}{\partial x_i} \in V',\ \forall q \in H$$

$$(4.9)\qquad \sigma_{i,z}(z,v)z_1 = z_1(x)h_i(x)$$

$$(4.10)\qquad \sigma^*_{i,z}(z,v)q = q(x)h_i(x)$$

$$(4.11)\qquad h_z = m$$

$$(4.12)\qquad \ell_z(z,v) = f_v$$

Therefore (3.4) can be written as follows

$$(4.13)\quad \begin{cases} p(x,T) = m(x) \\ -\,dp - \sum_{i,j} \frac{\partial}{\partial x_i}(a_{ij}p)dt = [-\sum_i a_i(x,u(t))\frac{\partial}{\partial x_i} + f(x,u(t)) + \sum_i h_i(x)K_i(x,t)]dt \\ \quad -\sum_i K_i(x,t)dw_i \end{cases}$$

Let us now write (3.1). We note that

$$\ell_v(z,v) = \int f_v(x,v)\, z(x)dx = (f_v,z)$$

$$g_v(z,v) = \sum_i \frac{\partial}{\partial x_i}(a_{i,v}(x,v)z)$$

$$\sigma_{i,v}(z,v) = 0$$

$$g^*_v(z,v)q = -\sum_i \langle \frac{\partial q}{\partial x_i}, a_{i,v}, z\rangle$$

hence (3.1) reads

$$(4.14)\quad \begin{cases} [\int_{R^n} f_v(x,u(t))y(x,t)dx - \sum_i \int_{R^n} \frac{\partial p}{\partial x_i}(x,t)\, y(x,t)\, a_{i,v}(x,u(t))dx] \\ \quad .\ (v - u(t)) \geq 0 \end{cases}$$

Remark 4.1. The relations (4.13), (4.14) are an improvement of the form given in A. BENSOUSSAN [3]. We get here a more explicit form of the martingale term, and a regularity property. Moreover we do not need substantial regularity properties of h which represents the observation, unlike in the previous approach based on the robust form.

REFERENCES

[1] BENSOUSSAN A. (1971). Filtrage optimal des systèmes linéaires, Dunod, Paris

[2] BENSOUSSAN A. (1978). Control of stochastic partial differential equations, in Distributed Parameters Systems, edited by W.H. Ray and D.G. Lainiotis, Marcel Dekker, N.Y.

[3] BENSOUSSAN A. (to be published in Stochastics). Maximum principle and dynamic programming approaches of the optimal control of partially observed diffusions.

[4] BENSOUSSAN A. (to be published in LSSTA, North Holland). Editor S.G. Tzafestas.

[1] KWAKERNAAK H. (July 1981). A minimum principle for stochastic control problems with output feedback, Systems and Control Letters, Vol. 1, n° 1.

[1] PARDOUX E. (1979). Stochastic partial differential equations and filtering of diffusion processes. Stochastic, Vol. 3, pp. 127-167.

SOME COMMENTS ON CONTROL AND ESTIMATION PROBLEMS FOR DIFFUSIONS IN BOUNDED REGIONS

René K. Boel
Research Fellow, NFWO
Rijksuniversiteit Gent, Belgium

and

Department of Systems Engineering
Research School of Physical Sciences
Australian National University
Canberra, ACT, Australia

Abstract

In two earlier papers [1,2] the use of diffusion processes with general boundary conditions was suggested as a method for analysing queueing networks. The distinguishing features compared to classical diffusion approximation were the delayed reflections in a random direction. This short paper first discusses advantages and disadvantages of these generalisations. Then it is shown that the backward and forward Kolmogorov equations for this model can be derived. This allows writing down explicitly problems of optimal recursive estimation and of optimal control.

INTRODUCTION

Consider a queueing network with finite buffers. The classical diffusion approximation [3] saturates all queues at the same time. This leads to instantaneously reflected diffusion processes. However at the same time the buffer sizes are renormalized to zero in the limit. The instantaneous reflection assumption implies in particular that the diffusion process spends zero time at the boundary, with probability one.

On the other hand, for a finite buffer queue, saturation cannot reasonably be assumed, and hence there is a positive probability that the queue will be empty. Therefore it was suggested in [1,2] that a delayed (slow) reflection be used in the model, as defined in [1,4]. Such processes have a positive probability that the set $\{X_t^{(i)}=0\}$ has non-zero Lebesque measure. In order to be consistent with this non-saturated model one then also has to make the angle of reflection a random process. This leads to the following diffusion equations (as in [1]), which, as in the remainder of this paper is written for X_t in the quadrant $[0,\infty)^2$:

$$dX_t = a(X_t)dt + b(X_t)dW_t + \sum_{i=1}^{2} 1_{X_{i,t}=0}\cdot(\alpha_i(X_t)d\ell_{i,t} + \beta_i(X_t)d\tilde{W}_{i,\ell_{i,t}})$$

$$\int_0^t 1_{X_{i,s}=0}\cdot ds = \gamma.\ell_{i,t} \tag{1}$$

where: $(a(x),b(x))$ describe average drift and variance of a diffusion on $(0,\infty)^2$.

$$\ell_{i,t} = \lim_{\varepsilon\downarrow 0} \varepsilon.\ \#\{\text{downcrossings of } X_{i,t} \text{ from } \varepsilon \text{ to } 0\}$$

$$= \lim_{\varepsilon\downarrow 0} \frac{1}{\varepsilon}\ \text{Leb}\{0 \leq s \leq t : 0 \leq X_{i,s} \leq \varepsilon\}$$

represents the local time at the boundary $x_i=0$; this is a measure of the time X_t spends near the boundary $X_{i,t}=0$.

γ, the stickyness of the process, is a measure of how long X_t sticks to a boundary once it has hit the boundary. The case $\gamma=0$ is called instantaneous reflection. Then $\{s : X_{i,s} = 0\}$ has Lebesque measure zero, but has no isolated points; $\ell_{i,t}$ is singular w.r.t. Lebesque measure. For the delayed reflection case, $\gamma > 0$, $\{s : X_{i,s} = 0\}$ has positive Lebesque measure a.s. but nevertheless does not contain any interval; $\ell_{i,t}$ now is absolutely continuous w.r.t. Lebesque measure. The generalisation to $\gamma_i(x)$ dependent on x and i is possible, as long as $\lim_{x_1 \downarrow 0} \gamma_2(x_1) = \lim_{x_2 \downarrow 0} \gamma_2(x_2)$.

$$\alpha_1(x) = \begin{bmatrix} 1 \\ \alpha_{12}(x_2) \end{bmatrix} \qquad \beta_1(x) = \begin{bmatrix} 0 \\ \beta_{12}(x_2) \end{bmatrix} \qquad x_1 = 0$$

$$\alpha_2(x) = \begin{bmatrix} \alpha_{21}(x_1) \\ 1 \end{bmatrix} \qquad (x) = \begin{bmatrix} \beta_{21}(x_1) \\ 0 \end{bmatrix} \qquad x_2 = 0$$

W_t, $\tilde{W}_{i,t}$ are independent Brownian motions. $\alpha_i(x)$ specifies the average angle of reflection of X_t after hitting the boundary $X_{i,t} = 0$. Notice that the normalization $\alpha_{ii}(x) = 1$ insures that there always is a normal component.

$(\alpha_{12}(x), \beta_{12}(x))$ describes a one-dimensional diffusion process in the time scale $\ell_{1,t}$ determining the tangential, random component of the reflection at boundary $x_1 = 0$. Assume $x_j^{-1}\beta_{ij}(x_j) \downarrow 0$ as $x_j \downarrow 0$ and $\alpha_{ij}(x_j) > 0$ for x_j small.

In section 2 the backward operator A of the diffusion defined by (1) will be derived, together with its domain, using Ito's differentiation rule. This allows formulation of a Bellman-Hamilton-Jacobi equation for an optimal control problem including a cost at the rate of the local times. Using partial integration one obtains the dual operator A*, the forward, Fokker-Planck operator of X_t. This is done in section 3, where it is used in the derivation of an optimal recursive estimator for X_t, given only noisy observations of it.

Whether it is sensible to use this rather complicated process as a model for heavy traffic queueing networks, is not yet clear. The boundary conditions of operator A and A* (obtained in sections 2 and 3) are very complicated. In some special cases (mostly one-dimenional, except [7,10]) it is known [5,6,7,10] that there exist unique solutions to the corresponding partial differential equations. Some explicit solutions in the one-dimensional case can be found in [9]. However, in general, the explicit solutions which one would like for analysis of queueing networks will be either impossible or very complicated.

One purpose of introducing delayed reflection in [2] was the suggestion that "ℓ_t is larger for a delayed reflection process" corresponding to more time spent by the queueing process at the boundary. However it has since been found that $E\ell_t$ is independent of γ for large t. Indeed, consider one-dimensional slowly reflected Brownian motion $dX_t = \text{sgn}W.dW_t + d\ell_t$; $1_{X_t=0}.dt = \gamma.d\ell_t$, and let T be the stopping time $T = \inf\{t > 0 : X_t = 1\}$ of first hitting the upper boundary (say full buffer), then $X_t - \ell_t$ is a martingale and optional sampling gives $E\ell_T = 1$ for all γ. It has been argued by J. Groh [8] that the distribution of ℓ_T is also independent of γ (unit-mean exponential). However, the microscopic (small increments in t) behaviour of ℓ_t will strongly depend on γ. Increasing γ smoothens ℓ_t, and $\gamma \downarrow 0$ makes ℓ_t nondifferentiable with probability 1.

While it is still possible that the correlation (experimentally measurable) of ℓ_T and T will depend on γ, the suggestion of [5] that γ be chosen on the basis of experimental values of ℓ_T and T will obviously be hard to carry out. Probably the choice of γ, whether in diffusion approximations to queueing networks, or in other physical models, will depend on the mathematical tractability on the one hand and the robustness of the results on the other hand. Indeed the results obtained in the following sections should agree with those obtained via boundary layer analysis (i.e. consider a layer of width ε with special properties, and study asymptotics for $\varepsilon \downarrow 0$).

The author would like to thank Dr M. Kohlmann and Dr J. Groh for many helpful discussions on the topic of this paper.

2. Stochastic control and backward operators

Consider the bounded diffusion process of equation (1):

$$dX_t = a(X_t)dt + b(X_t)dW_t + \sum_{i=1}^{2} (\alpha_i(X_t)d\ell_{i,t} + \beta_i(X_t)d\tilde{W}_{i,\ell_t}$$

$$1_{X_{i,t}=0}.dt = \gamma.d\ell_{i,t}$$

Note that γ determines the type of reflection. If $\gamma=0$, instantaneous reflection, then $\ell_{i,t}$ is singular w.r.t. Lebesque measure; if $\gamma>0$, delayed reflection, then $\ell_{i,t}$ is absolutely continuous, with however a zero derivative unless $X_{i,t}=0$.

The backward operator A is easily obtained by applying Ito's differentiation rule to any $f\varepsilon C_0^2([0,\infty)^2)$:

$$f(X_{t+dt}) - f(X_t) = \nabla f(X_t)^T(a(X_t)dt + \sum_{i=1}^{2} 1_{X_{i,t}=0}.\alpha_i(X_t).d\ell_{i,t}$$

$$+ (\nabla f(X_t))^T(b(X_t)dW_t + \sum_{i=1}^{2} 1_{X_{i,t}=0}.\beta_i(X_t).d\tilde{W}_{i,\ell_{i,t}})$$

$$+ \frac{1}{2}\sum_{i,j=1}^{2} (\frac{\partial^2 f(X_t)}{\partial x_i \partial x_j})((b(X_t)b^T(X_t))_{ij}dt$$

$$+ 1_{X_{i,t}=0}(\beta_{i,j}^2(X_t).d\ell_{i,t})$$

Then, if it exists and is continuous on $[0,\infty)^2$ (with continuity on the boundaries specifying the boundary condition and hence D(A)), A is defined by the limit:

$$\lim_{dt\downarrow 0} E\,(\frac{f(X_{t+dt})-f(X_t)}{dt} \mid X_t = x) = Af(x)$$

Two cases have to be distinguished:

i) $\gamma = 0$

$$Af(x) = a^T(x)\nabla f(x) + \frac{1}{2}\sum_{i,j} ((b(x)b^T(x))_{ij} \frac{\partial^2 f(x)}{\partial x_i \partial x_j}$$

$$x\varepsilon(0,\infty)^2 \qquad (2)$$

$$D(A) = \{f\varepsilon C_0^2([0,\infty)^2) \ : \ L_i f(x) = 0 \text{ for } x_i = 0\} \tag{2'}$$

where

$$L_i f(x) = \alpha_i^T(x)\nabla f(x) + \frac{1}{2}\beta_{ij}^2(x)\ \frac{\partial^2 f}{\partial x_i \partial x_j}(x)\ ,\ j\neq i$$

ii) $\gamma \neq 0$, $Af(x)$ unchanged

$$D(A) = \{f\varepsilon C_0^2([0,\infty)^2) \ : \ \gamma.\lim_{x_i\downarrow 0} Af(x) = L_i f(x)\ ,\ x_i=0\} \tag{2''}$$

Generalize equations (1) now to include a full information controller, that is consider a control law $u_t(\omega)\varepsilon U$ which is $F_t = \sigma(X_s, s\leq t)$ measurable. The choice of the control value u at time t makes the average drift at t equal to $a(X_t,u)$ and, if $X_{i,t}=0$, the average angle of reflection $\alpha_i(X_t,u)$. We can then use the measure transformation approach of [12] to stochastic control (b,β_i,γ independent of u). The backward operator A^u of the controlled diffusion, described by [2], is obvious, from the following equation:

$$dX_t = a(X_t,u_t)dt + b(X_t)dW_t + \sum_{i=1}^{2} (\alpha_i(X_t,u_t)d\ell_{i,t} + \beta_i(X_t)dW_{i,\ell_t}$$

$$1_{X_{i,t}=0}\ .\ dt = \gamma\ .\ d\ell_{i,t} \tag{3}$$

We consider this system on [0,T] and choose the control law u (i.e. the probability measure P_u corresponding to A_u) so as to minimize the cost

$$E_u\left[\int_0^T c(s,X_s,u_s)ds + \sum_{i=1}^{2}\int_0^T \eta_i(s,X_s,u_s)d\ell_{i,s} + C(X_T)\right]$$

In [1] the need in queueing applications for a cost running at the rate of local time, i.e. penalizing being close to the boundary, is explained. Using [12], as in [1], we obtain the optimality condition in function of the value function

$$V(t,x) = \inf_{u\varepsilon U_t^T} E_u[\int_t^T c(s,X_s,u_s)ds + \sum_{i=1}^{2}\int_t^T \eta_i(s,X_s,u_s)d\ell_{i,s}$$

$$+ C(X_T)\,|\,X_t = x]$$

<u>Optimality condition</u>: for all admissible control laws $u, \forall t, h>0$

$$V(t,X_t) \leq E_u[\int_t^{t+h} c(s,X_s,u_s)ds + \sum_{i=1}^{2} \int_t^{t+h} \eta_i(s,X_s,u_s)d\ell_{i,s}$$

$$+V(t+h,X_{t+h})|X_t] \qquad (4)$$

with equality holding if and only if u is optimal. Moreover, for all u, $V(T,x) = C(x)$. Letting h tend to 0 after dividing by h yields the Bellman-Hamilton-Jacobi equation satisfied by the value function:

$$0 \leq \frac{\partial V}{\partial t}(t,x) + \inf_{u\varepsilon U}\{a^T(x,u)\nabla V(t,x) + c(t,x,u)\}$$

$$+ \frac{1}{2}\sum_{i,j} (b(x)b^T(x))_{ij} \frac{\partial^2 V}{\partial x_i \partial x_j}(t,x) \qquad (5)$$

with the boundary conditions at $x_i=0$:

$$0 \leq \frac{\partial V}{\partial t}(t,x) + \gamma_0[\inf_{u\varepsilon U}\{\alpha_i^T(x,u)\nabla V(t,x) + \eta_i(t,x,u)\}$$

$$+ \frac{1}{2}\beta_{ij}^2(x) \cdot \frac{\partial^2 V}{\partial x_i \partial x_j}(t,x)]$$

and $V(T,x) = C(x)$

Remarks: 1. $\frac{\partial V}{\partial t}(t,x)$ continuous at boundaries insures $V\varepsilon D(A^u)$, i.e. $\lim_{x_i\downarrow 0} \gamma_i . A^u V = L_i^u V$.

2. The existence and uniqueness of solutions to complicated partial differential equations as (5),(6) is not obvious. See [5,6,7,10] for some results in this direction.

3. Combining with the results of section 3, it is quite feasible to formulate a partial observation control problem, and to write down the corresponding optimality condition.

3. Recursive optimal estimation and forward operators

Consider the diffusion process X_t in a bounded region, defined by (1), as an unobserved signal, to be estimated through its influence on the observations with independent noise (V_t Brownian motion, ind. of W_t):

$$dY_t = h(X_t)dt + dV_t \tag{7}$$

The standard non-linear estimation theory [12] can be applied to this problem. This has already been done by Pardoux [10] for slightly more restrictive assumptions.

Consider the reference measure P_0 defined such that X_t and Y_t are independent:

$$L_t = \frac{dP}{dP_0}\Big|_{F_t} = \exp\left(\int_0^t h(X_s)dY_s - \frac{1}{2}\int_0^t h^2(X_s)ds\right)$$

or

$$dL_t = h(X_t)L_t\, dY_t$$

Then (see [12]) for any $\phi \varepsilon D(A)$:

$$E(\phi(X_t)|\sigma(Y_s, s \leq t)) = \frac{E_0(L_t.\phi(X_t)|\sigma(Y_s, s \leq t))}{E_0(L_t|\sigma(Y_s, s \leq t))}$$

$$= \frac{\sigma_t(\phi)}{\sigma_t(1)} \tag{8}$$

and $\sigma_t(\phi)$ can be represented as

$$d_t\sigma_t(\phi) = \sigma_t(A\phi).dt + \sigma_t(h.\phi).dY_t \tag{9}$$

Notice that since $\phi \varepsilon D(A)$ implies

$$\lim_{x_i \downarrow 0} \gamma.Af(x) = L_i f(x)$$

(9) is equivalent to

$$d_t\sigma_t(\phi) = \sigma_t(A.\phi(x).1_{x\varepsilon(0,\infty)^2})dt + \frac{1}{\gamma}\sum_i \sigma_t(L_i\phi(x).1_{x_i=0}).dt$$
$$+ \sigma_t(h(x)\phi(x)).dY_t$$

In order to transform (9) into a recursive equation consider the unnormalized conditional density $q(t,x)$ on $(0,\infty)^2$, and the unnormalized densities $q_i(t,x_j)$ $(j \neq i)$ on the boundary $x_i=0$. It turns out that these will exist as soon as the corresponding (unconditional) transition densities exist for the diffusion process X_t solving (1). The need for $q_i(t,x_j)$ is obvious since X_t can spend, with non-zero probability, a non-zero fraction of time on the boundaries, if $\gamma > 0$.

Consider hence

$$\sigma_t(\phi) = \int_{(0,\infty)^2} \phi(x_1,x_2)q_t(x_1,x_2)dx_1dx_2 + \int_{(0,\infty)} \phi(0,x_2)q_1(t,x_2)dx_2 + \int_{(0,\infty)} \phi(x_1,0)q_2(t,x_1)dx_1 \tag{10}$$

and try to determine recursive equations for $q(t,x_1,x_2), q_1(t,x_2)$, $q_2(t,x_1)$. Note that the assumptions on α and β near the origin insure that there is no positive amount of time spent at the origin, and hence no term $\phi(0,0)q_0(t)$ is required.

Combining (9) and (10) and then integrating by parts, one obtains for all $\phi \in C_0^2$ such that $\gamma . \lim_{x_i \downarrow 0} A\phi(x) = L_i\phi(x)$

$$\int_{(0,\infty)^2} \phi(x_1,x_2)d_tq_1(t,x_1,x_2)dx_1dx_2 + \int_{(0,\infty)} \phi(0,x_2)d_tq_1(t,x_2)dx_2 + \int_{(0,\infty)} \phi(x_1,0)d_tq_2(t,x_1)dx_1$$

$$= \int_{(0,\infty)^2} \phi(x_1,x_2)[A^*q(t,x_1,x_2)dt + h(x_1,x_2)q(t,x_1,x_2)dY_t]dx_1dx_2$$

$$+ \int_{(0,\infty)} \phi(x_1,0)[A_2^*q(t,x_1,0)dt + \frac{1}{\gamma} L_2^*q_2(t,x_1)dt + h(x_1,0)q_2(t,x_1)dY_t]dx_1$$

$$+ \int_{(0,\infty)} \phi(0,x_2)[A_1^*q(t,0,x_2)dt + \frac{1}{\gamma} L_1^*q_1(t,x_2)dt + h(0,x_2)q_1(t,x_2)dY_t]dx_2$$

$$+ \int_{(0,\infty)} \frac{\partial\phi}{\partial x_2}(x_1,0)\,[-\tfrac{1}{2}(bb^T)_{22}(x_1,0)q(t,x_1,0) + \frac{1}{\gamma} q_2(t,x_1)]dtdx_1$$

$$+ \int_{(0,\infty)} \frac{\partial\phi}{\partial x_1}(0,x_2)\,[-\tfrac{1}{2}(bb^T)_{11}(0,x_2)q(t,0,x_2) + \frac{1}{\gamma} q_1(t,x_2)]dtdx_2$$

$$+ \phi(0,0)\,[(bb^T)_{12}(0,0)q(t,0,0) - \frac{1}{\gamma}\alpha_{12}(0)q_1(t,0) - \frac{1}{\gamma}\alpha_{21}(0)q_2(t,0)]$$

It can be verified that the class of allowable ϕ is sufficiently rich to conclude that the different coefficients multiplying ϕ and its derivatives have to be zero (it was assumed that q, q_1 and q_2 do not have a singularity near the origin, but this assumption can probably be justified). One then obtains the following system of linear partial differential equations driven by the observations. Together with (8) and (10) they provide a recursive optimal estimator:

$$d_t q(t,x_1,x_2) = A^* q(t,x_1,x_2)dt + h(x_1,x_2)q(t,x_1,x_2)dY_t \tag{11a}$$

$$d_t q_1(t,x_2) = (A_1^* q(t,0,x_2) + \frac{1}{\gamma} L_1^* q_1(t,x_2))dt + h(0,x_2)q_1(t,x_2)dY_t \tag{11b}$$

$$d_t q_2(t,x_1) = (A_2^* q(t,x_1,0) + \frac{1}{\gamma} L_2^* q_2(t,x_1))dt + h(x_1,0)q_2(t,x_1)dY_t \tag{11c}$$

with boundary conditions

$$\lim_{x_1 \downarrow 0} \tfrac{1}{2}(bb^T)_{11}(x_1,x_2)q(t,x_1,x_2) = \frac{1}{\gamma} q_1(t,x_2)$$

$$\lim_{x_2 \downarrow 0} \tfrac{1}{2}(bb^T)_{22}(x_1,x_2)q(t,x_1,x_2) = \frac{1}{\gamma} q_2(t,x_1)$$

$$\lim_{\substack{x_1 \downarrow 0 \\ x_2 \downarrow 0}} [(bb^T)_{12}(x_1,x_2)q(t,x_1,x_2) - \frac{1}{\gamma}(\alpha_{12}(x_2)q_1(t,x_2) + \alpha_{21}(x_1)q_2(t,x_1))] = 0$$

The operators used in the equations are defined below ($\underline{x} = (x_1,x_2)$):

$$A^*q(t,x_1,x_2) = -\frac{\partial}{\partial x_1}(a_1(\underline{x})q(t,\underline{x}) - \frac{\partial}{\partial x_2}(a_2(\underline{x})q(t,\underline{x}))$$

$$+ \frac{1}{2}\sum_{i,j=1}^{2} \frac{\partial^2}{\partial x_i \partial x_j}((bb^T)_{ij}(\underline{x})q(t,\underline{x}))$$

$$L_i^* q_i(t,x_j) = -\frac{\partial}{\partial x_j}(\alpha_{ij}(x_j)q_i(t,x_j))$$

$$+ \frac{1}{2}\frac{\partial^2}{\partial x_j^2}(\beta_{ij}^2(x_j)q_i(t,x_j)) \quad , \quad j\neq i$$

$$A_i^* q(t,x_i,x_j) = -a_i(\underline{x})q(t,\underline{x}) + \frac{1}{2}\frac{\partial}{\partial x_i}((bb^T)_{ii}(\underline{x})q(t,\underline{x}))$$

$$+ \frac{\partial}{\partial x_j}((bb^T)_{ij}(\underline{x})q(t,\underline{x})) \quad , \quad j\neq i$$

Remarks: 1. (A^*, L_i^*) and the associated boundary conditions of course specify the forward operator corresponding to the diffusion process X_t.

2. For the case of instaneous reflection, $\gamma=0$, $q_i(t,x_j) = 0$ and (11) reduces to (11a) but with the extra boundary conditions $A_i^* q(t,x) = 0$ for $x_i=0$.

4. Conclusions

This short paper illustrates how known results on stochastic control and nonlinear estimation apply to the diffusion processes with boundary conditions, which were introduced in [1,2] as models of a queueing network. Undoubtedly, these results could be extended to even fancier and more complicated theorems, dealing with partial observation control, robust estimation, etc. However, the basic question of how useful these equations are is still unanswered. First one should be able to find explicit, analytical solutions in simple cases to predict system behaviour dependent on a small number of parameters. Second, a robust method for choosing the parameters should be found. This should be consistent with the basic philosophy of these models, viz. the fact that one considers different time scales near boundaries (critical levels such as full or empty buffers) and away from boundaries. This seems justified because one should observe the system most carefully near these critical states.

References

1. R. Boel - M. Kohlmann: A control problem on a manifold with nonsmooth boundary, Proc. of the 2nd Bad Honnef Workshop on Stochastic Dynamical Systems, Springer Verlag Lecture Notes in Economics and Mathematical Systems Theory, 1982.

2. R. Boel: Boundary conditions for diffusion approximations to queueing problems, Proc. of the International Seminar on Modelling and Performance Evaluation Methodology, INRIA, Paris, 1983.

3. M. Reiman: Queueing networks in heavy traffic, Ph.D. Dissertation, Dept. of Op. Res., Stanford University, 1977.

4. I. Gihman - A. Skorohod: Stochastic Differential Equations, Springer, 1972.

5. H.P. McKean: Elementary solutions for certain parabolic partial differential equations, Trans. Amer. Math. Soc., 82 (1956), pp. 519-548.

6. H. Langer, L. Partzsch and D. Schütze: Uber verallgemeinerte gewöhnliche Differential Operatoren mit nichtlokalen Randbedingungen and die von ihnen erzeugten Markov-Prozesse, Kyoto University, Research Institute for Mathematical Sciences, Ser. A, 7 (1971-72), pp. 659-702.

7. I. Karatzas: Diffusions with reflections on an orthant and associated initial-boundary value problems, preprint 1981.

8. J. Groh, private communication, March 1983.

9. F. Knight: Essentials of Brownian motion and diffusion, Mathematical Surveys, no. 18, Am. Math. Soc., 1981.

10. E. Pardoux: Stochastic partial differential equations for the density of the conditional law of a diffusion process with boundary, in: "Stochastic Analysis", A. Friedman & M. Pinsky, eds., Academic Press, 1978, pp. 239-269.

11. R. Boel - M. Kohlmann: Semi-martingale models of stochastic optimal control, with applications to double martingales, SIAM J. Control and Optimization, 18 (1980), pp. 511-533.

12. M. Davis and S. Marcus: An introduction to nonlinear filtering, in: Stochastic Systems: The Mathematics of Filtering and Identification and Applications, pp. 53-75, M. Hazewinkel & J. Willems, eds., Reidel Publishing Cy., 1981.

THE SEPARATION PRINCIPLE FOR PARTIALLY OBSERVED LINEAR CONTROL SYSTEMS: A GENERAL FRAMEWORK

N. Christopeit
Institut für Ökonometrie
und Operations Research
Universität Bonn
Adenauerallee 24-42
5300 Bonn 1
West Germany

K. Helmes
Institut für Angewandte
Mathematik
Universität Bonn
Wegelerstraße 6
5300 Bonn 1
West Germany

1. Introduction and Problem Formulation

This note is intended to provide a framework for the separation approach to the question of existence of optimal controls for partially observed linear systems with Gaussian initial distribution, convering both the case of bounded as well as unbounded controls. It is based on ideas underlying certain results which were recently obtained for specific control problems ([2],[6]) and on tools from the general theory of optimal control of (non-linear) partially-observed diffusions ([9]). In order to set forth the character of such a synthesis we formulate assumptions about the model in terms of a general hypothesis, whose validity has to be checked in particular cases of interest. This way we avoid to impose in advance unnecessarily restrictive conditions on the data (cf. [8],[12]). To give just one example, neither the diffusion matrix of the state equation nor the matrix F (see (1.1) below) which relates the observation process to the state process has (in general) to be nondegenerate; instead, weaker conditions which are sufficient for the existence of an optimal control can be found, cf. Section 3. To be more specific, we are concerned with control of partially-observed diffusion processes of the form

$$dx_t = [A(t)x_t + b(t,u_t)]dt + C(t)dw_t^{(1)}, \tag{1.1}$$

$$dy_t = F(t)x_t dt + G(t)dw_t^{(2)}, \tag{1.2}$$

$$x(o) = x_o,\ y_o = o .$$

The n-dimensional state x_t evolves according to (1.1) from an initial value x_o, assumed to be Gaussian and independent of the $(d+k)$-dimensional Brownian motion $(w_t^{(1)}, w_t^{(2)})$. The control u_t, which takes values in a given set $\mathcal{U} \subset \mathbb{R}^m$ and which is 'some' functional of the observations $(y_s, s \leq t)$ (a precise formulation of this dependence is given below), has to be chosen so that the cost functional

$$(1.3) \qquad J[u] = E[\int_o^T \ell(s,x_s,u_s)ds + k(x_T)], \; T > 0 \; \text{fixed},$$

is minimized.

Throughout, the matrices $A(t)$, $C(t)$, $F(t)$ and $G(t)$ will be assumed to be of size $n\times n$, $n\times d$, $k\times n$, and $k\times k$, respectively, and continuous in t with $G(t)G'(t)$ being uniformly (in t) positive definite. The functions $b(t,u)$, $\ell(t,x,u)$ and $k(x)$ are measurable mappings defined on $\mathbb{R}^+ \times \mathcal{U}$, $\mathbb{R}^+ \times \mathbb{R}^n \times \mathcal{U}$ and $\mathbb{R}^n$, respectively, and satisfy the following conditions:

(i) $|b(t,u)| \leq K(1 + |u|)$ for all $t \geq 0$, $u \in \mathcal{U}$;

(ii) $|\ell(t,x,u)| \leq K(1 + |u|^{2p} + |x|^{2p})$ for some integer $p \geq 1$ and all $t \geq 0$, $u \in \mathcal{U}$, $x \in \mathbb{R}^n$;

(iii) $|k(x)| \leq K(1 + |x|^{2p})$ for all $x \in \mathbb{R}^n$ (K a constant).

Let us now give a precise formulation of the control problem. To this end, let $C^r = C([0,T];\mathbb{R}^r)$ denote the space of continuous functions $[0,T] \to \mathbb{R}^r$, equipped with the natural filtration $(\mathcal{C}_t^r)$. As our basic measure space we shall take

$$\Omega^o = \Omega_o \times \Omega_1 \times \Omega_2$$

where $\Omega_i = C^r$ with $r=d$, n and k, respectively, with generic element $\omega = (w^{(1)}, x, y)$ and endowed with the filtration

$$\mathcal{F}_t^o = \mathcal{C}_t^d \times \mathcal{C}_t^n \times \mathcal{C}_t^k$$

and $\mathcal{F}^o = \mathcal{F}_T^o$. Let $(\mathcal{G}_t)$ denote the filtration on Ω^o generated by the last k components of ω, i.e. $\mathcal{G}_t = \sigma\{y_s, s \leq t\}$.

Let now $(\tilde{\Omega}, \tilde{F}, \tilde{P})$ be any probability space carrying a $(d+k)$-dimensional Brownian motion $\tilde{w} = (\tilde{w}_t^{(1)}, \tilde{w}_t^{(2)})$. Then, for every $v \in L^2([0,T];\mathbb{R}^m)$, there exists a (pathwise) unique solution to the linear stochastic differential equation

(1.4) $\quad d\tilde{x}_t = A(t)\tilde{x}_t dt + b(t,v_t)dt + C(t)d\tilde{w}_t^{(1)}, \quad 0 \leq t \leq T,$

with initial value $\tilde{x}_o$, $\tilde{x}_o$ independent of $\tilde{w}^{(1)}$ and distributed as x_o. Let $\tilde{P}^v = \tilde{P} \circ (\tilde{w}^{(1)},\tilde{x})$ denote the (unique) distribution measure of $(\tilde{w}_t^{(1)},\tilde{x}_t)$ on $\Omega_o \times \Omega_1$ and take any function $u : [0,T] \times C^k \to \mathcal{U}$ which is measurable and adapted to $(\mathcal{C}_t^k)$ and satisfies $v(\cdot) = u(\cdot,y) \in L^2([0,T];\mathcal{U})$ for μ-almost all $y \in C^k$, where $\mu(dy)$ denotes the distribution measure of the process $(\int_o^t G(s)d\tilde{w}_s^{(2)})$. Then, on a set of μ-measure 1, the mapping $y \to \tilde{P}^{u(\cdot,y)}(A)$ is well defined and measurable for all Borel sets A in $\Omega_o \times \Omega_1$ (cf. Appendix, Lemma (A.0)), so that we may define a probability measure $\mathring{P}^u$ on Ω^o by

$$\mathring{P}^u(A \times B) = \int_B \tilde{P}^{u(\cdot,y)}(A)\mu(dy) .$$

This definition is in the spirit of [9] for wide-sense admissible controls. Note that under $\mathring{P}^u$

(1.5) $\quad dx_t = A(t)x_t dt + b(t,u(t,y))dt + C(t)dw_t^{(1)},$

(1.6) $\quad dy_t = G(t)d\hat{w}_t^{(2)}$

for some Brownian motion $\hat{w}^{(2)}$ independent of $w^{(1)}$.

Let now

$\mathring{\mathcal{A}} := \{u : [0,T] \times C^k \to \mathcal{U} \mid$

(i) u is measurable and $(\mathcal{C}_t^k)$-adapted;

(ii) $u(\cdot,y) \in L^2([0,T];\mathcal{U})$ for μ-almost all $y \in C^k$;

(iii) $\mathring{E}^u \zeta_o^T = 1\}$,

denote the class of "pre-admissible" controls, where we have put

$$\zeta_o^T = \exp\{\int_o^T [G(t)^{-1}F(t)x_t]'d\hat{w}_t^{(2)} - \frac{1}{2}\int_o^T |G(t)^{-1}F(t)x_t|^2 dt\} .$$

For every $u \in \mathring{\mathcal{A}}$ we define a measure P^u by

$$dP^u = \zeta_o^T d\mathring{P}^u .$$

Then, by Girsanov's theorem, P^u is a probability measure on $(\Omega^o, \mathcal{F}^o)$ and under P^u the coordinate process $t \to (x_t, y_t)$ is a solution to the stochastic differential equation (1.1), (1.2) with some new Brownian motion $w^{(2)}$ independent of $w^{(1)}$ and with $u_t(\omega) = u(t,y)$. Note that (u_t) is adapted to the observation σ-fields $(\mathcal{G}_t)$. We are now in the position to specify the class of admissible controls for the partially observed problem, viz.

$$\mathcal{A} := \{u \in \mathring{\mathcal{A}} \mid E^u \int_o^T |u_t|^{2p} dt < \infty\}.$$

(same p as in (ii) and (iii), but at least p=2).

The (linear) partially observed control problem now reads as follows:

($\mathcal{P}$) Choose $u \in \mathcal{A}$ so as to minimize

$$J[u] = E^u\{\int_o^T \ell(t,x_t,u_t)dt + k(x_T)\} .$$

Remark 1. Finiteness of J[u] for $u \in \mathcal{A}$ follows from (a straightforward extension of) Lemma (A.1), in which the stochastic equation (A1) contains an additional drift term $u_t dt$ such that $E\int_o^T |u|^{2p} dt < \infty$. (A2) is then valid for all integers $1 \leq m \leq p$. Moreover, the moment condition (for p=2) is needed to ensure that the process (x_t, y_t) is conditionally Gaussian (cf. Section 2).

Remark 2. A somewhat different formulation of the (linear) partially observed control problem is given in [12]. The main difference between the two formulations is the fact that in our approach the set of admissible controls is operationally defined whereas in [12] 'admissible' controls are defined in an implicit way.

2. The separated problem

It follows from Theorem 12.7 in [15] that for each $u \in \mathcal{A}$ the process $z_t = (x_t, y_t)$ is conditionally Gaussian under P^u. This means in particular that the conditional distribution of x_t under P^u given $\mathcal{G}_t$ is Gaussian with mean $\hat{x}_t^u$ and covariance matrix $R(t)$. The conditional mean $\hat{x}_t^u$ obeys the stochastic differential equation (under P^u)

$$(2.1)\qquad \begin{aligned} d\hat{x}_t^u &= A(t)\hat{x}_t^u dt + b(t,u_t)dt + K(t)d\nu_t^u , \\ \hat{x}_o^u &= E^o x_o . \end{aligned}$$

E^o denotes expection w.r. to P^o, the probability measure corresponding to $u \equiv 0$, but note that $E^u x_o$ is the same for all admissible u; the matrix $K(t) = R(t)F'(t)[G(t)G'(t)]^{-1}$, and the innovation process ν_t^u is given by

$$(2.2)\qquad d\nu_t^u = dy_t - F(t)\hat{x}_t^u dt = G(t)d\hat{w}_t^u, \quad \nu_o^u = 0,$$

with a k-dimensional Brownian motion $(\hat{w}_t^u)$ (under P^u) adapted to $(\mathcal{G}_t)$. The error covariance $R(t)$ is independent of u and satisfies the Riccati equation

$$\dot{R} = AR + RA' - RF'[GG']^{-1}FR + CC', \quad R(o) = \mathrm{cov}(x_o) .$$

Let $N(t,x;d\xi)$ denote the n-dimensional normal distribution with mean value x and covariance $R(t)$ and introduce the functions

$$\hat{\ell}(t,x,u) = \int_{\mathbb{R}^n} \ell(t,\xi,u)N(t,x;d\xi)$$

and

$$\hat{k}(x) = \int_{\mathbb{R}^n} k(\xi)N(T,x;d\xi) .$$

Note that $\hat{\ell}$ and $\hat{k}$, too, satisfy a polynomial growth condition since the moments of N are polynomials in x and $R(t)$. Then, for every admissible u,

$$(2.3)\qquad \begin{aligned} J[u] &= E^u\{\int_o^T E^u[\ell(t,x_t,u_t)\,|\,\mathcal{G}_t]dt + E^u[k(x_T)\,|\,\mathcal{G}_T]\} \\ &= E^u\{\int_o^T \hat{\ell}(t,\hat{x}_t^u,u_t)dt + \hat{k}(\hat{x}_T^u)\} . \end{aligned}$$

With this in mind, let us formulate what we shall call the separated control problem associated with $(\mathcal{P})$. To this end, let

$\hat{\mathcal{A}} := \Big\{ u = (u_t)$ process on $(\Omega^o, \mathcal{F}^o)$ such that

(i) u is measurable and $(\mathcal{G}_t)$-adapted; $u_t \in \mathcal{U}$ for all t;

(ii) there exists on $(\Omega^o, \mathcal{F}^o)$ a probability measure P such that $(\Omega^o, \mathcal{F}^o, P)$ carries a k-dimensional Wiener process $(w_t, \mathcal{G}_t)$ and

$$E \int_o^T |u_t|^{2p} dt < \infty \Big\} .$$

For each such $u \in \hat{\mathcal{A}}$ there exists then (under P) a unique $(\mathcal{G}_t)$-adapted solution (obtained by Picard iteration) to the equation

$$d\xi_t = A(t)\xi_t dt + b(t,u_t)dt + K(t)G(t)dw_t ,$$

$$\xi_o = E^o x_o . \tag{2.4}$$

The separated problem $(\hat{\mathcal{P}})$ is now defined as follows.

$(\hat{\mathcal{P}})$ Minimize

$$\hat{J}[u] = E\{\int_o^T \hat{\ell}(t,\xi_t,u_t)dt + \hat{k}(\xi_T)\}$$

in the class $\hat{\mathcal{A}}$ of admissible controls u subject to the constraint (2.4).

Remark 3. Note that, by the definition of $\hat{\mathcal{A}}$, corresponding to one admissible u there may be many different probability measures P and hence different joint probability distributions of (ξ_t, u_t). Hence, to be rigorous, we should write $\hat{J}[u,\xi,P]$ to take care of this ambiguity. We shall, however, confine ourselves to situations where the infimum is attained by Markov feedback control laws $u_t = \upsilon(t,\xi_t)$ for which (ξ,u) is unique in law (see below); therefore, meanwhile, we shall continue to use the somewhat loose notation $\hat{J}[u]$.

The values of $(\mathcal{P})$ and $(\hat{\mathcal{P}})$ are related in the following way:

Lemma 1. $\inf_{u \in \hat{\mathcal{A}}} \hat{J}[u] \leq \inf_{u \in \mathcal{A}} J[u]$.

Proof. For $u \in \mathcal{A}$, let $P = P^u$ be the probability measure constructed in Section 1 and let $w = \hat{w}^u$ be the Wiener process from (2.2). Then, since $E^u \int_0^T |u_t|^{2p} dt < \infty$ for $u_t(\omega) = u(t,y)$ by the definition of $\mathcal{A}$, $u \in \hat{\mathcal{A}}$. Moreover, $\xi_t := \hat{x}_t^u$ is then the unique solution of (2.4) (cf. (2.1), (2.2)), from which $\hat{J}[u] = J[u]$ by virtue of (2.3).

□

As it should be, $(\hat{\mathcal{P}})$ is a stochastic control problem with complete observation of the state, which evolves according to the dynamics governing the conditional mean of the state in $(\mathcal{P})$.

Note that one might have started with a broader class of admissible controls for the separated problem, in that also the probability space and the filtration could have been varied. Under the hypothesis (H) to be made below these subtleties will, however, make no difference.

In accordance with the concept of this note as outlined in the Introduction we now make the following *hypothesis*:

(H) There exists a Markov feedback control law $\hat{\upsilon}^*: [0,T] \times \mathbb{R}^k \to \mathcal{U}$ with the following properties:

(i) $|\hat{\upsilon}^*(t,\xi)| \leq K(1 + |\xi|)$, $(t,\xi) \in [0,T] \times \mathbb{R}^k$, for some constant K;

(ii) there exist unique strong solutions to the equation

$$d\xi_t = L(t)\xi_t dt + b(t, \hat{\upsilon}^*(t,\xi_t))dt + K(t)G(t)dw_t, \tag{2.5}$$

$$\xi_0 = E^0 x_0 ,$$

for $L(t) = A(t)$ and $L(t) = A(t) - K(t)F(t)$.

(iii) $\hat{\upsilon}^*$ determines (by the method described below) an optimal control $\tilde{u}$ for $(\hat{\mathcal{P}})$, i.e.

$$\hat{J}[\tilde{u}] = \inf_{u \in \hat{\mathcal{A}}} \hat{J}[u] .$$

Strong solutions are understood here in the sense of [13],[17] (cf. Appendix). As is shown in Lemma (A.2) this means that (ξ_t) can be represented as a causal functional of the Brownian paths (w_t).

Let $\tilde{\Xi}(t,y)$, $y \in C^k$, denote this representation for the solution of (2.5) with $L(t) = A(t)$, i.e.

$$(2.6) \qquad \tilde{\xi}_t = \tilde{\Xi}(t,w), \; 0 \leq t \leq T, \qquad P\text{-a.e.}$$

for any solution $(\tilde{\xi}_t)$ to (2.5) (with $L \equiv A$) defined on some probability space $(\Omega, \mathcal{F}, P)$ carrying a k-dimensional Brownian motion w. Define $\tilde{u}_t(\omega) := \hat{\mathcal{U}}^*(t, \tilde{\Xi}(t,w(\omega))) = \hat{\mathcal{U}}^*(t, \tilde{\xi}_t(\omega))$. Then, taking $(\Omega, \mathcal{F}) = (\Omega^o, \mathcal{F}^o)$ and, for example, $(w_t) = (y_t)$ under the Wiener measure P on $(\Omega^o, \mathcal{F}^o)$, it is easily verified that $\tilde{u} = (\tilde{u}_t) \in \hat{\mathcal{A}}$, the moment condition following from Lemma (A.1). Moreover, the distribution of $(\tilde{\xi}_t, \tilde{u}_t)$ is unique (independent of the probability space and the Brownian motion chosen) for all t; it follows again from Lemma (A.1) and Fubini's theorem that $\hat{J}[\tilde{u}]$ is well defined, finite and unique.

A detailed discussion of (H) for various cases of interest will be postponed to the next section.

Let us now see how an optimal control for the original partially observed problem $(\mathcal{P})$ can be obtained from such an optimal Markov feedback control law (satisfying (H)) for the separated problem. To this end, let the unique strong solution (ξ_t) of (2.5) for $L(t) = A(t) - K(t)F(t)$ have the representation

$$(2.7) \qquad \xi_t = \Xi(t, \int_0^{\cdot} G(s)\,dw_s) \qquad P\text{-a.e.}$$

as a causal functional of the paths $(\eta_t) = (\int_0^t G(s)\,dw_s)$ (cf. Appendix), where (w_t) is any k-dimensional Brownian motion on some probability space $(\Omega, \mathcal{F}, P)$. Then by virtue of (h) (i) and Lemma (A.1),

$$(2.8) \quad E\int_0^T |\xi_t|^2 dt = E\int_0^T |\Xi(t,\eta)|^2 dt = E^\mu \int_0^T |\Xi(t,y)|^2 dt < \infty ,$$

where E^μ denotes expectation with respect to μ, the distribution measure of the process (η_t) (cf. Section 1).
Now, put

$$(2.9) \qquad u_t^*(\omega) = u^*(t,y) = \hat{\mathcal{U}}^*(t, \Xi(t,y))$$

for $\omega = (w^{(1)},x,y) \in \Omega^o$. Then $u^* : [0,T] \times C^k \to \mathcal{U}$ is measurable, $(\mathcal{C}_t^k)$-adapted, and, by virtue of (2.8) and (H) (i), $u^*(\cdot,y) \in L^2([0,T];\mathcal{U})$ for μ-almost all y; i.e. u^* satisfies conditions (i) and (ii) in the definition of $\mathcal{A}$ (cf. section 1) and we can define the probability measure $\mathring{P}^* = \mathring{P}^{u^*}$. Condition (iii) (of $\mathring{\mathcal{A}}$) follows from the following result.

Lemma 2. $\mathring{E}^*\zeta_o^T = 1$.

Proof. Let (x_t, y_t) be the coordinate process on Ω^o (as in Section 1) and put $\xi_t^o = \Xi(t,y)$ (Ξ as in (2.7)). Then, under $\mathring{P}^*$,

$$d\xi_t^o = [A(t) - K(t)F(t)]\xi_t^o dt + b(t, \hat{\upsilon}^*(t,\xi_t^o))dt + K(t)G(t)d\hat{w}_t^{(2)} ,$$

$$dx_t = A(t)x_t dt + b(t, \hat{\upsilon}^*(t,\xi_t^o))dt + C(t)dw_t^{(1)}$$

(cf. (1.5), (1.6)). By virtue of (H) (i) and the growth condition imposed on $b(t,u)$ Lemma (A.1) (ii) applies, showing that for the process $\zeta_t = (\xi_t^o, x_t)$

$$\mathring{E}^* \exp[c\|\zeta\|_T^2] < \infty$$

for some positive constant c. Since $\|G^{-1}F\|_T^2 = \bar{L} < \infty$, we obtain for $\phi_t = G(t)^{-1}F(t)x_t$ the estimate $\|\phi\|_T^2 \leq \bar{L}\|x\|_T^2$. Hence, with $c_1 = c/\bar{L}$,

$$\mathring{E}^* \exp[c_1\|\phi\|_T^2] < \infty ,$$

whence $\mathring{E}^*\zeta_T^o(\phi) = 1$ follows by Corollary 7.2.2 in [14].

□

Thus u* is pre-admissible, and we can define the probability measure $P^* = P^{u^*}$ by

$$dP^* = \zeta_o^T d\mathring{P}^* .$$

As explained in Section 1, the coordinate process $(t,\omega) \to (x_t, y_t)$ on Ω^o then solves (1.1), (1.2) (with $u = u^*$) under P^*. Let $\hat{x}_t^* = \hat{x}_t^{u^*} = E^*\{x_t|\mathcal{G}_t\}$ (it will be shown below that the conditional expectation w.r. to P^* does indeed exist).

Theorem 1. *The control* u* *(cf.* (2.9)) *is optimal for* ($\mathcal{P}$) *and admits a feedback representation in terms of the estimated state* $\hat{x}^*$ *which is given by*

$$u_t^* = \hat{\upsilon}^*(t,\hat{x}_t^*) .$$

Proof. Let ξ^* be the unique solution to the equation

$$(2.10) \quad d\xi^*_t = [A(t) - K(t)F(t)]\xi^*_t dt + b(t,u^*_t)dt + K(t)dy_t$$

(with initial value $E^o x_o$) under P^*. Note that this solution is adapted to $(\mathcal{G}_t)$ and given by

$$\xi^*_t = \Phi(t)[E^o x_o + \int_o^t \Phi(s)^{-1} b(s,u^*_s)ds + \int_o^t \Phi(s)^{-1} K(s)dy_s]$$

(with Φ the transition matrix corresponding to $L = A - KF$, i.e. $\dot{\Phi}(t) = L(t)\Phi(t)$, $\Phi(O) = \mathrm{id}$), where the stochastic integral on the right hand side is indeed well defined since (under P^*) (y_s) is the Ito Process (1.2) and the paths $b(t,u^*_t)$ are in $L^2[O,T]$ $\mathring{P}^*$ - and hence P^* - a.e.

Put now $\xi^o_t = \Xi(t,y)$. Then, by the definition of Ξ,

$$(2.11) \quad d\xi^o_t = [A(t) - K(t)F(t)]\xi^o_t dt + b(t, \hat{\upsilon}^*(t,\xi^o_t))dt + K(t)dy_t$$

(and $\xi^o_o = E^o x_o$) under $\mathring{P}^*$. Moreover, from (2.9),

$$(2.12) \quad u^*_t = \hat{\upsilon}^*(t,\xi^o_t).$$

Since, by the mutual absolute continuity of $\mathring{P}^*$ and P^*, the stochastic integral $\int_o^t K(s)dy_s$ is the same (as a functional of y) under both measures, it follows from (2.10) - (2.12) that ξ^* and ξ^o coincide pathwise with ($\mathring{P}^*$ - and P^* -) probability one. Hence, under P^*,

$$dx_t = A(t)x_t dt + b(t, \hat{\upsilon}^*(t,\xi^*_t))dt + C(t)dw^{(1)}_t,$$

$$d\xi^*_t = [A(t) - K(t)F(t)]\xi^*_t dt + K(t)F(t)x_t dt +$$

$$+ b(t, \hat{\upsilon}^*(t,\xi^*_t))dt + K(t)G(t)dw^{(2)}_t,$$

whence, by virtue of (H) (i) and Lemma (A.1), $E^* \sup_{O<t<T} |\xi^*_t|^{2m} < C_m$ and, consequently,

$$E^* \int_o^T |u^*_t|^{2m} dt < K_m < \infty .$$

This shows that u^* is indeed admissible ($\in \mathcal{A}$), and $z_t = (x_t,y_t)$ is conditionally Gaussian under P^* with $\hat{x}^*$ satisfying

$$d\hat{x}^*_t = A(t)\hat{x}^*_t dt + b(t,u^*_t)dt + K(t)d\nu^*_t \ ,$$

$$d\nu^*_t = dy_t - F(t)\hat{x}_t dt = G(t)d\hat{w}^*_t$$

with some $(\mathcal{G}_t)$-adapted k-dimensional Brownian motion $\hat{w}^*$(cf. (2.1), (2.2)). Comparing with (2.10) we find that $\hat{x}^* = \xi^*$ P^*-a.e., hence

$$u^*_t = \hat{\mathcal{U}}^*(t,\hat{x}^*_t) \ .$$

Finally, since $\hat{x}^* = \xi^*$ is a solution under P^* to the equation

$$d\xi^*_t = A(t)\xi^*_t dt + b(t,\hat{\mathcal{U}}^*(t,\xi^*_t))dt + K(t)G(t)d\hat{w}^*_t$$

(with initial value $E^o x_o$), it follows that

$$\xi^*_t = \tilde{\Xi}(t,\hat{w}^*), \quad 0\leq t\leq T, \quad P^*\text{-a.e.}$$

(cf. (2.6)), hence $(\xi^*_t,u^*_t) \sim (\tilde{\xi}^*_t,\tilde{u}^*_t)$ for all t (where $\tilde{\xi}^*$ and $\tilde{u}^*$ are obtained from (2.6) by substituting $w = \hat{w}^*$) and, by (2.3) together with the considerations following (2.6) and (H) (iii),

$$J[u^*] = E^*\{\int_o^T \hat{\ell}(t,\xi^*_t,u^*_t)dt + \hat{k}(\xi^*_T)\}$$

$$= E^*\{\int_o^T \hat{\ell}(t,\tilde{\xi}^*_t,\tilde{u}^*_t)dt + \hat{k}(\tilde{\xi}^*_T)\}$$

$$= \hat{J}[\tilde{u}^*] = \inf_{u\in\hat{\mathcal{A}}} \hat{J}[u] \ .$$

Lemma 1 concludes the proof.

□

Remark 4. We close this section with some remarks on the classes $\mathcal{A}$ and $\hat{\mathcal{A}}$. If $\mathcal{U}$ is *bounded*, $\mathcal{A}$ is simply given by

$$\mathcal{A} = \{u:[0,T]\times C^k \to \mathcal{U} \mid u \text{ is measurable and } (\mathcal{C}^k_t)\text{-adapted}\}$$

and $\hat{\mathcal{A}}$ by

$$\hat{\mathcal{A}} = \{u = (u_t) \text{ measurable and } (\mathcal{G}_t)\text{-adapted process on } (\Omega^o,\mathcal{F}^o) \text{ with values in } \mathcal{U}\},$$

since condition (ii) may always be verified by taking as P the Wiener measure on $(\Omega^o,\mathcal{F}^o)$ and as (w_t) the coordinate process (y_t). Hence we may identify $\mathcal{A}$ and $\hat{\mathcal{A}}$ by means of the relation $u_t(\omega) =$

$u(t,y), \omega = (w^{(1)}, x, y) \in \Omega^o$ (note that, with this identification, $\mathcal{A} \subset \hat{\mathcal{A}}$ is always true, cf. proof of Lemma 1).

In any case, $\mathcal{U}$ bounded or not, both $\mathcal{A}$ and $\hat{\mathcal{A}}$ contain the class $\tilde{\mathcal{A}}$ of controls $u_t(\omega) = u(t,y)$ with linear growth in y, i.e.

$$|u(t,y)| \leq K(1 + \|y\|_t), \quad 0 \leq t \leq T.$$

Condition (ii) in $\hat{\mathcal{A}}$ follows from Lemma (A.1) applied to equation (2.4). Condition (ii) in $\mathcal{A}$ is obvious, while (iii) and the moment condition follow from Lemma (A.3) applied to the system (1.5), (1.6) or (1.1), (1.2), respectively.

3. Applications

In this section we shall discuss three problems for which hypothesis (H) (cf. Section 2) can be verified. In addition, we shall state a general existence result for an optimal control of problem ($\mathcal{P}$) under additional conditions imposed on the matrices C,F and the functions ℓ and k. The necessity of adding further conditions in the general case is due to the fact that unlike to the case of the problems mentioned above, this time the validity of (H) is guaranteed by general results on the existence of strong solutions to stochastic differential equations ([18]) and on the existence of (optimal) Markov controls ([4],[10]). For those specific examples one can do better, i.e. weaken some of these conditions, simply because it is possible to solve the separated control problem and to exploit structural properties of the optimal Markov control while checking (H) (ii).

The examples which we are going to consider are the three 'classics', the

(I) 'LQG-problem' (see, for instance, [10], ch. 6),

(II) 'predicted-miss' problem ([6] and the literature cited therein),

(III) 'minimum distance problem' ([1],[7]).

The state- and observation-equation for each one of the problems is given by (1.1) and (1.2) when

$$b(t,u) = B(t)u$$

and B(t) is a deterministic $n \times m$-matrix valued function. The problems differ from each other only in the kind of restrictions which are put on $\mathcal{U}$ and in the kind of cost functionals which are chosen:

(I) $\mathcal{U} = \mathbb{R}^m$, $\ell(t,x,u) = x'M(t)x + u'N(t)u$, $k(x) = x'Qx$, where the $n \times n$ matrices M(t) and Q are symmetric, nonnegative definite and the $n \times m$ matrices N(t) are symmetric, positive definite.

(II) $\mathcal{U} = [-1,1]^m$, $\ell \equiv 0$, $k(x) = g(\langle \zeta, x \rangle)$, where ζ is a given n-vector, $\langle \cdot, \cdot \rangle$ denotes scalar product in $\mathbb{R}^n$ and g is an even functional, increasing on $\mathbb{R}^+$ and satisfying a polynomial growth condition.

(III) $\mathcal{U}$ is the (Euclidean) unit ball in $\mathbb{R}^m$, $\ell(t,x,u) \equiv k(x) \equiv \|x\|$ is the (Euclidean) norm in $\mathbb{R}^n$ and, moreover, the system (1.1), (1.2) is 'rotational invariant'; for the precise meaning, see (3.1) below.

Lemma 3. Hypothesis (H) *holds for problem* (I).

Proof. It is well known, e.g. [10], p. 166, that the optimal Markov feedback control law for the separated problem ($\hat{\mathcal{P}}$) associated with (I) is linear in ξ, viz.

$$\hat{U}^*_I(t,\xi) = -S(t)\xi,$$

where $S(t) = N(t)^{-1}B'(t)\Delta(t)$ and Δ satisfies the matrix Riccati equation, $0<t<T$,

$$\dot{\Delta} = -\Delta A - A'\Delta + \Delta B N^{-1} B'\Delta - M,$$

$$\Delta(T) = Q.$$

Therefore (H) (i) is trivially satisfied; but (H) (i) is also fulfilled since (2.5) is now a linear stochastic differential equation. Optimality of $\hat{U}^*_I$ together with (H) (ii) yields (H) (iii).

□

Lemma 3 combined with Theorem 1 justifies the separation principle for the partially observed linear regulator, when the class of admissible controls is chosen to be $\mathcal{A}$ (cf. Section 1,2). Thus, the following result extends Theorem 11.1 in chapter 6 of [10].

Theorem 2. An optimal control for the partially observed linear regulator problem (I) *within the class* $\mathcal{A}$ *is*

$$u_I^*(t) = - S(t)\hat{x}_t^* ;$$

$\hat{x}^*$ *satisfies the stochastic differential equation*

$$dx_t^* = [A(t) - B(t)S(t)]\hat{x}_t^* dt + K(t)d\nu_t^*$$

with initial data $\hat{x}_o^* = E^o[x_o]$.

The multidimensional predicted miss problem with partial observations has been analyzed in [6]. Here we shall merely comment on how the results proved in [6] can be used to verify (H). Since the set of control values $\mathcal{U}$ for example (II) is bounded there is no need to bother about (H) (i). As in the previous case (H) (iii) follows from (H) (ii) and optimality of the Markov control. So let us concentrate on (H) (ii). We can only verify this property when the matrices A,C,F and G satisfy the following condition: Let $\Phi(T,t)$ denote the transition matrix corresponding to A and put $\zeta_t := \Phi'(T,t)\zeta$. We shall assume that $(\mathcal{G}_t)$ always contains information on $\langle\zeta_t, x_t\rangle$, i.e.

(B) $$\zeta_t' K(t)G(t)G'(t)K'(t)\zeta_t > 0 \text{ for all } 0 \leq t \leq T.$$

Lemma 4. Let (B) *be satisfied and put*

$$\hat{\upsilon}_{II}^{*\,i}(t,\xi) := - \operatorname{sign}\,(\langle\zeta_t,\xi\rangle b_i(t)), \quad i = 1,\ldots,m,$$

where $b(t) := B'(t)\zeta_t$. *Then* (H) (ii) *holds for* $\hat{\upsilon}_{II}^*$.

Lemma 5. Let (B) *hold. Then* $\hat{\upsilon}_{II}^*$ *is an optimal control for the separated problem associated with example* (II).

The proof of Lemma 4 can be found in [6], while Lemma 5 is proved in [7]. (Note that in [7] the assumption of continuity of k can be dispensed with.)

Theorem 1 combined with the Lemmas 4 and 5 yields:

Theorem 3. Let (B) *hold. An optimal control for the partially observed 'predicted miss problem'* (II) *within the class* $\mathcal{A}$ *is*

$$u^*_{II}(t) = \hat{\mathcal{U}}^*_{II}(t,\hat{x}^*_t);$$

$\hat{x}^*$ *satisfies the stochastic differential equation*

$$d\hat{x}^*_t = [A(t)\hat{x}^*_t + B(t)\,\hat{\mathcal{U}}^*_{II}(t,\hat{x}^*_t)]dt + K(t)d\nu^*_t$$

with initial data $\hat{x}^*_o = E^o[x_o]$.

The minimum distance problem with full information has been analyzed by many authors; see, for instance, [7] and the literature cited therein. For the partially observed problem it remains to specify the property of (1.1), (1.2) being rotational invariant. To this end, put $\tilde{B}(t) := \Phi(T,t)B(t)$ and $\tilde{\Gamma}(t) := \Phi(T,t)K(t)G(t)$ where $\Phi(T,t)$ denotes the transition matrix corresponding to A. We call the system (1.1), (1.2) 'rotational invariant' iff there are positive, continuous functions $p(t),q(t),r(t)$ and orthogonal matrices $\bar{B}(t),\bar{\Gamma}(t)$ and $\bar{R}(t)$ such that

(3.1) $\tilde{B}(t) = p(t)\bar{B}(t)$, $\tilde{\Gamma}(t) = q(t)\bar{\Gamma}(t)$ and $R(t) = r(t)\bar{R}(t)$.

Proceeding as we have done when analyzing example (II) and now using results proved in [7], Section 4, one shows:

Theorem 4. Let (1.1), (1.2) *be rotational invariant. An optimal control for the partially observed 'minimum distance problem'* (III) *within the class* $\mathcal{A}$ *is*

$$u^*_{III}(t) := - \bar{B}(t)' \frac{\hat{x}^*_t}{\|\hat{x}^*_t\|};$$

$\hat{x}^*_t$ *satisfies the stochastic differential equation*

$$d\hat{x}^*_t = [A(t)\hat{x}^*_t - B(t)\bar{B}(t)'\hat{x}^*_t/\|\hat{x}^*_t\|]dt + K(t)d\nu^*_t$$

with initial data $\hat{x}^*_o = E^o[x_o]$.

Proof. It suffices to comment on (H) (ii). Noting that rotational invariance forces K(t)G(t) to be nondegenerate for all $0<t<T$, existence of a unique strong solution follows from [18].

□

Finally, we formulate a general existence result which, unfortunately, is too weak to imply Theorems 2,3 or 4 although we are not aware of a particular minimum distance problem (example (III)) which is rotational invariant but does not satisfy the conditions assumed in our general result.

Theorem 5. Let $C(t)$, $F(t)$ *be* $n \times n$*-matrices which are invertible for all* $0 < t < T$. *Let the initial value* x_o *be non degenerate and the set* $\mathcal{U}$ *of control values be compact. Assume either that* ℓ, k *are bounded or that* $\ell(t,x,u)$ *is continuously differentiable in* t *and* u *for each fixed* x *(in addition to assumptions* (ii) *and* (iii) *in section* 1).

Then there exists an optimal separated control for problem $(\mathcal{P})$.

Proof. For bounded ℓ, k existence of an optimal Markov control for the separated problem follows from Theorem IV - 2 in [4]. For the second set of conditions, note that the error covariance $R(t)$ is positive definite for all $0 < t < T$ and hence the normal distributions $N(t,x;d\xi)$ possess densities. As a consequence, it follows that $\hat{\ell}(t,x,u) \in C^1([0,T] \times \mathbb{R}^n \times \mathcal{U})$, $\hat{k} \in C^2(\mathbb{R}^n)$ and $\hat{\ell}, \hat{\ell}_x, \hat{k}, \hat{k}_x$ satisfy a polynomial growth condition. Hence, by Theorem 6.3 in [10], there exists an optimal Markov feedback control law for $(\hat{\mathcal{P}})$. In both cases, (H) (ii) follows from Veretennikov's result in [18].

□

Remark 5. The condition (H) (ii) formalizes, in a sense, the basic difficulty underlying the separation approach to stochastic control of (linear) partially observed systems, viz. the existence of strong solutions to nonlinear stochastic differential equations with nonsmooth coefficients. If only the classical Itô solution theory were available, the scope of the separation approach would be confined to the setting which was considered by Wonham [19], viz. "natural" controls (cf. [10], chapter 6) which are Lipschitzian in the observations $(\mathcal{A}_{\text{Wonham}})$. Then under nondegeneracy conditions, in particular on F, he is able to show existence of an optimal Lipschitz-continuous Markov control $\hat{\upsilon}^*$ for $(\hat{\mathcal{P}})$ which thus (trivially) satisfies hypothesis (H). Hence $u_t^* = \hat{\upsilon}^*(t,\hat{x}_t^*) \in \mathcal{A}$. Differentiability of the functions F,G (assumed by Wonham) then implies that $u^* \in \mathcal{A}_{\text{Wonham}}$.

Remark 6. Another class of problems for which a separation principle holds without the very restrictive non-degeneracy condition on F is the class of (linear) partially observed stopping time problems (cf. [16]).

APPENDIX

Lemma (A.O) *Let* $u : [0,T] \times C^k \to \mathcal{U}$ *be a measurable,* $\mathcal{C}_t^k$*-adapted function such that* $v(\cdot) := u(\cdot, y) \in L^2([0,T]; \mathbb{R}^m)$ *for* μ*-almost all* $y \in C^k$. *Let* P^y *denote the measure* $P^{u(\cdot,y)}$ *defined at the beginning of Section* 1. *Then for any Borel set* A *in* $\Omega_o \times \Omega_1$ *the function* $y \to P^y[A]$ *is well defined on a set of* μ*-measure zero and is* $\mathcal{C}^k/\mathcal{B}$*-measurable.*

Proof. Fix $y \in C^k$ and let us denote by $(\tilde{x}_{t;y})$ the unique solution to the stochastic differential equation (1.4) for $v = u(\cdot, y)$. Let $\Phi(t,s)$ denote the transition matrix corresponding to A. Then $(\tilde{x}_{t;y})$ is given by the equation

$$\tilde{x}_{t;y} = \Phi(t,o)x_o + \int_o^t \Phi(t,s)b(s,u(s,y))ds + \int_o^t \Phi(t,s)C(s)d\tilde{w}_s^{(1)} .$$

The representation formula for the solution to equation (1.4) implies that the mapping

$$(\tilde{w}^{(1)}, \tilde{x}) : \tilde{\Omega} \times C^k \to \Omega_o \times \Omega_1$$

$$(\tilde{\omega}, y) \to (\tilde{w}^{(1)}(\tilde{\omega}), \tilde{x}_{\cdot;y}(\tilde{\omega})) =: \pi_{\cdot;y}(\tilde{\omega})$$

is $\tilde{\mathcal{F}} \times \mathcal{C}^k / \mathcal{C}^d \times \mathcal{C}^n$-measurable. Hence, the function

$$(\tilde{\omega}, y) \to I_A[(\pi_{\cdot;y}(\tilde{\omega})]$$

is $\tilde{\mathcal{F}} \times \mathcal{C}^k / \mathcal{B}$ -measurable for any Borel set $A \subset \Omega_o \times \Omega_1$.

Since

$$P^y[A] = \int I_A[\pi_{\cdot;y}(\tilde{\omega})]d\tilde{P}(\tilde{\omega})$$

the assertion follows.

□

Lemma (A.1) *Let* $\xi = (\xi_t)$ *be a solution (on some probability space* $(\Omega,\mathcal{F},P)$*) to the stochastic differential equation*

$$\text{(A1)} \qquad d\xi_t = f(t,\xi)dt + \sigma(t,\xi)dw_t, \quad 0 \leq t \leq T,$$

with initial value ξ_0 *independent of the Brownian motion* (w_t). *Suppose that* $f : [0,T] \times C^n \to \mathbb{R}^n$ *and* $\sigma : [0,T] \times C^n \to \mathbb{R}^{n\times r}$ *are measurable,* $(\mathcal{C}^n_t)$*-adapted and satisfy the linear growth condition*

$$|f(t,x)| + |\sigma(t,x)| \leq K(1 + \|x\|_t), \quad 0 \leq t \leq T,$$

where $\|x\|_t = \sup_{0\leq s\leq t} |x_s|$.

(i) *If* $E|\xi_0|^{2m} < \infty$, $m \geq 1$, *there exists a constant* C_m *such that*

$$\text{(A2)} \qquad E \sup_{0\leq t\leq T} |\xi_t|^{2m} \leq C_m$$

and

$$\text{(A3)} \qquad \lim_{\Delta t \downarrow 0} E \sup_{|s-t|\leq \Delta t} |\xi_s - \xi_t|^{2m-1} = 0 .$$

(ii) *If* $E \exp[c_0|\xi_0|^2] < \infty$ *for some* $c_0 > 0$, *then*

$$E \exp[c\|\xi\|^2_T] < \infty \quad \textit{for some } c > 0.$$

Proof. (i) The estimate (A2) is standard (cf. [11]). As to (A3), we obtain from a Hölder estimate (for $s \leq t \leq T$)

$$|\xi_t - \xi_s|^4 \leq 8\{2K^4(t-s)^2 \int_s^t (1 + \|\xi\|^2_r)dr + |\int_s^t \sigma(r,\xi)dw_r|^4\},$$

from which, by virtue of (A2) and standard estimates from stochastic calculus (for $|t-s| \leq 1$)

$$\begin{aligned} E|\xi_t - \xi_s|^4 &\leq 8\{2K^4(t-s)^2 \int_s^t (1 + E\|\xi\|^2_r)dr \\ &\quad + 48K^4n^3r^2(t-s) \int_s^t (1 + E\|\xi\|^4_r)dr\} \\ &\leq \text{const. } |t-s|^2 . \end{aligned}$$

As is shown in the proof of Theorem 12.3 in [3] this implies that for every $\varepsilon > 0$, $\eta > 0$, there exists a Δt such that

$$P\{\sup_{|s-t|<\Delta t}|\xi_s-\xi_t|>3\varepsilon\}<\eta\ ,$$

which means that

$$\sup_{|s-t|<\Delta t}|\xi_s-\xi_t|\to 0$$

in probability. But by virtue of (A2) the $\sup_{|s-t|<\Delta t}|\xi_s-\xi_t|^{2m-1}$ are uniformly (in Δt) integrable, whence (A3) follows.

(ii) Basically, the proof is the same as that of Theorem 5.7.1 in [14]. But, for completeness, and since the multidimensional case requires some slight modification, we reproduce it here. Let δ be any positive number. Then, for $u\in[t,t+\delta]$,

$$\begin{aligned}|\xi_u-\xi_t|^2 &< 2[(\int_t^u|f_v|dv)^2+|\int_t^u\sigma_v dw_v|^2]\\ &< 2[2\ \delta K^2\int_t^u(1+\|\xi\|_v^2)dv+|\int_t^u\sigma_v dw_v|^2]\ ,\end{aligned}$$

where we have put $f_v=f(v,\xi)$ and $\sigma_v=\sigma(v,\xi)$. Since, for $v>t$,

$$\|\xi\|_v=\max[\|\xi\|_t,\ \sup_{t<s<v}|\xi_s|],$$

we obtain

$$\begin{aligned}|\xi_u|^2 &< 2|\xi_t|^2+2|\xi_u-\xi_t|^2\\ &< 2|\xi_t|^2+8\delta^2K^2(1+\|\xi\|_t^2+\sup_{t<u<t+\delta}|\xi_u|^2)\\ &\qquad +4\sup_{t<u<t+\delta}|\int_t^u\sigma_v dw_v|^2\ .\end{aligned}$$

Taking the supremum over $[t,t+\delta]$ of the left hand side, we obtain

$$\begin{aligned}(1-8\delta^2K^2)\sup_{t<u<t+\delta}|\xi_u|^2 &< 8\delta^2K^2+(2+8\delta^2K^2)\|\xi\|_t^2\\ &\qquad +4\sup_{t<u<t+\delta}|\int_t^u\sigma_v dw_v|^2\ .\end{aligned}$$

Choose now $\delta=T/N$ (N a positive integer) so small that $\alpha=1-8\delta^2K^2>0$ and put $\alpha_1=8\delta^2K^2/\alpha$, $\alpha_2=(2+8\delta^2K^2)/\alpha$, $\alpha_3=4/\alpha$.

Then, for every $\varepsilon > 0$, a Hölder estimate yields

(A6) $$E \exp[\varepsilon \sup_{t\le u\le t+\delta} |\xi_u|^2] \le e^{\alpha_1\varepsilon} (E \exp[2\alpha_2\varepsilon \|\xi\|_t^2])^{1/2}$$

$$\cdot (E \exp[2\alpha_3\varepsilon \sup_{t\le u\le t+\delta} |\int_t^u \sigma_v dw_v|^2])^{1/2} .$$

But

$$|\int_t^u \sigma_v dw_v|^2 \le r \sum_{i=1}^n \sum_{k=1}^r (\int_t^u \sigma_v^{ik} dw_v^k)^2 ,$$

whence

$$\sup_{t\le u\le t+\delta} |\int_t^u \sigma_v dw_v|^2 \le r \sum_{i=1}^n \sum_{k=1}^r \sup_{t\le u\le t+\delta} |\int_t^u \sigma_v^{ik} dw_v^k|^2 .$$

By choosing ε_o so small that $\lambda = 2\alpha_3\varepsilon_o r$ satisfies

$$16KTrn\lambda < 1$$

and

$$2\alpha_2\varepsilon_o \le c_o ,$$

we can achieve that

$$E \exp[rn\lambda \sup_{t\le u\le t'} |\int_t^u \sigma_v^{ik} dw_v^k|^2] \le B < \infty$$

for all $i = 1,\dots,n$, $k = 1,\dots,r$ and $0 \le t < t' \le T$ (cf. [14], Lemma 5.7.2.). Hence, by the (generalized) Hölder inequality,

$$E \exp[2\alpha_3\varepsilon_o \sup_{t\le u\le t+\delta} |\int_t^u \sigma_v dw_v|^2] \le B$$

for all $t \in [0,T-\delta]$, and (A6) becomes

(A7) $$E \exp[\varepsilon \sup_{t\le u\le t+\delta} |\xi_u|^2] \le e^{\alpha_1\varepsilon_o}\{B\cdot E \exp[2\alpha_2\varepsilon \|\xi\|_t^2]\}^{1/2} ,$$

for all $\varepsilon \le \varepsilon_o$.

Take now $t = 0$. Then, for $\varepsilon = \varepsilon_o$,

$$E \exp[\varepsilon_o \|\xi\|_\delta^2] \le e^{\alpha_1\varepsilon_o}(B\cdot Ee^{c_o|\xi_o|^2})^{1/2}$$

$$= e^{\alpha_1\varepsilon_o}(BK_o)^{1/2} < \infty .$$

Proceeding by induction, suppose that positive numbers ε_i, $i = 0,1,\dots,\nu-1$, have been found such that $\varepsilon_i \leq \varepsilon_o$ and

(A8) $$E \exp[\varepsilon_{i-1} \|\xi\|^2_{i\delta}] < \infty, \quad i = 1,\dots,\nu .$$

Then, putting $\varepsilon_\nu = \varepsilon_{\nu-1}/2\alpha_2$, we have $\varepsilon_\nu < \varepsilon_{\nu-1} \leq \varepsilon_o$ and, from (A7)

(A9) $$E \exp[\varepsilon_\nu \sup_{\nu\delta \leq u \leq (\nu+1)\delta} |\xi_u|^2] \leq e^{\alpha_1 \varepsilon_o} \{B \cdot E \exp[\varepsilon_{\nu-1} \|\xi\|^2_{\nu\delta}]\}^{1/2} ,$$

which is finite because of (A8). Since $\|\xi\|^2_{(\nu+1)\delta} = \max(\|\xi\|^2_{\nu\delta}, \sup_{\nu\delta \leq u \leq (\nu+1)\delta} |\xi_u|^2)$, it follows from (A8) and (A9) that

$$E \exp[\varepsilon_\nu \|\xi\|^2_{(\nu+1)\delta}] < \infty .$$

Hence the assertion holds for $\nu = N-1$ (i.e. $(\nu+1)\delta = T$) if we take $c = \varepsilon_{N-1}$.

□

Suppose now that equation (A1) admits a unique strong solution (in the sense of [13]) corresponding to some fixed initial value ξ_o. I.e., the exists a mapping $F : C^r \to C^n$, $\bar{\mathcal{C}}^r_t/\bar{\mathcal{C}}^n_t$-measurable for every $t \in [0,T]$ (where $(\bar{\mathcal{C}}^r_t)$ is the canonical filtration on C^r completed with respect to r-dimensional Wiener measure P^W), such that for any r-dimensional standard Brownian motion $w = (w_t)$, defined on some probability space $(\Omega, \mathcal{F}, P)$, $\xi = F(w)$ is a solution to (A1) and, converseley, for any solution (ξ,w) to (A1), $\xi = F(w)$ holds a.e.

<u>Lemma (A.2)</u>. *Under the assumptions of Lemma* (A.1), *there exists a process* $\tilde{\Xi} : [0,T] \times C^r \to \mathbb{R}^n$ *such that*

(i) $\tilde{\Xi}$ *is progressively measurable* (*with respect to the natural filtration* $(\mathcal{C}^r_t)$) *and has* P^W - *a.e. continuous trajectories;*

(ii) $\tilde{\Xi}(\cdot,z) = F(z)$ *for* P^W*-almost all* z.

In particular, $\tilde{\xi} = \tilde{\Xi}(\cdot,w)$ is a solution to (A1) for every r-dimensional standard Brownian motion, and any solution $(\tilde{\xi},w)$ of (A1) satisfies $\tilde{\xi} = \tilde{\Xi}(\cdot,w)$ a.e.

<u>Proof</u>. Let $w = (w_t)$ be any r-dimensional standard Brownian motion on some probability space $(\Omega, \mathcal{F}, P)$. Let $0 = t_1^{(N)} < \dots < t_N^{(N)} = T$, N =

1,2,..., be a sequence of partitions of [0,T] with mesh tending to 0. For $z \in C^r$ put

$$F^N(z)(t) = \tilde{F}^{(N)}_{j-1}(z) \quad \text{for} \quad t^{(N)}_{j-1} < t \leq t^{(N)}_j ,$$

(and $F^N(z)(0) = F^N(z)(0+)$), where $\tilde{F}^{(N)}_{j-1}$ is any $\mathcal{C}^r_{t^{(N)}_{j-1}}$-measurable random vector such that

$$\tilde{F}^{(N)}_{j-1}(z) = F(z)(t^{(N)}_{j-1}) \qquad P^W\text{-a.e.}$$

Then the process $(t,z) \to F^N(z)(t)$ is progressively measurable. Let $\xi = F(w)$ and denote by E^W (E) expectation with respect to P^W (P). Then

$$\begin{aligned}
E^W \|F^N - F\|_T &= E^W \sup_{\substack{t^{(N)}_{j-1} < t \leq t^{(N)}_j \\ j=2,\dots,N}} |F(\cdot)(t^{(N)}_{j-1}) - F(\cdot)(t)| \\
&= E \sup_{\substack{t^{(N)}_{j-1} < t \leq t^{(N)}_j \\ j=2,\dots,N}} |F(w(\cdot))(t^{(N)}_{j-1}) - F(w(\cdot))(t)| \\
&= E \sup_{\substack{t^{(N)}_{j-1} < t \leq t^{(N)}_j \\ j=2,\dots,N}} |\xi_{t^{(N)}_{j-1}} - \xi_t| \\
&\leq E \sup_{|s-t| \leq \text{mesh}\,\{t^{(N)}_j\}} |\xi_s - \xi_t| \\
&\to 0
\end{aligned}$$

by virtue of Lemma (A.1). Hence $\|F^N - F\|_T \to 0$ in $L^1(dP^W)$. Let (N') be a subsequence such that

(A5) $$\|F^{N'} - F\|_T \to 0 \text{ on } \Gamma ,$$

where Γ is a $\mathcal{C}_T$-measurable set of P^W-measure 1. Define

$$\tilde{\Xi}(t,z) = \begin{cases} \lim\limits_{N'\to\infty} F^{N'}(z)(t) & \text{where the (finite) limit exists,} \\ 0 & \text{else.} \end{cases}$$

Then, $\tilde{\Xi}$ is a progressively measurable process and, by virtue of (A5),

$$||\tilde{\Xi}(\cdot,z) - F(z)||_T = 0 \quad \text{for } z \in \Gamma \ ,$$

which shows in particular that $\tilde{\Xi}$ has P^W - a.e. continuous trajectories.

<u>*Remark*</u> 7. If we are dealing with stochastic differential equations of the type (2.5) with $G(t)G'(t)$ uniformly positive definite, then the strong solution may equally well be expressed in terms of the paths $(\int_0^t G(s)dw_s)$:

$$\xi_t = \tilde{\Xi}(t,w) = \Xi(t, \int_0^{\cdot} G(s)dw_s), \quad 0 \leq t \leq T, \qquad \text{a.e.,}$$

where $\Xi(t,y)$ is measurable and adapted to $(\mathcal{C}_t^k)$.

References

[1] V.E. Beneš: Composition and invariance methods for solving some stochastic control problems, *Adv. Appl. Prob.* 7 (1975), 299-329.

[2] V.E. Beneš, I. Karatzas: Examples of optimal control for partially observable systems: Comparison, classical and martingale methods, *Stochastics* 5 (1981), 43-64.

[3] P. Billingsley: *Convergence of Probability Measure*, Wiley, New York, 1968.

[4] J.M. Bismut: Théorie probabiliste du contrôle des diffusions, *Mem. AMS* 1967, vol. 4 (1976).

[5] N. Christopeit: The separation principle for partially observed linear control systems: a general framework. To appear in Springer Lect. Notes in Control and Inf. Sciences.

[6] N. Christopeit, K. Helmes: Optimal control for a class of partially observable systems, *Stochastics* 8 (1982), 17-38.

[7] N. Christopeit, K. Helmes: On Beneš' bang-bang control problem, *Appl. Math. Optimization* 9 (1982), 163-176.

[8] M.H.A. Davis: The separation principle in stochastic control via Girsanov solutions, *SIAM J. Control and Optimization* 14 (1976), 176-188.

[9] W.H. Fleming, E. Pardoux: Optimal control for partially observed diffusions, *SIAM J. Control and Optimization* 20 (1982), 261-285.

[10] W.H. Fleming, R. Rishel: *Deterministic and Stochastic Optimal Control*, Springer-Verlag, New York, 1975.

[11] I.I. Gihman, A.V. Skorohod: *The Theory of Stochastic processes III*, Springer-Verlag, Berlin, 1979.

[12] U.G. Haussmann: Optimal control of partially observed diffusions via the separation principle, in *Lect. Notes in Control and Inf. Sciences*, vol. 43 (1982), 302-311.

[13] N. Ikeda, S. Watanabe: *Stochastic Differential Equations and Diffusion Processes*, North-Holland Publ. Company, Amsterdam, 1981.

[14] G. Kallianpur: *Stochastic Filtering Theory*, Springer-Verlag, New York, 1980.

[15] R.S. Liptser, A.N. Shiryayev: *Statistics of Random Processes II*, Springer-Verlag, New York, 1977.

[16] J.-L. Menaldi: Le principle de séparation pour le problème de temps d'arrêt optimal, *Stochastics* 3 (1979), 47-59.

[17] T. Yamada, S. Watanabe: On the uniqueness of solutions of stochastic differential equations, *J. Math. Kyoto Univ.* 11 (1971), 155-167.

[18] A.J. Veretennikov: On strong solutions and explicit formulas for solutions of stochastic integral equations, *Math. USSR Sbornik* 39 (1981), 387-4O3.

[19] W.M. Wonham: On the separation theorem of stochastic control, *SIAM J. Control* 6 (1968), 312-326.

APPROXIMATIONS FOR DISCRETE-TIME PARTIALLY OBSERVABLE STOCHASTIC CONTROL PROBLEMS

Giovanni B. Di Masi
CNR-LADSEB and
Istituto di Elettrotecnica
Università di Padova
Padova, Italy

Wolfgang J. Runggaldier
Seminario Matematico
Università di Padova
Padova, Italy

Abstract: We consider a discrete-time stochastic control problem with partial observation. The usual approach of transforming such problem into an equivalent one with complete observation requires the solution of a generally difficult nonlinear filtering problem. We first define a particular class of problems, for which the filtering problem can be explicitly solved and δ-optimal controls computed. We then show that, under suitable assumptions, given any $\varepsilon > 0$, such controls lead to ε-optimal controls for the original problem.

INTRODUCTION

On a given probability space $(\Omega, \mathcal{F}, P)$ consider the partially observed process (x_t, y_t), $x_t, y_t \in \mathbb{R}$, defined for $t=0,1,\dots,T$ by

$$x_{t+1} = a(x_t, u_t) + v_t \quad ; \quad x_o \sim \mathcal{N}(0,1) \tag{1.a}$$

$$y_t = c(x_t) + w_t \quad ; \quad y_o = w_o \tag{1.b}$$

where $\{v_t\}$ and $\{w_t\}$ are independent white Gaussian noises, both independent of x_o, and $u = \{u_t\}$ is a sequence of admissible controls, namely such that, for each t, u_t takes values in a given set $U \subset \mathbb{R}$ and depends only on past and present observations $y^t := \{y_o, y_1, \dots, y_t\}$ and past controls $u^{t-1} := \{u_o, \dots, u_{t-1}\}$.
For each admissible control sequence u and given cost functions $r(x,u)$ and $b(x)$, which we assume bounded from below, define

$$v(u) = E\left\{ \sum_{s=0}^{T-1} r(x_s, u_s) + b(x_T) \right\} \tag{2}$$

The problem is to find an admissible control u^* such that

$$v(u^*) = v := \inf_u v(u) \tag{3}$$

The sequence u^*, if it exists, will be called optimal, and v the optimal value. For problem (1)-(2) an optimal control does not exist in general. Defining however as ε-optimal a control u^ε such that

$$v(u^\varepsilon) \leqslant v + \varepsilon \qquad (4)$$

it can be shown [5, Sec.8.4. and p.83] that an ε-optimal control exists for all $\varepsilon > 0$. The purpose of this study is to derive a method for computing such a control.

A first step in approaching a stochastic control problem with incomplete observation as (1)-(2) is (see f.e. [1,5]) to transform it into an equivalent problem with complete observation by taking as new state at time t the conditional density of x_t, given y^t and u^{t-1}. One of the difficulties that arise with this approach is to determine the transition law for the new state, as this amounts to solving for each given control sequence u a nonlinear filtering problem for model (1).

Nonlinear filtering problems of this type have been studied by the authors in [3]. the approach used there, which will be synthesized in Section 1, consists in determining a particular class of filtering problems for which an explicitly computable finite-dimensional filter can be obtained and then approximating the optimal filter for model (1) by those obtained for the particular class.

For the stochastic control problem (1)-(2) in general only ε-optimal controls exist and their computation is formally equivalent to determining an approximate optimal control. Motivated by the filtering results in [3] our approach consists in the following two steps:

First we define a particular class of problems of type (1)-(2) for which the associated filtering problem admits an explicit finite-dimensional solution and the equivalent complete-observation problem admits, for each $\delta > 0$, an explicitly computable δ-optimal control. This will be the subject of Section 2.

Second, for each problem (1)-(2) we define a sequence of problems in the particular class with the property that, given $\varepsilon > 0$, it is possible to find an element in the sequence and a $\delta > 0$ so that a δ-optimal control for the latter is ε-optimal for the former. This will be the subject of the last Section 3.

1. APPROXIMATIONS TO THE FILTERING PROBLEM

Consider now the filtering problem for model (1), where we drop the dependence on the control u. It consists in evaluating $g(x_t|y^t)$, namely the conditional density of x_t given y^t, where

$$x_{t+1} = a(x_t) + v_t \tag{5.a}$$

$$y_t = c(x_t) + w_t \tag{5.b}$$

Instead of calculating the entire conditional density $g(x_t|y^t)$, we may also consider the problem of determining for a given Borel function h the conditional expectation $E\{h(x_t)|y^t\}$ which represents the least squares estimate of $h(x_t)$ given y^t.
The solution to a discrete-time nonlinear filtering problem may most conveniently be obtained by means of the following unnormalized version of the recursive Bayes formula (discrete-time Zakai equation [2])

$$q_t(x_t|y^t) = \psi(t,x_t,y_t)\int p(t,x_t;t-1,x_{t-1})q_{t-1}(x_{t-1}|y^{t-1})dx_{t-1} \tag{6}$$

where for model (5)

$$\psi(t,x_t,y_t) = \exp\left[c(x_t)y_t - \frac{1}{2}c^2(x_t)\right] \tag{7}$$

$$p(t,x_t;\ t-1,x_{t-1}) = \exp\left[-\frac{1}{2}(x_t-a(x_{t-1}))^2\right] \tag{8}$$

and $q_t(x_t|y^t)$ is such that

$$g(x_t|y^t) = \frac{q_t(x_t|y^t)}{\int q_t(x_t|y^t)dx_t} \tag{9}$$

Notice that equation (6) not only involves an integral, which may be difficult to compute, but this integral is also parametrized by x_t. Consider now the following particular class of problems of type (5)

$$x_{t+1} = \sum_{i=1}^{n} a_i\, I_{B_i}(x_t) + v_t \tag{10.a}$$

$$y_t = \sum_{i=1}^{n} c_i\, I_{B_i}(x_t) + w_t \tag{10.b}$$

where $\{B_i\}$ is a finite partition of the real line into n intervals $B_i=(b_{i-1},b_i]$, $(b_o=-\infty, b_n=+\infty)$ and I_{B_i} denote their indicator functions. It is not difficult to show [3] that for model (10) the relation (6) admits the following finite-dimensional solution

$$q_t(x_t\ y^t) = \exp\left[\sum_{j=1}^{n} I_{B_j}(x_t)(c_j y_t - \frac{1}{2} c_j^2)\right] \cdot$$

$$\cdot \sum_{i=1}^{n} d_i(y^{t-1}) \exp\left[-\frac{1}{2}(x_t-a_i)^2\right] \tag{11}$$

where $d_i(y^{t-1})$ are computed recursively as

$$d_i(y^{t-1}) = \sum_{j=1}^{n} d_j(y^{t-2}) \exp\left[c_i y_{t-1} - \frac{1}{2} c_i^2\right] \cdot$$

$$\cdot \left[\operatorname{erf}\left(\frac{1}{\sqrt{2}}(b_i-a_j)\right) - \operatorname{erf}\left(\frac{1}{\sqrt{2}}(b_{i-1}-a_j)\right)\right] \tag{12}$$

for $t>1$, where $\operatorname{erf}(x) = \frac{2}{\pi} \int_0^x \exp(-t^2)dt$, and

$$d_i(y_o) = \int_{b_{i-1}}^{b_i} \exp\left[-\frac{1}{2} x_o^2\right] dx_o \tag{13}$$

We now approximate the optimal filter for the general model (5) by that for a model in the particular class (10). In [3] this is done when the filtering problem consists in determining $E\{h(x_t) \mid y^t\}$ under the assumptions:

A.1.1. The functions a, c in (5) and the function h are Lipschitz-continuous.

A.1.2. $\lim_{x\to\pm\infty} a(x) < +\infty$, $\lim_{x\to\pm\infty} c(x) < +\infty$.

A.1.3. The function h is bounded.

Furthermore, the approximating problem is constructed in such a way that the observation process y_t is the same as in the original problem (notice that y_t provides the only available data).
The approach used consists in first constructing sequences of step functions $\{a^{(n)}\}$ and $\{c^{(n)}\}$ converging uniformly to a and c respectively and defining processes $\{x_t^{(n)}\}$ by

$$x_{t+1}^{(n)} = a^{(n)}(x_t^{(n)}) + v_t \quad ; \quad x_o^{(n)} = x_o \tag{14}$$

Then define probability measures P_o and $P^{(n)}$ $(n \in \mathbb{N})$ on $(\Omega, \mathcal{F})$ by means of their Radon-Nikodym derivatives

$$\frac{dP_o}{dP} = \prod_{s=1}^{T} \exp\left[-c(x_t)y_t + \frac{1}{2} c^2(x_t) \right] \tag{15}$$

$$\frac{dP^{(n)}}{dP_o} = \prod_{s=1}^{T} \exp\left[c^{(n)}(x_t^{(n)})y_t - \frac{1}{2} c^{(n)^2}(x_t^{(n)}) \right] \tag{16}$$

It can be shown that under P_o the distributions of $\{x_t\}$ and $\{x_t^{(n)}\}$ remain the same as under P, and $\{y_t\}$ becomes Gaussian white noise independent of $\{x_t\}$ and $\{x_t^{(n)}\}$. Furthermore, under $P^{(n)}$ the distributions of $\{x_t\}$ and $\{x_t^{(n)}\}$ still remain the same, while $\{y_t\}$ satisfies

$$y_t = c^{(n)}(x_t^{(n)}) + \tilde{w}_t \; ; \quad y_o = \tilde{w}_o \tag{17}$$

where $\tilde{w}_t$ is white Gaussian noise.
On the spaces $(\Omega, \mathcal{F}, P^{(n)})$ then consider the sequence of models of the particular type (10)

$$x_{t+1}^{(n)} = a^{(n)}(x_t^{(n)}) + v_t \; ; \quad x_o^{(n)} = x_o \tag{18.a}$$

$$y_t = c^{(n)}(x_t^{(n)}) + \tilde{w}_t \; ; \quad y_o = \tilde{w}_o \tag{18.b}$$

For each of the models (18) it is then possible to explicitly compute

$$E^{(n)}\{ h(x_t^{(n)}) \mid y^t \} \tag{19}$$

where $E^{(n)}$ denotes expectation with respect to the measure $P^{(n)}$.
Under the assunptions A.1.1.-A.1.3. it is shown in [3] that

$$\left| E^{(n)}\{ h(x_t^{(n)}) \mid y^t \} - E\{ h(x_t) \mid y^t \} \right| \leqslant H^{(n)}(y^t) \tag{20}$$

where the bound $H^{(n)}(y^t)$ can be explicitly computed and $\lim_{n\to\infty} H^{(n)}(y^t) = 0$ for all y^t.
For the proof of (20) both filters are expressed by means of the so-called Kallianpur-Striebel formula, namely

$$E\{ h(x_t) \mid y^t \} = \frac{E_o\{ h(x_t) L_t \mid y^t \}}{E_o\{ L_t \mid y^t \}} \tag{21}$$

$$E^{(n)}\left\{ h(x_t^{(n)}) \mid y^t \right\} = \frac{E_o\left\{ h(x_t^{(n)}) L_t^{(n)} \mid y^t \right\}}{E_o\left\{ L_t^{(n)} \mid y^t \right\}} \tag{22}$$

where P_o is the common reference probability measure defined in (15) and

$$L_t = \prod_{s=1}^{t} \exp\left[c(x_t) y_t - \frac{1}{2} c^2(x_t) \right] \quad ; \quad L_o = 1 \tag{23}$$

$$L_t^{(n)} = \prod_{s=1}^{t} \exp\left[c^{(n)}(x_t^{(n)}) y_t - \frac{1}{2} c^{(n)^2}(x_t^{(n)}) \right] \; ; \; L_o^{(n)} = 1 \tag{24}$$

2. A PARTICULAR CLASS OF CONTROL PROBLEMS

Consider the following particular case of problem (1)-(2)

$$x_{t+1} = \sum_{i=1}^{n} \sum_{k=1}^{n} a_i^k I_{B_i}(x_t) I_{U_k}(u_t) + v_t \tag{25.a}$$

$$y_t = \sum_{i=1}^{n} c_i I_{B_i}(x_t) + w_t \tag{25.b}$$

$$v(u) = E\left\{ \sum_{t=0}^{T-1} \left[\sum_{i=1}^{n} \sum_{k=1}^{n} r_i^k I_{B_i}(x_t) I_{U_k}(u_t) \right] + \right.$$

$$\left. + \sum_{i=1}^{n} b_i I_{B_i}(x_T) \right\} \tag{26}$$

where a_i^k, r_i^k, b_i, c_i $(i,k=1,\dots,n)$ are given real numbers and $\{B_i\}$, $\{U_k\}$ are intervals which constitute finite partitions of the real line and of U respectively.

Denoting by $\pi_t(k)$ and π_t the n-vectors whose components are

$$\pi_t^i(k) = P\{ x_t \in B_i \mid y^t, u^{t-2}, u_{t-1} \in U_k \} \quad \text{and}$$

$\pi_t^i = P\{ x_t \in B_i \mid y^t, u^{t-1} \}$ respectively, then, using the recursive Bayes formula (6) one has the following transition law

$$\pi_{t+1}^j(k) = \frac{\sum_{i=1}^{n} \pi_t^i p_{ij}(k) \exp\left[-\frac{1}{2} (y_{t+1} - c_j)^2 \right]}{\sum_{h=1}^{n} \sum_{i=1}^{n} \pi_t^i p_{ih}(k) \exp\left[-\frac{1}{2} (y_{t+1} - c_h)^2 \right]} \tag{27}$$

where

$$p_{ij}(k) = \frac{1}{\sqrt{2\pi}} \int_{B_j} \exp\left[-\frac{1}{2}(x-a_i^k)^2\right] dx \qquad (28)$$

It is clear from the particular structure of problem (25)-(26) that the vector π_t is a sufficient statistic in the sense that (see f.e. [1]) it contains all the information on the past history (y^t, u^{t-1}) which is relevant for control purposes. The vector π_t can therefore be considered as the solution to the filtering problem associated to (25)-(26), which by (27) can be explicitly computed by a finite dimensional procedure. Such vector π_t will now be taken as the state variable of the corresponding completely observed control problem, hereafter denoted by (P), thus obtaining as state space, denoted by Π, a compact subset of $\mathbb{R}^n$.

Notice that the denominator in (27) is the conditional density $g(y_{t+1}|y^t, u^{t-1}, u_t \in U_k)$ of y_{t+1} given y^t and u^t, which may also be interpreted as the density of y_{t+1} given π_t and u_t, so that the notations $g(y_{t+1} | y^t, u^{t-1}, u_t \in U_k)$ and $g(y_{t+1} | \pi_t, k)$ will be used alternatively.

Problem (P) is now characterized by the state transition law

$$\pi_{t+1}(k) = G(\pi_t, k, y_{t+1}) \qquad (29)$$

obtained from (27), by admissible control sequences $u = \{u_t\}$ such that u_t depends only on π_t, and cost functions given by

$$r(\pi_t, u_t) = \sum_{i=1}^{n} \pi_t^i \sum_{k=1}^{n} r_i^k \, I_{U_k}(u_t) \qquad (30.a)$$

$$b(\pi_T) = \sum_{i=1}^{n} \pi_T^i \, b_i \qquad (30.b)$$

In the rest of this Section it will be shown that, given any $\delta > 0$, it is possible to determine a δ-optimal control for (P). For this purpose we shall adapt to our situation the approximate dynamic programming approach of Hinderer [6] (see also [7]). It consists essentially in using the dynamic programming algorithm over a finite subset of the state space and then extending the thus obtained control law to the entire state space.

The control problem corresponding to such dynamic programming algorithm will be denoted by $(\tilde{P})$. For its characterization consider a finite partition $\{\Pi_j\}$ (j=1,...,m) of the state space Π and a set of representative vectors $\tilde{\Pi} = \{\tilde{\pi}_j \mid \tilde{\pi}_j \in \Pi_j,\ j=1,\dots,m\}$ such that $\pi_o \in \tilde{\Pi}$ where π_o is the n-vector whose components are given by $P\{x_o \in B_i\}$. The set $\tilde{\Pi}$ will be the state space of $(\tilde{P})$. Defining for arbitrary k, j (k=1,...,n; j=1,...,m) and $\pi \in \Pi$

$$Y_j^{\pi k} := \{ y \mid G(\pi,k,y) \in \Pi_j \}$$

we have for (P) the transition function

$$p(\pi,k,\Pi_j) = P\{\pi_{t+1} \in \Pi_j \mid \pi_t=\pi,\ u_t \in U_k\} = \int_{Y_j^{\pi k}} g(y \mid \pi,k)dy \qquad (31)$$

and it is natural to take for $(\tilde{P})$ the transition probabilities given by (h, j=1,...,n)

$$p(\tilde{\pi}_h,k,\tilde{\pi}_j) = P\{\tilde{\pi}_{t+1} = \tilde{\pi}_j \mid \tilde{\pi}_t=\tilde{\pi}_h, u_t \in U_k\} = \int_{Y_j^{\tilde{\pi} k}} g(y \mid \tilde{\pi}_h,k)dy \qquad (32)$$

Problem $(\tilde{P})$ is then characterized by the transition probabilities (32), admissible control sequences $u=\{u_t\}$ such that u_t depends only on $\tilde{\pi}_t$ and cost functions $r(\tilde{\pi}_t, u_t)$, $b(\tilde{\pi}_T)$ as given in (30). Since the state space of $(\tilde{P})$ is finite and the choice of a control $u_t \in U$ reduces to the choice of a k (k=1,...,n), an optimal control for $(\tilde{P})$, hereafter denoted by $\tilde{u}=\{\tilde{u}_t\}$, can actually be computed via dynamic programming.

Following Hinderer [6] we shall show in Theorem 1 below that $\tilde{u}$, when suitably extended to all of Π , is δ-optimal for (P) with δ tending to zero as the partition $\{\Pi_j\}$ becomes finer and finer. To this end define

$$\tilde{\pi}(\pi) = \sum_{j=1}^{m} \tilde{\pi}_j\, I_{\Pi_j}(\pi) \qquad (33)$$

and notice that any mapping h defined on $\tilde{\Pi}$, in particular $\tilde{u}$ and the optimal cost-to-go for $(\tilde{P})$, can be extended to all of Π by setting for all $\pi \in \Pi, h(\pi)=h(\tilde{\pi}(\pi))$. Furthermore, for a given partition $\{\Pi_j\}$ let

$$\Delta := \sup_{\pi \in \Pi} \max_{1 \leq i \leq n} |\pi^i - \tilde{\pi}^i(\pi)| \qquad (34)$$

and, for any partition $\{\Pi_j\}$ having the same Δ define the following quantities, where $r(\pi,k)=r(\pi,u)$ for $u \in U_k$,

$$K_{r+}^{\Delta} := \sup_{\pi,k} \{ r(\pi,k) - r(\tilde{\pi}(\pi),k) \} \qquad (35)$$

$$K_{r-}^{\Delta} := \inf_{\pi,k} \{ r(\pi,k) - r(\tilde{\pi}(\pi),k) \} \qquad (36)$$

$$K_{p}^{\Delta} := \sup_{\pi,k} \{ \sum_{j=1}^{m} | p(\pi,k,\Pi_j) - p(\tilde{\pi}(\pi),k,\Pi_j) | \} \qquad (37)$$

It then follows immediately that

$$K_{r+}^{\Delta} \leqslant \sup_{\pi,k} | r(\pi,k) - r(\tilde{\pi}(\pi),k) | \leqslant \sup_{\pi,k} \{ \sum_{i=1}^{n} |\pi^i - \tilde{\pi}^i(\pi)| \, |r_i^k| \} \leqslant nR\Delta \qquad (38)$$

where

$$R = \max_{1\leqslant i\leqslant n} \max_{1\leqslant k\leqslant n} |r_i^k| \qquad (39)$$

and analogously it follows that

$$- K_{r-}^{\Delta} \leqslant nR\Delta \qquad (40)$$

It can also be shown [4] that

$$K_p^{\Delta} \to 0 \quad \text{as} \quad \Delta \to 0 \qquad (41)$$

More precisely, the proof given in [4] implicitly suggests an upper bound on K_p^{Δ} that goes to zero as Δ goes to zero.
Defining, for any mapping $h(z)$,

$$\mathrm{sp}\{h(z)\} := \sup_z h(z) - \inf_z h(z) \qquad (42)$$

and denoting by $\tilde{v}_t(\tilde{\pi}_t)$ the optimal cost-to-go for $(\tilde{P})$ at time $t (t\leqslant T-1)$, namely

$$\tilde{v}_t(\tilde{\pi}_t) = \inf_u E \{ \sum_{s=t}^{T-1} r(\tilde{\pi}_s,u_s) + b(\tilde{\pi}_T) \mid \tilde{\pi}_t \} \quad , \qquad (43)$$

we now have
<u>Theorem 1</u>: If $(\tilde{P})$ corresponds to a partition $\{\Pi_j\}$ of Π with a given $\Delta>0$, and $\tilde{u}$ is the optimal control for $(\tilde{P})$ extended to Π, then

$$|v(\tilde{u}) - v| \leqslant n\Delta\,(B+2TR) + TK_p^{\Delta}\,\mathrm{sp}\{b(\tilde{\pi})\} + \frac{T(T-1)}{2} K_p^{\Delta}\,\mathrm{sp}\{\tilde{v}_{T-1}(\tilde{\pi}) - b(\tilde{\pi})\} \qquad (44)$$

where $B := \max_{1\leqslant i\leqslant n} |b_i|$

The Proof of Theorem 1 is based on [6; Thms 4.3, 4.4, 4.8] and can be found in [4].

Taking into account the comments following (41), the important consequence of Theorem 1 is that, given any $\delta > 0$, we can determine a partition $\{\Pi_j\}$ such that

$$|v(\tilde{u}) - v| \leq \delta \tag{45}$$

i.e. $\tilde{u}$ is δ-optimal for problem (P).

3. APPROXIMATIONS TO THE CONTROL PROBLEM

The main purpose of this Section is to construct for each problem (1)-(2) a problem in the particular class (25)-(26) such that a δ-optimal control for the latter is ε-optimal for the former. Analogously to the approach used in Section 1 we shall actually construct an entire sequence of problems in the particular class (25)-(26) and call the n-th element in the sequence the n-th approximating problem.

We shall use the following assumptions:

A.3.1.: There exist sequences of step functions $a^{(n)}(x,u)$, $c^{(n)}(x)$, $b^{(n)}(x)$, $r^{(n)}(x,u)$ such that $\|a^{(n)}(x,u)-a(x,u)\|$, $\|c^{(n)}(x)-c(x)\|$, $\|b^{(n)}(x)-b(x)\|$, $\|r^{(n)}(x,u)-r(x,u)\| \to 0$ as $n \to \infty$ where $\|\cdot\|$ denotes the sup-norm.

A.3.2.: The functions $c(x)$ and $b(x)$ are Lipschitz-continuous with Lipschitz constants L_c and L_b respectively. Furthermore, $a(x,u)$ and $r(x,u)$ are Lipschitz-continuous in x, uniformly in u, with Lipschitz-constants L_a and L_b respectively.

Furthermore, let B and C denote two constants such that $|b(x)| \leq B$, $|r(x,u)| \leq B$, $|c(x)| \leq C$, and define $B^{(n)} := \|b^{(n)}(x)-b(x)\|$, $C^{(n)} := \|c^{(n)}(x)-c(x)\|$.

As is Section 1 we want the approximating problems to have the same observation process y_t as the original problem; this implies that also the admissible controls remain the same. To this end we again use the absolutely continuous transformations of probability measures (15), (16). Then, defining for a given control sequence $u = \{u_t\}$ the processes $\{x_t^{(n)}\}$ by

$$x_{t+1}^{(n)} = a^{(n)}(x_t^{(n)}, u_t) + v_t \quad ; \quad x_o^{(n)} = x_o, \tag{46}$$

it can be shown that under P_o the distributions of $\{x_t\}$ and $\{x_t^{(n)}\}$ remain the same as under P and $\{y_t\}$ becomes white Gaussian noise independent of $\{x_t\}$ and $\{x_t^{(n)}\}$. Furthermore, under $P^{(n)}$ the distributions of $\{x_t\}$ and $\{x_t^{(n)}\}$ still remain the same, while $\{y_t\}$ satisfies

$$y_t = c^{(n)}(x_t^{(n)}) + \tilde{w}_t \quad ; \quad y_o = \tilde{w}_o \tag{47}$$

where $\tilde{w}_t$ is white Gaussian noise.
Notice that, using A.3.1., one easily has for a given control u

$$|x_t^{(n)} - x_t| \leqslant A_t^{(n)} \quad , \quad t = 1,\ldots, T \tag{48}$$

where

$$A_t^{(n)} = \|a^{(n)} - a\| \left[\sum_{s=0}^{t-1} L_a^s \right] \tag{49}$$

which does not depend on u.
On the spaces $(\Omega, \mathscr{F}, P^{(n)})$ we now consider the following stochastic control problems with incomplete observation

$$x_{t+1}^{(n)} = a^{(n)}(x_t^{(n)}, u_t) + v_t \quad ; \quad x_o^{(n)} = x_o \tag{50.a}$$

$$y_t = c^{(n)}(x_t^{(n)}) + \tilde{w}_t \quad ; \quad y_o = \tilde{w}_o \tag{50.b}$$

$$v^{(n)}(u) = E^{(n)}\left\{ \sum_{t=0}^{T-1} r^{(n)}(x_t^{(n)}, u_t) + b^{(n)}(x_T^{(n)}) \right\} \tag{51}$$

where $E^{(n)}$ denotes expectation with respect to $P^{(n)}$. Furthermore, let $v^{(n)} = \inf_u v^{(n)}(u)$. Problems (50)-(51) are of the same type as in the previous Section 2 and can therefore be solved to obtain a δ-optimal control sequence. Such control, hereafter denoted by $u^{n,\delta}$, can then be applied to the original problem (1)-(2) and it remains to show that for $\varepsilon > 0$ given, an $n \in \mathbb{N}$ and a $\delta > 0$ can be found such that

$$v(u^{n,\delta}) \leqslant v + \varepsilon \tag{52}$$

We now have the following Proposition, whose proof is in [4].

Proposition 1 : If for all admissible controls u

$$|v^{(n)}(u) - v(u)| \leqslant \varepsilon/2-\delta \ , \quad (0<\delta<\varepsilon/2) \tag{53}$$

then

$$|v(u^{n,\delta})-v| < \varepsilon$$

□

Using Proposition 1, our problem now reduces to that of finding, for $\varepsilon > 0$ and $\delta\,(0<\delta<\varepsilon/2)$ given, an $n\in\mathbb{N}$ such that for any control u relation (53) holds. This is made possible by the results in Theorem 2 and its Corollary below.

For the statement of Theorem 2 let $v_t(u;y^t)$ and $v_t^{(n)}(u;y^t)$ denote, for the original problem (1)-(2) and the n-th approximation respectively, the cost-to-go at time t having fixed a control $u=\{u_t\}$ and observed y^t, namely

$$v_t(u;y^t) = E\{\sum_{s=t}^{T-1} r(x_s,u_s)+b(x_T) \mid y^t\} \tag{54}$$

$$v_t^{(n)}(u;y^t) = E^{(n)}\{\sum_{s=t}^{T-1} r^{(n)}(x_s^{(n)},u_s)+b^{(n)}(x_T^{(n)}) \mid y^t\} \tag{55}$$

Furthermore, let for t=0,...,T,

$$z_{1,t}^{(n)} := L_b A_t^{(n)} + B^{(n)} \tag{56}$$

$$z_{2,t}^{(n)} := 2(L_c A_t^{(n)} + C^{(n)}) \tag{57}$$

$$z_{3,t}^{(n)} := 2tC(L_c A_t^{(n)} + C^{(n)}) \tag{58}$$

$$Q_t := \exp\left[\frac{1}{2}\, t\, C^2\right] \tag{59}$$

and, for $q \in \mathbb{N}$,

$$\alpha_q := 2\exp\left[\frac{(qC)^2}{2}\right] \tag{60}$$

$$\beta_q := 2\exp\left[\frac{(qC+1)^2}{2}\right] \tag{61}$$

Theorem 2: Under A.3.1. and A.3.2., for t=0,...,T-1 and all u

$$|v_t^{(n)}(u;y^t)-v_t(u;y^t)| \leqslant \sum_{h=1}^{T-t} R_{h,t}^{(n)} \; H_t^{4h}(y^t)+ S_{h,t}^{(n)} \; K_t^{(4h)}(y^t) \qquad (62)$$

where

$$H_t(y^t) = \exp\left[c \sum_{s=1}^{t} |y_s|\right], \quad t=1,\dots,T-1; \; H_o(y_o)=1$$

$$K_t^{(q)}(y^t) = H_t^q(y^t)\left[\sum_{s=1} |y_s|\right], \quad t=1,\dots,T-1; \; K_o^{(q)}(y_o)=0$$

and $R_{h,t}^{(n)}$ and $S_{h,t}^{(n)}$ can be computed recursively as follows:

for t=T-1

$$R_{1,T-1}^{(n)} = \left[Z_{1,T}^{(n)} + Z_{1,T-1}^{(n)} + 4B\, Z_{3,T-1}^{(n)}\right] Q_{T-1}^2$$

$$S_{1,T-1}^{(n)} = 4B\, Z_{2,T-1}^{(n)}\, Q_{T-1}^2$$

for t < T-1, h=1

$$R_{1,t}^{(n)} = \Big[(T-t)B\;[\beta_1 Z_{2,t+1}^{(n)} + \alpha_1 Z_{3,t+1}^{(n)} + \alpha_1 Z_{3,t}^{(n)}] +$$

$$+ Z_{1,t}^{(n)} + 2BZ_{3,t}^{(n)}\Big] Q_t^2$$

$$S_{1,t}^{(n)} = \Big[(T-t)B\;[\alpha_1 Z_{2,t+1}^{(n)} + \alpha_1 Z_{2,t}^{(n)}] + 2BZ_{2,t}^{(n)}\Big] Q_t^2$$

for t < T-1, h > 1

$$R_{h,t}^{(n)} = \left[\alpha_{4h-3} R_{h-1,t+1}^{(n)} + \beta_{4h-3} S_{h-1,t+1}^{(n)}\right] Q_t^2$$

$$S_{h,t}^{(n)} = \alpha_{4h-3} S_{h-1,t+1}^{(n)}\, Q_t^2$$

Proof (Outline only) The proof involves lenghty calculations, so we refer to [4] mentioning here only the basic steps involved. We have for all u

$$|v_t^{(n)}(u;y^t)-v_t(u;y^t)| \leqslant |E^{(n)}\{ r^{(n)}(x_t^{(n)},u_t)| y^t\} -E\{r(x_t,u_t)|y^t\}| +$$

$$+ |E^{(n)}\{ v_{t+1}^{(n)}(u;y^{t+1})| y^t\} - E\{v_{t+1}(u;y^{t+1})| y^t\}| \qquad (63)$$

so that we only need a bound on each of the two expressions on the right. These correspond to the errors in the approximation to a filtering and prediction problem respectively. Using the common reference probability measure P_o, both filter and predictor can be expressed by means of the so-called Kallianpur-Striebel formula, so that a bound can be obtained as for the filtering problem in Section 1.

Remark. Notice that the quantities $R_{h,t}^{(n)}$ and $S_{h,t}^{(n)}$ go to zero with $\|a^{(n)}-a\|$, $\|c^{(n)}-c\|$, $\|b^{(n)}-b\|$, $\|r^{(n)}-r\|$, implying that the bound on the right of (62) goes to zero as $n \to \infty$.

Corollary: Under A.3.1. and A.3.2. for all u

$$|v^{(n)}(u) - v(u)| \leqslant \sum_{h=1}^{T} R_{h,0}^{(n)}$$

□

Using the previous Remark, it follows from Proposition 1 and the above Corollary to Theorem 2, that, given $\varepsilon > 0$ and $\delta > 0$ with $\delta < \varepsilon/2$, it is possible to find $n \in \mathbb{N}$ such that a δ-optimal control for the n-th approximating problem (50)-(51) is ε-optimal for the original problem (1)-(2).

REFERENCES

1. Bertsekas D.P., Shreve S.E. "Stochastic optimal control: the discrete-time case", Academic Press, 1978

2. Di Masi G.B., Runggaldier W.J. "On measure transformations for combined filtering and parameter estimation in discrete time", Systems & Control Letters 2, pp.57-62, 1982

3. Di Masi G.B., Runggaldier W.J. "Approximations and bounds for discrete-time nonlinear filtering" in "Analysis and Optimization of Systems" (Bensoussan A., Lions J.L., eds.) L.N. in Control and Information Sciences 44, pp.191-202, Springer-Verlag, 1982

4. Di Masi G.B., Runggaldier W.J. "An approach to discrete-time stochastic control problems under partial observation" CNR-LADSEB Int. Rept. 02-83, 1983

5. Dynkin E.B., Yushkevich A.A. "Controlled Markov processes" Springer-Verlag, 1979

6. Hinderer K. "On approximate solutions of finite-stage dynamic programs" in "Dynamic programming and its applications" (Puterman M. ed.), pp. 289-317, Academic Press, 1979

7. Whitt W. "Approximations of dynamic programs I" Math. Oper. Res. 3, pp. 231-243, 1978.

Nonexistence of finite dimensional filters for conditional statistics of the cubic sensor problem

by

M. Hazewinkel[*], S.I. Marcus[**] & H.J. Sussmann[***]

ABSTRACT

Consider the cubic sensor $dx = dw$, $dy = x^3dt + dv$ where w, v are two independent brownian motions. Given a function $\phi(x)$ of the state x let $\hat{\phi}_t(x)$ denote the conditional expection given the observations y_s, $0 \leq s \leq t$. This paper consists of a rather detailed discussion and outline of proof of the theorem that for nonconstant ϕ there can not exist a recursive finite dimensional filter for $\hat{\phi}$ driven by the observations.

CONTENTS

1. Introduction
2. System theoretic part I: Precise formulation of the theorem
3. System theoretic part II: The homomorphism principle and outline of the proof (heuristics)
4. Analytic part
5. System theoretic part III: realization theory
6. Algebraic part
7. Putting it all together and concluding remarks

KEY WORDS & PHRASES: *cubic sensor, recursive filter, robust filtering, Weyl Lie algebra*

---

*) Mathematical Centre, P.O.Box 4079, 1009 AB Amsterdam.

**) Dept. Electrical Engr. Univ. of Texas at Austin, Texas. Supported in past by the National Science Foundation under Grant ECS-8022033 and in past by the Joint Services Electronics Program under Contract F49620-77-c-0101.

***) Dept. Math., Rutgers Univ. New Brunswick, New Jersey.

1. INTRODUCTION

The cubic sensor problem is the problem of determining conditional statistics of the state of a one dimensional stochastic process $\{x_t: t \geq 0\}$ satisfying

$$(1.1) \qquad dx = dw, \quad x_0 = x^{in}$$

with w a Wiener process, independent of x^{in}, given the observation process $\{y_t: t \geq 0\}$ satisfying

$$(1.2) \qquad dy = x^3 dt + dv, \quad y_0 = 0$$

where v is another Wiener process independent of w and x^{in}. Given a smooth function $\phi: \mathbb{R} \to \mathbb{R}$ let $\hat{\phi}_t$ denote the conditional expection

$$(1.3) \qquad \hat{\phi}_t = \hat{\phi}(x_t) = E[\phi(x_t) \mid y_s, \ 0 \leq s \leq t]$$

By definition a smooth finite dimensional recursive filter for ϕ_t is a dynamical system on a smooth finite dimensional manifold M governed by an equation

$$(1.4) \qquad dz = \alpha(z)dt + \beta(z)dy, \quad z_0 = z^{in}$$

driven by the observation process, together with an output map

$$(1.5) \qquad \gamma: M \to \mathbb{R}$$

such that, if z_t denotes the solution of (1.4),

$$(1.6) \qquad \gamma(z_t) = \hat{\phi}_t \quad \text{a.s.}$$

Roughly speaking one now has the theorem that for nonconstant ϕ such filters cannot exist. For a more precise statement of the theorem see 2.10 below.

It is the purpox of this note to give a fairly detailed outline of the proof of this theorem and to discuss the structure of the proof. That is the general principles underlying it. The full precise details of the analytic and realization theoretic parts of the proof will appear in [Sussmann 1983a, 1983b], the details of the algebraic part of the proof can be found in [Hazewinkel - Marcus, 1982]. An alternative much better and shorter proof of the hardest bit of the algebraic part will appear in [Stafford, 1983].

2. SYSTEM THEORETIC PART. I: PRECISE FORMULATION OF THE THEOREM

2.1 The setting

The precise system theoretic - probabilistic setting which we shall use for the cubic sensor filtering problem is as follows

(i) (Ω,A,P) is a probability space

(ii) $(A_t: 0 \le t)$ is an increasing family of σ-algebras

(iii) (w,v) is a two-dimensional standard Wiener process adapted to the A_t.

(iv) $x = \{x_t: t \ge 0\}$ is a process which satisfies $dx = dw$, i.e.

$$x_t = x_0 + w_t \qquad \text{a.s. for each } t \tag{2.1}$$

(v) x_0 is A_0-measurable and has a finite fourth moment

(vi) $\{y_t: t \ge 0\}$ is a process which satisfies $dy = x^3 dt + dv$, i.e.

$$y_t = \int_0^t x_s^3 ds + v_t \qquad \text{a.s. for each } t \tag{2.2}$$

(vii) the processes v, w, x, y all have continuous sample paths, so that in particular (2.1) and (2.2) actually hold and not just almost surely. (More precisely one can always find if necessary modified versions of v, w, x, y such that (vii) (also) holds).

2.3. The filtering problem

Let y_t, $t \ge 0$ be the σ-algebra generated by the y_s, $0 \le s \le t$ and let $\phi: \mathbb{R} \to \mathbb{R}$ be a Borel measurable function. Then the filtering problem (for this particular ϕ) consists of determining $E[\phi(x_t) \mid Y_t]$.

2.4. Smooth finite dimensional filters

Consider a (Fisk-Stratonovič) stochastic differential equation

(2.5) $$dz = \alpha(z)dt + \beta(z)dy, \quad z \in M,$$

where M is a *finite dimensional* smooth manifold and α and β are smooth vectorfields on M. Let there also be given an initial state and a smooth output map

(2.6) $$z^{in} \in M, \quad \gamma\colon M \to \mathbb{R}.$$

The equation (2.5) together with the initial condition $z(0) = z^{in}$ has a solution $z = \{z_t : t \geq 0\}$ defined up to a stopping time T, which satisfies

(2.7) $$0 < T \leq \infty \text{ a.s.}, \quad \{\omega \mid T(\omega) > t\} \in Y_t, \text{ for } t \geq 0.$$

Moreover there is a unique maximal solution, i.e. one for which the stopping time T is a.s. $\geq T_1$ if T_1 is the stopping time of an arbitrary other solution z_1. In the following $z = \{z_t : t \geq 0\}$ denotes such a maximal solution.

The system given by (2.5), (2.6) is now said to be a smooth finite dimensional filter for the cubic sensor (2.1) (i) - (vii) if for y equal to the observation process (2.2) the solution z of (2.5) satisfies

(2.8) $$E[\phi(x_t) \mid Y_t] = \gamma(z_t) \text{ a.s. on } \{\omega \mid T(\omega) > t\}.$$

2.9. Statement of the theorem

With these notions the main theorem of this note can be stated as:

2.10. THEOREM. *Consider the cubic sensor 2.1. (i) - (vii); i.e. assume that these conditions hold. Let* $\phi\colon \mathbb{R} \to \mathbb{R}$ *be a Borel measurable function which satisfies for some* $\beta \geq 0$ *and* $0 \leq r < 4$

(2.11) $$|\phi(x)| \leq \exp(\beta |x|^r), \quad -\infty < x < \infty.$$

Assume that ϕ is not almost everywhere equal to a constant. Then there exists no smooth finite dimensional filter for the conditional statistic $E[\phi(x_t) \mid Y_t]$.

3. SYSTEM THEORETIC PART. II: THE HOMOMORPHISM PRINCIPLE AND OUTLINE OF THE PROOF (HEURISTICS)

3.1. The Duncan-Mortensen-Zakai equation

Consider a nonlinear stochastic dynamical system

$$dx_t = f(x_t)dt + G(x_t)dw_t, \quad x^{in} = x_0, \quad x_t \in \mathbb{R}^n, \quad w_t \in \mathbb{R}^m \tag{3.2}$$

where w_t is a standard Brownian motion independent of the initial random variable x^{in} and where f and G are appropriate vector valued and matrix valued functions. Let the observations be given by

$$dy_t = h(x_t)dt + dv_t, \quad y_t \in \mathbb{R}^p \tag{3.3}$$

where v_t is another standard Brownian motion independent of w and x^{in}. Let $\hat{x}_t$ denote the conditional expectation

$$\hat{x}_t = E[x_t \mid Y_t] = E[x_t \mid y_s, 0 \le s \le t] \tag{3.4}$$

where Y_t is the σ-algebra generated by the y_s, $0 \le s \le t$. Let p(x,t) be the density of $\hat{x}_t$ where it is assumed (for the purposes of this heuristic section) that p(x,t) exists and is sufficiently smooth as a function of x and t. Then an unnormalized version $\rho(x,t)$ satisfies the Duncan-Mortensen-Zakai equation

$$d\rho(x,t) = \Big(\frac{1}{2}\sum_{i,j}\frac{\partial^2}{\partial x_i \partial_j}((GG^T)_{ij} - \sum_i \frac{\partial}{\partial x_i} f_i - \frac{1}{2}\sum h_j^2\Big)\rho(x,t)dt \tag{3.5}$$

$$+ \sum h_j \rho(x,t)dy_{jt}, \quad \rho(x,0) = \text{density of } x^{in}$$

where $h_j = h_j(x)$ is the j-th component of h, $(GG^T)_{i,j}$ is the (i,j)-th entry of the product of the matrix G(x) with its transpose and $f_i = f_i(x)$ is the i-th component of f(x). The equation (3.5) is a stochastic partial differential equation in Fisk-Stratonovič form. In the case of the cubic sensor (2.1), (2.2) (or (1.1), (1.2)) the equation becomes

(3.6) $$d\rho(x,t) = (\frac{1}{2}\frac{d^2}{dx^2} - \frac{1}{2}x^6)\rho(x,t)dt + x^3\rho(x,t)dy$$

3.7. <u>The homomorphism principle</u>

Now assume for a given $\phi: \mathbb{R}^n \to \mathbb{R}$ we have a smooth finite dimensional filter

(3.8) $$dz = \alpha(z)dt + \sum_j \beta_j(z)dy_j, \quad z_0 = z^{in}, \quad \gamma: \mathbb{R}^n \to \mathbb{R}$$

to calculate the statistic $\hat{\phi}_t = E[\phi(x_t) \mid Y_t]$. I.e. $\hat{\phi}_t = \gamma(z_t)$ a.s. if z_t is the solution of (3.8). The equation (3.8) is to be interpreted in the Statonovič sense.

Then, very roughly, we have two ways to process an observation path $y^\omega: s \to y_s(\omega)$, $0 \le s \le t$ to give the same result. One way is by means of the filter (3.8), the other way is by means of the infinite dimensional system (3.5) (defined on a suitable space of functions) coupled with the output map

(3.9) $$\Phi: \psi \to (\int\psi(x)dx)^{-1}\int\psi(x)\phi(x)dx$$

Assuming that (3.8) is observable, deterministic realization theory [Sussmann 1977] then suggest that there exists a smooth map F from the reachable part (from $\rho(x,0)$) of (3.6) to the reachable part of (3.8), which takes the vectorfields of (3.6) to the vectorfields of (3.6) and which is compatible with the output maps γ and (3.9). The operators in (3.6) define linear vectorfields in the state space of (3.6) (a space of functions). Let $L_0, L_1, \ldots, L_p$ be the operators occuring in (3.5) so that $d\rho = L_0\rho dt + L_1\rho dy_1 + \ldots + L_p\rho dy_p$. The Lie algebra of differential operators generated by $L_0, \ldots, L_p$ is called the *estimation Lie algebra,* and is denoted $L(\Sigma)$. The idea of studying this Lie algebra to find out things about filtering problems is apparently due to both Brockett and Mitter, cf e.g. [Brockett 1981] and Mitter [1981] and the references in these two papers.

Let $L \mapsto \tilde{L}$ be the map which assigns to an operator the corresponding linear vectorfield (analogous to the map which assigns to an $n \times n$ matrix $A = (a_{ij})$ the linear vector field $\Sigma a_{ij} x_i \frac{\partial}{\partial x_j}$ as $\mathbb{R}^n$). Then $L \mapsto -\tilde{L}$ is a homomorphism of Lie algebras. Further F induces a homomorphism of Lie algebras dF: $\tilde{L}_0 \to \alpha$, $\tilde{L}_i \to \beta_i$, $i = 1, \ldots, p$. Thus the existence of a finite dimensional filter should imply the existence of a homomorphism of Lie algebras $L(\Sigma) \to V(M)$ where $V(M)$ is the Lie algebra of smooth vectorfields on a smooth finite dimensional manifold M. This principle, originally enunciated by Brockett, has come to be called the homomorphism principle.

3.10. Pathwise filtering (robustness)

As it stands to remarks in 3.7 above are quite far from a proof of the homomorphism principle. First of all (3.6) and (3.8) are stochastic differential equations and as such they have solutions defined only almost everywhere. The first thing to do to remedy this situation is to show that these equations make sense and have solutions pathwise so that they can be interpreted as processing devices which accept an observation path $y: [0,t] \to \mathbb{R}^p$ and produce outputs $\hat{\Phi}_t(y)$ as a result. Another reason for looking for pathwise robust versions which is most important for actual applications, lies in the observation that actual physical observation paths will be piece-wise differentiable and that the space of all such paths is of measure zero in the probability space of paths underlying (3.6) and (3.8). Cf. [Clark, 1978].

Another difficulty in using the remarks of 3.7 above to establish a general homomorphism principle lies in the fact that (3.6) evolves on an infinite dimensional state space. A different approach to the establishing of homomorphism principles (than the one used in this paper) is described in [Hijab, 1982].

3.11. On the proof of theorem 2.10

In this paper the following route is followed to establish the homomorphism principle for the case of the cubic sensor. First for suitable $\phi: \mathbb{R} \to \mathbb{R}$ it is established that there exists a robust pathwise version of the functional $\hat{\phi}_t$. More precisely if C_t is the space of continuous functions $[0,t] \to \mathbb{R}$ then it is shown that there exists a functional $\Delta_t^\phi: C_t \to \mathbb{R}$ such that (proposition 4.15)

$$(3.12) \qquad \hat{\phi}_t = \frac{\Delta_t^\phi(y)}{\Delta_t^1(y)} \qquad \text{a.s. if } \; y = y^\omega.$$

The next step is to show that $\Delta_t^\phi(y)$, $y \in C_t$ is given by a density $n_t(y)(x)$ so that $\Delta_t^\phi(y) = \int n_t(y)(x)\,\phi(x)dx$ and to show that $n_t(y)(x)$ is smooth (as a function of x).

The next step is to use that there exist (up to a stopping time) pathwise and robust solutions of stochastic differential equations like (3.8). Robustness of both (3.6) and (3.8) then gives the central equality (4.24) *anywhere* (not just a.s.), that is:

$$(3.13) \qquad \frac{\Delta_t^\phi(y)}{\Delta_t^1(y)} = \gamma(z_t(y)), \quad y \in C_t.$$

The next step is to prove results about the smoothness properties of the density $n_t(y)$ as a function of $t_1, \ldots, t_m$ for paths y such that $u = \dot{y}$ is of the bang-bang type: $u(s) = \bar{u}_m \in \mathbb{R}$ for $0 \le t < t_m$, equal to $\bar{u}_{m-1}$ for $t_m \le t < t_m + t_{m-1}$, etc. ... and to observe that $(t,x) \mapsto n_t(y)(x)$ satisfies the DMZ equation (3.6). This permits to write down and calculate the result of applying $\frac{\partial^m}{\partial t_1 \ldots \partial t_m}\Big|_{t_1 = \ldots = t_m = 0}$ to both sides of (3.13) and gives a relation of the type

$$(3.14) \qquad (A(\bar{u}_m) \ldots A(\bar{u}_1)\gamma)(z) = \tilde{L}(\bar{u}_m) \ldots \tilde{L}(\bar{u}_1)\Phi(\psi_z)$$

where $A(\bar{u})$ is the vectorfield $\alpha + \bar{u}\beta$, $L(\bar{u})$ the operator $L_0 + \bar{u}L_1 = (\frac{1}{2}\frac{d^2}{dx^2} - \frac{1}{2}x^6) + \bar{u}x^3$ and $\tilde{L}(\bar{u})$ the linear vectorfield associated to $L(\bar{u})$, Φ the functional (3.9), and ψ_z a function corresponding to z, cf. 5.10.

A final realization theoretic argument having to do with reducing the filter dynamical system (3.8) to an equivalent observable and reachable system then establishes the homomorphism principle in the case of the cubic sensor and the fact that if the homomorphism is zero ϕ was a constant.

The remaining algebraic part of the proof consists of two parts:

(i) a calculation of $L(\bar{z})$ for the cubic sensor. It turns out that $L(\bar{z})$ is in this case equal to the Heisenberg-Weyl algebra W_1 of all differential operators (any order) in x with polynomial coefficients.

(ii) the theorem that if V(M) is the Lie algebra of smooth vectorfields on a smooth finite dimensional manifold and $\alpha: W_1 \to V(M)$ a homomorphism of Lie algebras then $\alpha = 0$.

4. ANALYTIC PART

4.1. The space of functions E

Let E denote the space of all Borel functions $\phi: \mathbb{R} \to \mathbb{R}$ such that there exists constants $C \in \mathbb{R}$, $\alpha \in \mathbb{R}$ and r, $0 \leq r < 4$ such that

$$|\phi(x)| \leq C \exp(\alpha|x|^r) \quad \text{for all } x \in \mathbb{R}. \tag{4.2}$$

The space $E(\alpha,r)$ is the normed space of all Borel functions for which

$$||\phi||_{(\alpha,r)} = \sup\{|\phi(x)| \exp(-\alpha|x|^r) ; x \in \mathbb{R}\} \tag{4.3}$$

is finite. The space E is the union of the $E(\alpha,r)$ and is topologized as such, i.e. as the inductive limit of the $E(\alpha,r)$.

4.2. Some bounds

Let C_t be the space of continuous functions $[0,t] \to \mathbb{R}$ such that $y(0) = 0$ and C_t^1 the functions of class C^1 in C_t, and $C_t^{1\#}$ the functions $y \in C_t$ which are piecewise C^1. Let H be the space of functions on Ω

$$H = \bigcap_{1 \leq p < \infty} L^p(\Omega, A, P). \tag{4.5}$$

For the processes x and w of 2.1, β, $r \in \mathbb{R}$, and a given $y \in C_t$ we define

$$(4.6) \qquad U(\beta,r,t,y) = \exp(\beta|x_t|^r + y(t)x_t^3 - 3\int_0^t y(s)x_s^2 dw_s - 3\int_0^t y(s)x_s ds - \frac{1}{2}\int_0^t x_s^6 ds).$$

The reason for considering this expression becomes clearer below. The $U(0,0,t,y)$ occur in a slightly modified version of the Kallianpur-Stiebel formula for $\hat{\Phi}_t$. The very slight modification (an integration by parts to remove dy_s) sees to it that the formula makes pathwise sense for continuous sample paths $y_s(\omega)$. The number 4 of 4.1 above gets into the picture as a result of wanting to keep $E[\phi(x_t)U|0,0,t,y)]$ bounded for bounded y and t.

4.7. <u>Lemma</u>. $U(\beta,r,t,y) \in H$ *if* $0 \leq \beta$, $0 \leq r < 4$, $0 < t$, $y \in C_t$.

This is proved by straightforward estimates (and the Ito formula) using each time $-\frac{1}{8}\int_0^t x_s^6 ds$ to keep the contributions of each of the other terms in check.

4.8. <u>A robust version for $\hat{\Phi}_t = E[\phi(x_t) \mid Y_t]$</u>

Now let B denote the space of bounded Borel functions $\phi: \mathbb{R} \to \mathbb{R}$ endowed with the sup norm. Define for $\phi \in B$

$$(4.9) \qquad \langle N_t(y),\phi\rangle = E[\phi(x_t)U(0,0,t,y)].$$

This is well defined because $U(0,0,t,y) \in L^1(\Omega,A,P)$ and $\phi(x_t)$ is bounded. One now shows using 4.7 that the finite positive measure $N_t(y)$ has a density $n_t(y)(x) \in L^1(\mathbb{R})$. Now define for all $\phi \in E$

$$(4.10) \qquad \langle N_t(y),\phi\rangle = \int_{-\infty}^{+\infty} \phi(x)n_t(y)(x)dx$$

(which agrees (4.9) for $\phi \in B$). The functional $N_t(y)$: $E \to \mathbb{R}$ satisfies

$$(4.11) \qquad ||\langle N_t(y),\phi\rangle|| \leq C||\phi||_{(\beta,r)} \quad \text{for } \phi \in \varepsilon(\beta,r)$$

with C independent of y, t for bounded $||y||_t$ and t in a compact subset of $(0,\infty)$.

4.12. Proposition. $N_t(y)$ *depends continuously on* y, *for each fixed* t, $\phi \in E$.

Now define $\Delta_t^\phi : C_t \to \mathbb{R}$ by

$$\Delta_t^\phi(y) = \langle N_t(y), \phi\rangle. \tag{4.13}$$

Then $\Delta_t^\phi(y)$ is continuous and one shows that $\Delta_t^1(y) > 0$ for all $t > 0$ so that we can define

$$\delta_t^\phi(y) = \frac{\Delta_t^\phi(y)}{\Delta_t^1(y)}. \tag{4.14}$$

4.15. Proposition. *If* $\phi \in E$, $\omega \mapsto \delta_t^\phi(y_t(\omega))$, *where* $y_t(\omega)$ *is the* ω*-path of the process* y *of (2.1) is a version of* $\hat{\phi}_t = E[\phi(x_t) \mid y_t]$ *for* $t > 0$. *(So* δ_t^ϕ *is a robust pathwise version of* $\hat{\phi}_t$.*)*

This is proved via the Kallianpur-Stiebel formula which says in our case that if

$$k_t(\phi) = \int_\Omega \phi(x_t(\omega'))\exp\left(\int_0^t x_s^3(\omega')dy_s(\omega) - \frac{1}{2}\int_0^t x_s^6(\omega')ds\right)dP(\omega')$$

then

$$\hat{\phi}_t = \frac{k_t(\phi)}{k_t(1)} .$$

Indeed rewrite the term $\int_0^t x_s^3(\omega')dy_s(\omega)$ by means of partial integration as

$$\int_0^t x_s^3(\omega')dy_s(\omega) = x_t^3(\omega')y_t(\omega) - \int_0^t y_s(\omega)d(x_s^3(\omega')) =$$

$$= x_t^3(\omega')y_t(\omega) - 3\int_0^t y_s(\omega)x_s^2(\omega')dw_s(\omega') -$$

$$- 3\int_0^t y_s(\omega)x_s(\omega)ds$$

and it readily follows that $\int U(0,0,t,y_t(\omega))\phi(x_t(\omega'))dP(\omega') = \Delta_t^{\phi}(y)$ is a version of $k_t(\phi)$.

4.16. Smoothness properties of $n_t(y)(x)$

Let F be the space of all C^{∞}-functions $\phi\colon \mathbb{R} \to \mathbb{R}$ for which $\exp(\beta|x|^r)|\phi^{(k)}(x)|$ is bounded for all $\beta \geq 0$, $0 \leq r < 4$, $k \in \mathbb{N} \cup \{0\}$ and give F the topology defined by the family of norms

$$(4.17) \qquad ||\phi||_{F(\beta,r,k)} = \sup\{\exp(\beta|x|^r)\ |\phi^{(k)}(x)|\}.$$

4.18. Lemma. *If* $x_0 \in H$ *then* $n_t(y) \in F$ *for all* $t > 0$, $y \in C_t$ *and* $n_t(y)(x) > 0$ *for all* $x \in \mathbb{R}$, $t > 0$ *for* y *differentiable*.

This is approached by considering the derivatives of $N_t(y)$ defined by $\langle N_t(y)',\phi\rangle = \langle N_t(y),\phi'\rangle$ for smooth $\phi : \mathbb{R} \to \mathbb{R}$.

4.19. Robustness for the filter

Now consider a stochastic differential equation with output map and initial condition driven by the observation process y_t

$$(4.20) \qquad dz = \alpha(z)dt + \beta(z)dy_t, \quad z(0) = z^{in}, \quad z \mapsto \gamma(z), \quad z \in M$$

as we would have for a filter for $\hat{\phi}$ cf. 2.4 above. Equation (4.20) is to be interpreted in the Stratonovič sense. Let T be the stopping time for a maximal solution. Then, as was shown in [Sussman, 1978] these equations admit robust solutions in the following sense.

Consider the equation for $y \in C_t$

$$(4.21) \qquad dz = \alpha(z)dt + \beta(z)dy, \quad z(0) = z^{in}.$$

A curve $z\colon \tau \to z(\tau)$, $0 \leq \tau \leq t$ is said to be a solution of (4.21) if there exists a neighborhood U of y in C_t with the property that there is a continuous map $U \to C([0,t],M)$ $\tilde{y} \to z(\tilde{y})$ to the space of continuous curves in M such that $z(\tilde{y})$ is a solution of (4.21) in the usual sense for all $\tilde{y} \in U \cap C_t^1$ (so that the equation can be written as a usual differential equation) and $z(y) = z$.

With this notion of solutions the robustness result is:

4.22. THEOREM [Sussman, 1978]. *(i) Given any continuous* $y\colon [0,\infty) \to \mathbb{R}$, $y(0) = 0$, *there exists a time* $T(y) > 0$ *such that there is a unique solution* $\tau \to z(y)(\tau)$ *of (4.21). If* $T(y) < \infty$ *then* $\{z(y)(t)\colon 0 \le t < T(y)\}$ *is not relatively compact on* M.

(ii) If y *is a Wiener process with continuous sample paths defined on* (Ω,A,P) *and if* $y^{\omega}(t) = y_t(\omega)$, *then* $\omega \to T(y^{\omega})$ *is a version of the stopping time up to which the Stratonovič solution of (4.20) is defined and* $\omega \to z(y^{\omega})(t)$, $0 \le t < T(y^{\omega})$ *is a version of the solution* z_t *for each* $t > 0$.

In our setting y_t is not a Wiener process, but the same techniquess apply, and the same results hold.

In other words up to a stopping time, solutions of (4.20) exist pathwise, they are continuous as a function of the path and hence can be calculated as limits of solutions to the corresponding nonstochastic differential equations (4.21) for (piecewise) differential continuous y.

4.23. Everywhere equality of the robust filter output and the robust DMZ output and consequences

Now let (4.20) be a smooth filter for $\hat{\phi}$ in the sense of section 2.4 above and let $\phi \in E$. Choose the robust version of $\hat{\phi}_t$, i.e. the map $\omega \mapsto \delta_t^{\phi}(y^{\omega})$ and choose the robust solution $\omega \mapsto z(y^{\omega})(t)$ of (4.20). The fact that (4.20) is a filter for $\hat{\phi}_t$ says by definition that $\delta_t^{\phi}(y^{\omega}) = \gamma(z(y^{\omega})(t))$ for almost all ω such that $T(y^{\omega}) > t$. The robustness of the two versions now readily implies that

$$(4.24) \qquad \delta_t^{\phi}(y) = \gamma(z(y)(t))$$

holds everywhere whenever $t > 0$, $y \in C_t$, $T(y) > t$.

4.25. Smoothness properties of the family of densities $n_t(y)$

When y is piecewise C^1 and the initial probability density ν^{in} is in F the study of the measure $N_t(y)$ is much easier. (By modifying the data (Ω, A, P), etc. it can actually be arranged that ν^{in} is in F essentially by replacing ν^{in} with the density at a slightly later time $\tau < t$). Now integration by parts gives

$$(4.26) \qquad U(0,0,t,y) = \exp\left(\int_0^t x_s^3 \dot{y}(s)ds - \tfrac{1}{2}\int_0^t x_s^6 ds\right)$$

Now let $\phi: \mathbb{R} \to \mathbb{R}$ be of class C^2, then the differential of $\phi(x_t)U(0,0,t,y)$ can easily be computed to be

$$(4.27) \qquad [\phi'(x_t)dw_t + \tfrac{1}{2}\phi''(x_t)dt + (x_t^3\,\dot{y}(t) - \tfrac{1}{2}x_t^6)\phi(x_t)]U(0,0,t,y)$$

so that (if, say, ϕ has compact support), y_τ denoting the restriction of y to $[0,\tau]$

$$(4.28) \qquad E[\phi(x_t)U(0,0,t,y) - \phi(x_\tau)U(0,0,\tau,y_\tau)] =$$

$$\int_\tau^t \left(\frac{1}{2}\frac{d^2}{dx^2} + x^3\dot{y}(t) - \frac{x^6}{2}\right)\phi(x))\big|_{x=x_s} U(0,0,s,y_s)ds$$

and this in turn says that the densities $n_t(y)$ of $N_t(y)$, i.e. the functions $(t,x) \to n_t(y)(x)$ satisfy the partial differential equation

$$(4.29) \qquad \frac{\partial\rho}{\partial t} = \frac{1}{2}\frac{\partial^2\rho}{\partial x^2} + (x^3u(t) - \frac{x^6}{2})\rho, \quad \rho(0,x) = n_0(x)$$

where n_0, the initial density is in F and $u = \dot{y}$. One has

4.30. LEMMA. *Let* u *be piecewise continuous an* $[0,T]$, *and for each* $n_0 \in F$ *let* $\rho_{n_0,t}$ *be the function* $x \to \rho(t,x)$ *where* ρ *solves (4.29) then* $(n_0,t) \to \rho_{n_0,t}$: $F \times [0,T] \to F$ *is continuous.*

Now let

(4.31) $$L_0 = \frac{1}{2}\frac{d^2}{dx^2} - \frac{x^6}{2}, \quad L_1 = x^3,$$

considered as (differential) operators $F \to F$. For each constant $\bar{u}$ let $L(\bar{u}) = L_0 + \bar{u}L_1$ and let $\exp(tL(\bar{u}))\psi$ for $\psi \in F$ denote the solution of (4.29) with $u(\cdot) \equiv \bar{u}$, $n_0 = \psi$.

Let $K \subset \mathbb{R}^n$ be a convex subset with nonempty interior. A family $\{\psi(v): v \in K\}$ of elements of F is said to depend smoothly on v if $(x,v) \to \psi(v)(x)$ is a C^∞ function on $K \times \mathbb{R}$ and for each $\underline{m} = (m_1,\ldots,m_n)$, $v \mapsto (\frac{\partial^{\underline{m}}}{\partial v^{\underline{m}}}\psi)(v)$, $v \in K$ takes values in F and is a continuous map $K \to F$. One then has

4.32. LEMMA. $\{\exp(tL(\bar{u}))\psi(v)\}$ *depends smoothly on* (v,t) *if* $\psi(v), v \in K$ *is a smooth family.*

4.33. Corollary. *Let* $\bar{u}_1,\ldots,\bar{u}_m \in \mathbb{R}$. *Then if* $\psi \in F$ *the family* $\{\exp(t_1L(\bar{u}_1)) \ldots \exp(t_mL(\bar{u}_m))\psi : (t_1,\ldots,t_m) \in [0,\infty]^m\}$ *depends continuously on* $t_1,\ldots,t_m$; *moreover for each* $\underline{\mu} = (\mu_1,\ldots,\mu_m)$ *we have*

(4.34) $$\frac{\partial^{\underline{\mu}}}{\partial t^{\underline{\mu}}}(\exp(t_1L(\bar{u}_1)) \ldots \exp(t_mL(\bar{u}_m))\psi) =$$

$$L(\bar{u}_1)^{\mu_1}\exp(t_1L(\bar{u}_1))L(\bar{u}_2)^{\mu_2}\exp(t_2L(\bar{u}_2)) \ldots L(\bar{u}_m)^{\mu_m}\exp(t_mL(\bar{u}_m))\psi.$$

5. SYSTEM THEORETIC PART III: REALIZATION THEORY

5.1. Some differential topology on F

Let U be an open subset of the space of smooth functions F. A map $\Phi: U \to \mathbb{R}$ is said to be of class C^∞ if the function $\nu \to \Phi(\psi(\nu))$ is C^∞ in the usual sense for every family $\{\psi(v) : v \in K\}$ depending smoothly on v in the sense described in section 4 above. This class of functions is denoted $C^\infty(U)$. If λ is a continuous linear functioned on F then λ (restricted to any U) is of class C^∞. Note that $C^\infty(U)$ is closed under pointwise multiplication and division by functions in $C^\infty(U)$ which are everywhere nonzero.

Let L be a continuous linear operator on F, then L defines a "linear vectorfield"

$\tilde{L}: C^\infty(F) \to C^\infty(F)$ (and $C^\infty(U) \to C^\infty(U)$ for each U) defined by

$$(5.2) \qquad (\tilde{L}\Phi)(\psi) = \frac{d}{dt}\Big|_{t=0} \Phi(\psi + tL\psi)$$

This is completely analogous to the map which assigns to an $n \times n$ matrix $A = (a_{ij})$ the "linear vector field" $\Sigma a_{ij} x_i \frac{\partial}{\partial x_j}$. It is totally routine to check that

$$(5.3) \qquad [L_1, L_2]^\sim = -[\tilde{L}_1, \tilde{L}_2]$$

5.4. LEMMA. *Let* $\{\psi(t) : 0 \le t < \varepsilon\} \subset U$ *depend smoothly on* t *and let* $\dot{\psi}$ *be the t-derivative of* ψ. *Then for all* $\Phi \in C^\infty(U)$

$$(5.5) \qquad \frac{d}{dt}\Big|_{t=0} \Phi(\psi(t)) = \frac{d}{dt} \Phi(\psi(0) + t\dot{\psi}(0)).$$

In particular if L *is a continuous linear operator on* F *such that* $L\psi(0) = \dot{\psi}(0)$, *then*

$$(5.6) \qquad (\tilde{L}\Phi)(\psi(0)) = \frac{d}{dt}\Big|_{t=0} \Phi(\psi(t)).$$

Now let $U \subset F$ be the set of all $\psi \in F$ such that $\int_{-\infty}^{+\infty} \psi(x)dx > 0$ and let $\Phi: U \to \mathbb{R}$ be given by the kind of formula occuring in our conditional expectation expressions

$$(5.7) \qquad \Phi(\psi) = \frac{\int \phi(x)\psi(x)dx}{\int \psi(x)dx}.$$

For the smooth families $e^{tL(\bar{u})}\psi$, where $L(\bar{u})$ is as in 4.32 above, one finds

$$(5.8) \qquad (\tilde{L}(\bar{u})\Phi)(\psi) = \frac{d}{dt} \Phi(e^{tL(\bar{u})}\psi)$$

and repeating this

$$(5.9) \qquad \tilde{L}(\bar{u}_m) \ldots \tilde{L}(\bar{u}_1)\Phi(\psi) =$$

$$\frac{\partial^m}{\partial t_1 \partial t_2 \ldots \partial t_m}\Big|_{t_1 = \ldots = t_m = 0} \Phi(e^{t_1 L(\bar{u}_1)} \ldots e^{t_m L(\bar{u}_m)}\psi).$$

5.10. The Lie-algebraic inplications of the existence of a smooth filter

Now let us repeat these remarks for the more familiar case of vectorfields α, β on a smooth finite dimensional manifold M with for each $\bar{u} \in \mathbb{R}$, $A(\bar{u})$ the vectorfield $\alpha + \bar{u}\beta$. Let $\pi((\bar{u}_1,t_1),(\bar{u}_2,t_2),\ldots,(\bar{u}_m,t_m));z)$ be the result of letting z evolve on M along $A(\bar{u}_m)$ during time t_m, then along $A(\bar{u}_{m-1})$ during time t_{m-1}, ... Let $\gamma: M \to \mathbb{R}$ be a smooth function. We have of course

$$(5.11) \qquad (A(\bar{u})\gamma)(z) = \frac{d}{dt}\Big|_{t=0} \gamma(\pi((\bar{u},t);z))$$

and

$$(5.12) \qquad (A(\bar{u}_m)\ldots A(\bar{u}_1)\gamma)(z) =$$

$$\frac{\partial^m}{\partial t_1 \ldots \partial t_m}\Big|_{t_1=\ldots=t_m=0} \gamma(\pi((\bar{u}_1,t_1),\ldots,(\bar{u}_m,t_m);z).$$

Let $R \subset M$ be the set of all points in M which can be reached from z^{in} by means of these bang-bang-bang controls in time $< \bar{T}$ i.e. R is the set of all $\pi((\bar{u}_1,t_1),\ldots,(\bar{u}_m,t_m);z^{in})$ with $\Sigma\, t_i < \bar{T}$. Let $z \in R$ and choose a bang-bang control which steers z^{in} to z in time $\tau < \bar{T}$; let $\psi_z \in \mathcal{F}$ be the solution of the "control version" of the $\mathcal{DMZ}$ equation (4.29), with initial condition n_0, the density of ν^{in}. Then $\psi_z \in U$, because $\Delta_t^1(y) > 0$ (cf. just below proposition 4.12). Now let $\bar{u}_1,\ldots,\bar{u}_m$, $t_1,\ldots,t_m$ satisfy $|\bar{u}_i| = 1$, $|t_1,\ldots,t_m| < \bar{T} - \tau$, $t_i \geq 0$ and assume that (α,β,γ) on M define a smooth filter for a given $\phi \in \mathcal{E}$ in the sense of subsection 2.4. Let Φ be the corresponding functional (5.7). Then by (4.24) we have

$$(5.13) \qquad \Phi(e^{t_1 L(\bar{u}_1)} \ldots e^{t_m L(\bar{u}_m)} \psi_z) = \gamma(\pi((\bar{u}_1,t_1)),\ldots,(\bar{u}_m,t_m));z)$$

(and this was really the whole reason for establishing formula (4.24), that is the reason why we needed to prove the existence of a robust pathwise version $\hat{\phi}_t$).

Now let $\underline{\underline{A}}$ denote the free associative algebra on two generators a_-, a_+. Let $\underline{\underline{A}}_1$ be the associative algebra (under composition) of linear maps

$C^\infty(U) \to C^\infty(U)$ generated by $\tilde{L}(-1)$, $\tilde{L}(1)$ and let $\underline{\underline{A}}_2$ be the associative algebra (again under composition) of differential operators on M. Homomorphisms of associative algebras $\nu_1: \underline{\underline{A}} \to \underline{\underline{A}}_1$, $\nu_2: \underline{\underline{A}} \to \underline{\underline{A}}_2$ are defined by $\nu_1(a_-) = \tilde{L}(-1)$, $\nu_1(a_+) = \tilde{L}(1)$, $\nu_2(a_-) = A(-1)$, $\nu_2(a_+) = A(1)$.

Let $\underline{\underline{L}}$ denote the free Lie-algebra on the generators a_-, a_+ (viewed as a subalgebra of A) and let $\underline{\underline{L}}_1$, $\underline{\underline{L}}_2$ denote the Lie algebras generated respectively by $\tilde{L}(-1)$, $\tilde{L}(1)$ and $A(-1)$, $A(1)$ (as subalgebras of $\underline{\underline{A}}_1$ and $\underline{\underline{A}}_2$). Then of course we have induced homomorphisms $\nu_i: L \to L_i$

(5.14)

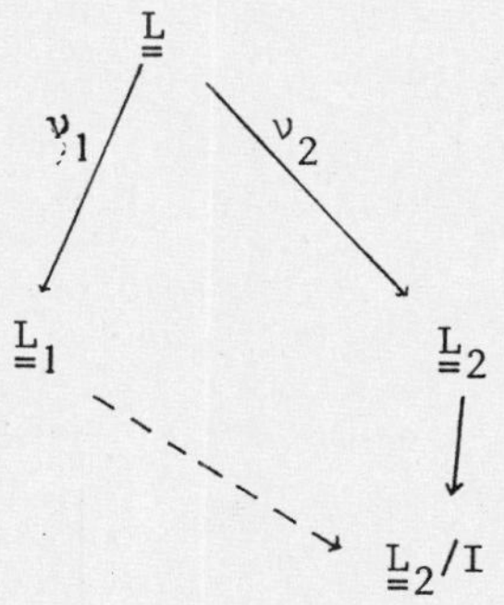

Let I denote the set of those vectorfields $V \in \underline{\underline{L}}_2$ such that

(5.15) $$([V_1,[V_2,[\ldots[V_m,V]\ldots]]\gamma)(z) = 0 \text{ for all } V_1,\ldots,V_m \in \underline{\underline{L}}_2,\quad m \in N \cup \{0\}$$

5.16. <u>LEMMA</u>. I *is an ideal and if* $a \in \underline{\underline{L}}$ *is such that* $\nu_1(a) = 0$ *then* $\nu_2(a) \in I$.

It follows that there is a homomorphism of Lie algebras $L_1 \to L_2/I$ making diagram (5.14) commutative.

The lemma is proved by combining (5.12) and (5.9) with (5.13) and this is why we needed to establish smoothness properties of families like $\exp(tL(\bar{u})\psi)$.

5.17. Foliations and such

The last step in this section is now to show that $\underline{\underline{L}}_2/I$, or more precisely a suitable quotient, is (isomorphic to) a subalgebra of a Lie algebra of vectorfields in a smooth finite dimensional manifold (a subquotient manifold of M) for suitable z. Let S be a set of smooth vectorfields on M. For each $z \in M$ consider $S(z) = \{V(z) : V \in S\}$. For $\underline{\underline{L}}_2$ and I, $\underline{\underline{L}}_2(z)$ and $I(z)$ are vectorspaces. Let k be the maximum of the dim $\underline{\underline{L}}_2(z)$ for $z \in R$ and k_0 the maximum of the dim $I(z)$, for $z \in R$ and dim $\underline{\underline{L}}_2(z) = k$. Choose a $\bar{z}$ in the relative interior of R such that dim $\underline{\underline{L}}_2(\bar{z}) = k$, dim $I(\bar{z}) = k_0$, and choose a neighborhood N of $\bar{z}$ (in M) such that dim $\underline{\underline{L}}_2(z) \geq k$, dim $I(z) \geq k_0$ for $z \in N$. (This can be done (obviously)). Suppose $\bar{z}$ can be reached from z^{in} in time $\tau < \bar{T}$. Now let M_0 be a connected submanifold of N of which all points are reachable from $\bar{z}$ in time $<\bar{T} - \tau - \delta$ for some $\delta > 0$ and which is maximal in dimension in the set of all such manifolds. Then α, β are tangent to M_0 and $\bar{z}$ (as is easily checked). Also (becaume $M_0 \subset N$ and $M_0 \subset R$) we have dim $(\underline{\underline{L}}_2(z)) = k$ for all $z \in M_0$ so that by Frobenius theorem there exists a submanifold M_1 of M_0 whose tangent space at each point $z \in M_1$ is precisely $\underline{\underline{L}}_2(z)$. One then also has that dim $I(z) = k_0$ for all $z \in M_1$ so that I has integral manifolds M_2 locally near $\bar{z}$. M_1 is then foliated by the integral manifolds of I so that M_1 locally near $\bar{z}$ looks like $M_1 \simeq M_2 \times M_3$. The Lie algebra of vectorfields of M_3 is then isomorphic to the quotient $\underline{\underline{L}}_2|_{M_1}/I|_{M_1}$.

So That by restriction to M_1 the dotted arrow is diagram 5.14 gives a homomorphism of Lie algebras

$$\underline{\underline{L}}_1 \to V(M_3) \tag{5.18}$$

5.19. Proposition. *Assume that the homomorphism of Lie algebras (5.18) is zero and assume moreover that* $\underline{\underline{L}}_1$ *contains all the operators* $L_k = \frac{d}{dx}x^k$, $k = 0,1,\dots$. *Then* ϕ *is a constant almost everywhere.*

This is seen as follows. This homomorphism is zero iff $k_0 = k$ so that for $z \in R$, $V\gamma(z) = 0$ for all $V \in \underline{\underline{L}}_2$, which gives $(\tilde{L}\Phi)(\psi_2) = 0$ for all $L \in \underline{\underline{L}}_2$. Now calculate $(\tilde{L}\Phi)(\psi_2)$ using formula (5.7) for $\Phi(\psi_2)$ to find that

(5.20) $\quad <\phi,L\psi_z> <1,\psi_z> = <\phi,\psi_z> <1,L\psi_z>$ for all $L \in \underline{\underline{L}}_2$.

As $<1,L_k\psi_z> = 0$ this gives

(5.21) $$\int \phi(x)[x^k\psi_z'(x) + kx^{k-1}\psi_z(x)]dx = 0, \quad k = 0,1,2,\ldots .$$

From this, using that $\phi(x)\psi_z'(x)$ and $\phi(x)\psi_2(x)$ are bounded by $e^{\beta|x|^r}$ for some $\beta > 0$, $r < 4$, one sees by considering the Fourier transforms of $\phi(x)\psi_z'(x)$ and $\phi(x)\psi_z(x)$ that $\phi'\psi_z = 0$ and as ψ_2 never vanishes that ϕ is constant.

6. ALGEBRAIC PART

6.1. The Weyl Lie algebras W_n

The Weyl Lie algebra W_n is the algebra of all differential operators (any order) in $\frac{\partial}{\partial x_1},\ldots,\frac{\partial}{\partial x_n}$ with polynomial coefficients. The Lie brackett operation is of course the commutator $[D_1,D_2] = D_1D_2 - D_2D_1$. A basis for W_1 (as a vector space over $\mathbb{R}$) constists of the operators

(6.2) $$x^i \frac{\partial^j}{\partial x^j}, \quad i,j = 0,1,2,\ldots$$

(where of course $x^i \frac{\partial^0}{\partial x^0} = x^i$, $x^0 \frac{\partial^j}{\partial x^j} = \frac{\partial^j}{\partial x^j}$, $x^0 \frac{\partial^0}{\partial x^0} = 1$). One has for example

$$[\frac{\partial^2}{\partial x^2},x^2] = 4x\frac{\partial}{\partial x} + 2$$

as is easily verified by calculating

$$[\frac{\partial^2}{\partial x^2},x^2]f(x) = \frac{\partial^2}{\partial x^2}(x^2f(x)) - x^2\frac{\partial^2}{\partial x^2}(f(x))$$

for an arbitrary test function (polynomial) $f(x)$.

Some easy facts (theorems) concerning the Weyl Lie algebras W_n are (cf. [Hazewinkel-Marcus, 1981] for proofs):

6.3. Proposition. *The Lie algebra* W_n *is generated (as a Lie algebra) by the elements* x_i, $\partial^2/\partial x_i^2$, $x_i^2(\partial/\partial x_i)$, $i = 1,\dots,n$; $x_i x_{i-1}$, $i = 2,\dots,n$. *In particular* W_1 *is generated by* x, $\partial^2/\partial x^2$, $x^2(\partial/\partial x)$.

6.4. Proposition. *The only nontrivial ideal of* W_n *is the one-dimensional ideal* $\mathbb{R}1$ *of scalar multiples of the identity operator.*

If M is a C^∞ differentable manifold let V(M) denote the Lie-algebra of all C^∞ vectorfields on M (i.e. the Lie algebra of all derivations on the ring of smooth functions on M). If $M = \mathbb{R}^n$, $V(\mathbb{R}^n)$ is the Lie algebra of all differential operators of the form

$$\sum_{i=1}^{n} g_i(x_1,\dots,x_n)\frac{\partial}{\partial x_i}$$

with $g_i(x_1,\dots,x_n)$ a smooth function on $\mathbb{R}^n$.

A deep fact concerning the Weyl Lie algebras W_n is now

6.5. Theorem. *Let* M *be a finite dimensional smooth manifold. Then there are no nonzero homomorphisms of Lie algebras* $W_n \to V(M)$ or $W_n/\mathbb{R}1 \to V(M)$ *for* $n \geq 1$.

The original proof of this theorem ([Hazewinkel-Marcus, 1981] is long and computational. Fortunately there now exists a much better proof (about two pages) of the main and most difficult part [Stafford, 1982], essentially based on the observation that the associative algebra W_1 cannot have left ideals of finite codimension. For some more remarks about the proof cf. 6.8 below.

6.6. The Lie algebra of the cubic sensor

According to section 2 above the estimation Lie algebra $L(\Sigma)$ of the cubic sensor is generated by the two operators

$$L_0 = \frac{1}{2}\frac{d^2}{dx^2} - \frac{1}{2}x^6, \quad L_1 = x^3.$$

Calculating $[L_0,L_1]$ gives $C = 3x^2\frac{d}{dx} + 3x$. Let $\mathrm{ad}_C(-) = [C,-]$. Then

$(\mathrm{ad}_C)^3 B = C^{st}. x^6$ which combined with A gives as that $(d^2/dx^2) \in L(\Sigma)$. To show that also $x^2 \frac{d}{dx} \in L(\Sigma)$ requires the calculation of some more bracketts (about 15 of them). For the details cf. [Hazewinkel-Marcus, 1981]. Then $x, x^2 \frac{d}{dx}, \frac{d^2}{dx^2} \in L(\bar{z})$ which by proposition 6.3 implies:

6.7. Theorem. *The estimation Lie algebra* $L(\Sigma)$ *of the cubic sensor is equal to the Weyl Lie algebra* W_1.

In a similar manner one can e.g. show that the estimation Lie algebra of the system $dx_t = dw_t$, $dy_t = (x_t + \varepsilon x_t^3)dt + dv_t$ is equal to W_1 for all $\varepsilon \neq 0$. It seems highly likely that this is a generic phenomenon i.e. that the estimation Lie algebra of a system of the form $dx_t = f(x_t)dt + G(x_t)dt$, $dy_t = h(x_t)dt + dv_t$ with $x \in \mathbb{R}^n$ and f, G and h polynomial is equal to W_n for almost all (in the Zariski topology sense) polynomials f, G, h.

6.8. Outline of the proof of the nonembedding theorem 6.5

Let $\hat{V}_n$ be the Lie algebra of all expressions

$$(6.9) \qquad \sum_{i=1}^{n} f_i(x_1,\ldots,x_n) \frac{\partial}{\partial x_i}$$

where $f_1(x),\ldots,f_n(x)$ are formal power series in $x_1,\ldots,x_n$. (No convergence properties are required). Suppose that

$$(6.10) \qquad \alpha: W_n \to V(M)$$

is a nonzero homomorphism of Lie algebras into some V(M) with M finite dimensional. Then there is a $D \in W_n$ and an $m \in M$ such that the tangent vector $\alpha(D)(m) \neq 0$. Now take formal Taylor series of the $\alpha(D)$ around m (with respect to local coordinates at m) to find a nonzero homomorphism of Lie algebra

$$(6.11) \qquad \hat{\alpha}: W_n \to \hat{V}_m$$

where $m = \dim(M)$.

Observe that W_1 is a sub-algebra of W_n (consisting of all differential operators not involving x_i, $i \geq 2$, and $\partial/\partial x_i$, $i \geq 2$) so that it suffices to prove theorem 6.5 for the case n = 1.

Because the only nontrivial ideal of W_1 is $\mathbb{R}1$ (cf. proposition 6.4) the existence of a nonzero $\hat{\alpha}: W_1 \to \hat{V}_m$ implies that W_1 or $W_1/\mathbb{R}1$ can be embedded in $\hat{V}_m$.

The Lie-algebra $\hat{V}_m$ carries a filtration $\hat{V}_m = L_{-1} \supset L_0 \supset L_1 \supset \ldots$ where the L_i are sub-Lie-algebras. This filtration has the following properties

(6.12) $$[L_i, L_j] \subset [L_{i+j}]$$

(6.13) $$\bigcap_{i=-1}^{\infty} L_i = \{0\}$$

(6.14) $$\dim(L_{-1}/L_i) < \infty, \quad i = -1, 0, 1, \ldots$$

where "dim" means dimension of real vectorspaces.

Indeed let

(6.15) $$f_i(x_1, \ldots, x_n) = \sum_{\nu} a_{i,\nu} x^{\nu}$$

$\nu = (\nu_1, \ldots, \nu_m)$, $\nu_i \in \mathbb{N} \cup \{0\}$ a multi index, be the explicit power series for $f_i(x)$. Then $L_j \subset \hat{V}_m$ consists of all formal vectorfields (6.15) for which

(6.16) $$a_{i,\nu} = 0 \quad \text{for all } \nu \text{ with } |\nu| \leq j$$

where $|\nu| = \nu_1 + \ldots + \nu_m$.

If there were an embedding $W_1 \to \hat{V}_m$ or $W_1/\mathbb{R}1 \to \hat{V}_m$ the Lie algebra W_1 or $W_1/\mathbb{R}1$ would interit a similar filtration satisfying (6.12) - (6.15). One can now show, essentially by brute force calculations that W_1 and $W_1/\mathbb{R}1$ do not admit such filtrations. Or much better one observes that (6.12) and (6.14) say that L_i i = 0,1,2,... is a subalgebra of finite codimension and applies Toby Stafford's result, loc. cit. that W_1 has no such sub-Lie-algebras.

7. PUTTING IT ALL TOGETHER AND CONCLUDING REMARKS

To condude let us spell out the main steps of the argument leading to theorem 2.10 and finish the proof together with some comments as to the generalizability of the various steps.

We start with a stochastic system, in particular the cubic sensor

$$(7.1) \qquad dx = dw, \quad x(0) = x^{in}, \quad dy = x^3 dt + dv$$

described more precisely in 2.1 and with a reasonable function ϕ of the state of which we want to compute the conditional expectation $\hat{\phi}_t$.

The first step now is to show that there exists a pathwise and robust version of $\hat{\phi}_t$. More precisely it was shown in section 4 that there exist a functional

$$(7.2) \qquad \delta_t^{\phi}(y) = \frac{\Delta_t^{\phi}(y)}{\Delta_t^{1}(y)}, \quad \Delta_t^{\phi}(y) = \langle N_t(y), \phi \rangle$$

such that the measures $N_t(y)$ depends continuously on the path $y: [0,t] \to \mathbb{R}$, such that $\Delta_t^1(y) > 0$ all $t > 0$, such that the density $n_t(y)$ is smooth and such that for $y(t) = y_t(\omega) =: y^{\omega}(t)$ a sample path of (7.1) then

$$(7.3) \qquad \hat{\phi}_t(\omega) = \delta_t^{\phi}(y^{\omega}).$$

From this we also obtained in the case of the cubic sensor that $n_t(y)(x)$ as a function of (t,x) satisfies the (control version) of the DMZ equation

$$(7.4) \qquad \frac{\partial}{\partial t} n_t(y)(x) = \left(\frac{1}{2}\frac{\partial^2}{\partial x^2} - \frac{1}{2}x^6\right) n_t(y)(x) + n_t(y)(x)\dot{y}(t)x^3$$

for piecewise differentiable functions $y: [0,t] \to \mathbb{R}$. And we showed that the family of densities $n_t(y)$, as a function of t, is smooth in the sense described in 4.25. Actually a more precise statement is needed, we need smoothness as a function of $t_1, \dots, t_m$ if $\dot{y} = u$ with u a bang-bang control of the type $u(t) = \bar{u}_i \in \mathbb{R}$ for $t_1 + \dots + t_{i-1} \le t < t_1 + \dots + t_i$, $|u_i| = 1$.

This whole bit is the part of the proof that seems most resistant to generalization. At present at least this requires reasonable growth bounds on the exponentials occurring in the Kallianpur-Stiebel formula (that is the explicit pathwise expressions for $\Delta_t^\phi(y)$). In particular let us call a family Φ_t of continuous maps $C_t \to \mathbb{R}$ a path-wise version of $\hat{\phi}_t$, if $\omega \mapsto \Phi_t(y^{\omega,t})$, $y^{\omega,t}(s) = y_s(\omega)$, $0 \leq s \leq t$ is a version of $\hat{\phi}_t$. Then it is not at all clear that path-wise versions exist for arbitrary nonlinear filtering problems.

Now suppose that there exists a smooth finite dimensional filter for $\hat{\phi}_t$. That is a smooth dynamical system

$$(7.5) \qquad dz = \alpha(z) + \beta(z)dy, \quad \gamma: M \to \mathbb{R}, \quad z(0) = z^{in}$$

such that if $z_y(t)$ denotes the solution of (7.5) then

$$(7.6) \qquad \gamma(z_y(t)) = \hat{\phi}_t = \delta_t^\phi(y)$$

almost surely. As described in 4.19 above up to a stopping time there also exists a robust pathwise version of the solutions of (7.5) so that $z_y(t)$ exists for all continuous y and so that (7.6) holds always. Now let $L_0 = \frac{1}{2}\frac{d^2}{dx^2} - \frac{1}{2}x^6$, $L_3 = x^3$, $L(\bar{u}) = L_0 + uL_1$. The next step is to show smoothness of

$$(7.7) \qquad e^{t_1 L(\bar{u}_s)} \cdots e^{t_m L(\bar{u}_m)} \psi$$

for smooth ψ as a function of $t_1,\ldots,t_m$, and to calculate $\partial^m/\partial t_1 \ldots \partial t_m$ of (7.7). The result being formula (4.34).

The next thing is to reinterpret a differential operator on F as a linear vectorfield $\tilde{L}$ on F by giving meaning to $\tilde{L}\Phi$ for Φ a functional $F \to \mathbb{R}$ for instance a functional of the form $\delta_t^\phi(y)$.

This permits us to give meaning to expressions like

$$(7.8) \qquad \frac{\partial^m}{\partial t_1 \ldots \partial t_m} \delta_t^\phi(y) \Big|_{t_1 = \ldots = t_m = 0}, \quad t = t_1 + \ldots + t_m$$

for $y \in C_t^1$ with $\dot{y} = u$ a bang-bang function. The same operator can be applied to the left handside of (7.6) and as both sides depend smoothly on $t_1,\dots,t_m$ there results from (7.6) an equality of the type

$$(7.9) \qquad (A(\bar{u}_m)\dots A(\bar{u}_1)\gamma)(z) = (\tilde{L}(\bar{u}_m)\dots\tilde{L}(\bar{u}_1)\Phi)(\psi_z)$$

where $z \in M$ and $\psi_z \in F$ are corresponding quantities in that they result from feeding in the same control function $\dot{y}(t)$ to the evolution equations for z and ψ respectively.

This relation in turn using some techniques familiar from nonlinear realization theory (essentially restriction to the completely reachable and observable subquotient of M) then implies that there is a homomorphism of Lie algebras from the Lie algebra $L(\Sigma)$ generated by L_0 and L_1 to a Lie algebra of smooth vectorfields. Moreover under the rather inelegant extra assumption that $L(\Sigma)$ contains the operator $\frac{d}{dx}x^k$ we showed that ϕ must have been constant if this homomorphism of Lie algebras is zero. (Proposition 5.19).

The final part is algebra and shows (i) that $L(\Sigma) = W_1$ so that in particular $\frac{d}{dx}x^k \in L(\Sigma)$ for all $k = 0,1,\dots$ and (ii) there are no nonzero homomorphisms of Lie algebras $W_1 \to V(M_1)$ for M_1 a smooth finite dimensional manifold. Thus both hypothesis of proposition 5.19 are fulfilled and ϕ is a constant. This proves the main theorem 2.10.

It seems by now clear [Hazewinkel-Marcus, 1981b] that the statement $L(\Sigma) = W_k$, $k = \dim$ (state space) will turn out to hold for a great many systems (though anything like a general proof for certain classes of systems is lacking). The system theoretic part of the argument is also quite general. The main difficulty of obtaining similar more general results lies thus in generalizing the analytic part or finding suitable subsitutes for establishing the homomorphism principle, perhaps as in [Hijab, 1982].

It should also be stressed that the main theorem 2.10 of this paper only says things about exact filters; it says nothing about approximate filters. On the other hand it seems clear that the Kalman-Bucy filter for $\hat{x}_t$ for

$$(7.10) \qquad dx = dw, \quad dy = xdt + dv$$

should for small ε give reasonable approximate results for

(7.11) $dx = dw, \quad dy = (x+\varepsilon x^3)dt + dv.$

Yet the estimation Lie algebra of (7.11) is for $\varepsilon \neq 0$ also equal to W_1 (a somewhat more tedious calculation cf. [Hazewinkel, 1981]) and the arguments of this paper can be repeated word for word (practically) to show that (7.11) does not admit smooth finite dimensional filters (for non-constant statistics). Positive results that the Kalman-Bucy filter of (7.10) does give an approximation to $\hat{x}_t$ for (7.11) are contained in loc. cit. [Sussmann, 1982], and [Blankenship - Liu - Marcus, 1983].

It is possible that results on approximate filters can be obtained by considering $L(\Sigma)$ not as a bare Lie algebra but as a Lie algebra with two distinguished generations L_0, L_1 which permits us to consider also the Lie algebra $L_s(\Sigma)$ generated by sL_0, sL_1 (where s is an extra variable) and to consider statements like $L_s(\Sigma)$ is close to $L_s(\Sigma')$ module s^t.

REFERENCES

BLANKENSHIP, G.L., C.-H. LIU, and S.I. MARCUS, 1983, *Asymptotic expansion and Lie algebras for some nonlinear filtering problems*, IEEE Trans. Aut. Control 28 (1983).

BROCKETT, R.W., 1981, *Nonlinear systems and nonlinear estimation theory*, In: [Hazewinkel-Willems, 1981], 441-478.

CLARK, J.M.C., 1978, *The design of robust approximations to the stochastic differential equations of nonlinear filtering*, In: J.K. Skwirzynski (ed.), Communication systems and random process theory, Sijthoff and Noordhoff, 1978.

DAVIS, M.H.A., 1982, *A pathwise solution to the equations of nonlinear filtering*, Teor. Verojatnost i. Prim. 27 (1982), 1, 160-167.

HAZEWINKEL, M., 1981, *On deformations, approximations and nonlinear filtering*, Systems and Control Letters 1 (1981).

HAZEWINKEL, M. & S.I. MARCUS, 1981, *Some facts and speculations on the role of Lie-algebras in nonlinear filtering*, In [Hazewinkel-Willems, 1981], 591-604.

HAZEWINKEL, M. & J.C. WILLEMS (eds), 1981, *Stochastic systems: the mathematics of filtering and identification and application,* Reidel Publ. Cy, 1981.

HAZEWINKEL, M. & S.I. MARCUS, 1982, *On Lie algebras and finite dimensional filtering,* Stochastics 7 (1982), 29-62.

HERMANN, R., 1963, *On the accessibility problem in control theory,* In: Int. Symp. on nonlinear differential equations and nonlinear mechanics, Acad. Pr., 1963, 325-332.

HIJAB, O., 1982, *Finite dimensional causal functionals of brownian motion,* To appear in Proc. NATO-ASI Nonlinear stochastic problems (Algarve, 1982), Reidel Publ. Cy.

MARCUS, S.I., S.K. MITTER & D. OCONE, 1980, *Finite dimensional nonlinear estimation for a class of systems in continuous and discrete time,* In: Proc. Int. Conf. on Analysis and Opt. of Stoch. Systems, Oxford 1978, Acad. Pr., 1980.

MITTER, S.K., 1981, *Nonlinear filtering and stochastic mechanics,* In: [Hazewinkel-Willems, 1981], 479-504.

NAGANO, S.K., 1966, *Linear differential systems with singularities and an application to transitive Lie algebras*, J. Math. Soc. Japan 18 (1966), 398-404.

STAFFORD, 1983, *On a result of Hazewinkel and Marcus,* to appear.

SUSSMANN, H.J., 1977, *Existence an uniqueness of minimal realizations of nonlinear systems*, Math. Systems Theory 10 (1977), 263-284.

SUSSMANN, H.J., 1978, *On the gap between deterministic and stochastic ordinary differential equations*, Ann. of Prob. 6 (1978), 19-41.

SUSSMANN, H.J., 1981, *Rigorous results on the cubic sensor problem*, In: [Hazewinkel-Willems, 1981], 637-648.

SUSSMANN, H.J., 1982, *Approximate finite-dimensional filters for some nonlinear problems,* Systems and control letters, to appear.

SUSSMANN, H.J., 1983a, *Rigorous results on robust nonlinear filtering,* to appear.

SUSSMANN, H.J., 1983b, *Nonexistence of finite dimensional filters for the cubic sensor problem,* to appear.

AN EXTENSION OF THE PROPHET INEQUALITY

D. P. Kennedy

1. INTRODUCTION

Let $X = \{X_r, r \geq 1\}$ be a sequence of non-negative random variables and let $F = \{F_r, r \geq 1\}$ be the filtration generated by the sequence, i.e., $F_r = \sigma(X_1, X_2, \ldots, X_r)$. Denote by T the set of (finite-valued) stopping times of F. The optimal stopping problem involves maximizing the expected reward, EX_T, over stopping times $T \in T$. It is well known (cf. [1]) that allowing randomized stopping times does not increase the maximal expected reward. In particular, if we denote by V the set of non-negative processes $V = \{V_r, r \geq 1\}$ adapted to F and satisfying $\sum_{r=1}^{\infty} V_r = 1$, a.s., we may regard the elements of V as randomized stopping times of F, with $E(\sum_1^{\infty} V_r X_r)$ as the expected reward associated with V. More usually, a randomized stopping time T is a (finite-valued) stopping time relative to a filtration $G = \{G_r, r \geq 1\}$ such that for each r, $F_r \subset G_r$ and the σ-fields $F_\infty = \sigma(\bigcup_r F_r)$ and G_r are conditionally independent given F_r, (cf. [8]). But, given such a randomized stopping time, setting $V_r = P\{T=r \mid F_r\}$ gives

$$\sum_1^{\infty} V_r = \sum_1^{\infty} P\{T=r \mid F_\infty\} = P\{T<\infty \mid F_\infty\} = 1 \text{, a.s.,}$$

i.e., $\{V_r, r \geq 1\} \in V$ and $EX_T = E(\sum_1^{\infty} V_r X_r)$. It may be shown ([3]) that

$$\sup_{V \in V} E(\sum_1^{\infty} V_r X_r) = \sup_{T \in T} EX_T \quad . \tag{1.1}$$

Here, we will consider a generalization of the optimal stopping problem. For $p \geq 1$, denoting by V_p the set of non-negative processes $V = \{V_r, r \geq 1\}$ adapted to F and satisfying $\sum_1^\infty V_r^p = 1$, a.s., we will consider the problem of maximizing $E(\sum_1^\infty V_r X_r)$ over $V \in V_p$. (Note that $V = V_1$, so the optimal stopping problem corresponds to the case $p = 1$.) This problem has been dealt with in [3] and an economic interpretation of the model and its solution has been given in [4]. A related sequential game has been treated in [5].

We will consider the case where the random variables $\{X_r, r \geq 1\}$ are independent. In this context the prophet inequality is

$$E(\sup_{r\geq 1} X_r) \leq 2 \sup_{T\in T} EX_T \quad . \tag{1.2}$$

This shows that the maximal expected reward of a gambler (i.e. a player using non-anticipating stopping rules) is at least half that of a prophet (a player with complete foresight). A form of this result was first given by Krengel and Sucheston [6] who established (1.2) with the constant 2 on the right-hand side replaced by 4. An argument attributed to Garling (cf. [7]) proved (1.2) and showed that the inequality is sharp. Hill and Kertz [2] have also dealt with the problem and have shown that strict inequality holds in (1.2) in all but trivial cases.

In this note we will prove a form of (1.2) for the extended optimal stopping problem. If we let W_p denote the set of sequences $W = \{W_r, r \geq 1\}$ of non-negative random variables satisfying $\sum_1^\infty W_r^p = 1$, a.s., (but not necessarily adapted) and if q is conjugate to p in the sense that $p^{-1} + q^{-1} = 1$, we have

$$\sup_{W\in W_p} E(\sum_1^\infty W_r X_r) = E[\sum_1^\infty X_r^q]^{1/q} \quad .$$

We will establish the following result.

Theorem If the random variables $\{X_r, r \geq 1\}$ are independent, then for each $p > 1$, $p^{-1} + q^{-1} = 1$,

$$E\left[\sum_{1}^{\infty} X_r^q\right]^{1/q} \leq 2 \sup_{V \in \mathcal{V}_p} E\left(\sum_{1}^{\infty} V_r X_r\right) . \tag{1.3}$$

Again, (1.3) shows that in the extended problem the maximal expected return for the gambler using non-anticipating rules (in $\mathcal{V}_p$) is at least half that of the prophet using rules with complete foresight (in $\mathcal{W}_p$). In the following, if we interpret expressions of the form $\left[\sum_{1}^{\infty} a_r^q\right]^{1/q}$ as $\sup_{r \geq 1} a_r$ when $q = \infty$, the argument also holds for $p = 1$.

2. PROOF OF THE THEOREM

The result (1.3) will be an immediate consequence of the following.

Proposition. For each $n \geq 1$, if the random variables $X_1, \ldots, X_n$ are independent then for $p > 1$, $p^{-1} + q^{-1} = 1$,

$$E\left[\sum_{1}^{n} X_r^q\right]^{1/q} \leq c_n \sup_{V \in \mathcal{V}_p} E\left(\sum_{1}^{n} V_r X_r\right) , \tag{2.1}$$

where $c_n \equiv c_n(p)$, satisfies $c_n^p = (2 - c_{n-1})^p + c_{n-1}^p$, $n > 1$, with $c_1 = 1$.

Proof. The proof proceeds by induction on n; the case $n = 1$ is immediate so assume that (2.1) holds for any $n - 1$ independent non-negative random variables. For $1 \leq i \leq n$, set

$$w_i = \sup_{V \in \mathcal{V}_p} E\left(\sum_{i}^{n} V_r X_r\right)$$

then it follows that (cf. [3]) $w_i = E[X_i^q + w_{i+1}^q]^{1/q}$, $i < n$, and $w_n = EX_n$. Without any loss of generality assume that $w_1 < \infty$. Observe that for $a \geq c \geq 0$ and $b \geq 0$ we have

$$a + (b^q + c^q)^{1/q} \geq (a^q + b^q)^{1/q} + c \quad . \tag{2.2}$$

Using (2.2), it follows that for any $x \geq w_3 \geq \ldots \geq w_n$,

$$\begin{aligned}
E[x^q + X_2^q + \ldots + X_n^q]^{1/q} &\leq E[x^q + X_2^q + \ldots + X_{n-1}^q]^{1/q} + EX_n \\
&= E[x^q + X_2^q + \ldots + X_{n-1}^q]^{1/q} + w_n \\
&\leq E[x^q + X_2^q + \ldots + X_{n-2}^q]^{1/q} + E[X_{n-1}^q + w_n^q]^{1/q} \\
&= E[x^q + X_2^q + \ldots + X_{n-2}^q]^{1/q} + w_{n-1} \\
&\leq E[x^q + X_2^q + \ldots + X_{n-3}^q]^{1/q} + w_{n-2} \\
&\;\vdots \\
&\leq x + w_2 \\
&\leq 2^{1/p}[x^q + w_2^q]^{1/q} \quad .
\end{aligned} \tag{2.3}$$

By the convexity of $[x^q + a^q]^{1/q}$, we see that for $0 \leq x \leq w_2$,

$$\begin{aligned}
E[x^q + X_2^q + \ldots + X_n^q]^{1/q} &\leq \frac{x}{w_2} E[w_2^q + X_2^q + \ldots + X_n^q]^{1/q} \\
&\quad + (1 - \frac{x}{w_2}) E[X_2^q + \ldots + X_n^q]^{1/q} \quad .
\end{aligned} \tag{2.4}$$

Using (2.3) the first term on the right-hand side of (2.4) is dominated by $2x$, while the inductive hypothesis gives

$$E[X_2^q + \ldots + X_n^q]^{1/q} \leq w_2 c_{n-1}$$

and so the second term is bounded above by $c_{n-1}(w_2 - x)$. For $0 \leq x \leq w_2$, the function,

$$f(x) = [(2 - c_{n-1})x + w_2 c_{n-1}]/[x^q + w_2^q]^{1/q}$$

is maximised at $\bar{x} = [(2 - c_{n-1})/c_{n-1}]^{p/q} w_2$, and

$$f(\bar{x}) = [(2 - c_{n-1})^p + c_{n-1}^p]^{1/p} = c_n \quad . \tag{2.5}$$

Observing that $2^{1/p} \le c_n$ and combining (2.3) and (2.5) shows that for all $x \ge 0$,

$$E[x^q + X_2^q + \ldots + X_n^q]^{1/q} \le c_n [x^q + w_2^q]^{1/q}$$

which implies that

$$\begin{aligned} E[X_1^q + X_2^q + \ldots + X_n^q]^{1/q} &\le c_n \; E[X_1^q + w_2^q]^{1/q} \\ &= c_n \, w_1 \; , \end{aligned}$$

completing the induction.

Since $c_n \uparrow 2$ as $n \to \infty$, we have for each n ,

$$\begin{aligned} E\,[\sum_1^n X_r^q]^{1/q} &\le c_n \sup_{V \in \mathcal{V}_p} E(\sum_1^n V_r X_r) \\ &\le 2 \sup_{V \in \mathcal{V}_p} E(\sum_1^\infty V_r X_r) \quad , \end{aligned}$$

and letting $n \to \infty$, monotone convergence gives (1.3).

In general, when the random variables are not independent then a result of the form (1.3) cannot hold. The best that can be said is that the inequality (2.1) holds with c_n replaced by $n^{1/p}$. This may be seen by considering the case when each $V_r = 1/n^{1/p}$, then

$$\begin{aligned} \sup_{V \in \mathcal{V}_p} E(\sum_1^n V_r X_r) &\ge E(\sum_1^n X_r)/n^{1/p} \\ &\ge E\,[\sum_1^n X_r^q]^{1/q} \,/n^{1/p} \quad . \end{aligned} \tag{2.6}$$

The inequality in (2.6) is sharp. For $1 \le i \le n$, if we set

$$W_i = \operatorname*{ess\,sup}_{V \in \mathcal{V}_p} E\,[\sum_i^n V_r X_r | \mathcal{F}_i]$$

then it has been shown in [3] that

$$W_i^q = X_i^q + E[W_{i+1} | \mathcal{F}_i]^q \text{ , a.s., } \quad i = 1,\ldots,n-1 \text{ ,}$$

with $W_n = X_n$, and furthermore with w_i defined as in the Proposition, $w_i = EW_i$ for each i. If X_r is a martingale it may be checked that $W_i = (n-i+1)^{1/q} X_i$, $i = 1,\ldots,n$ and so $w_1 = n^{1/q} EX_1$. But for $0 < \alpha < 1$ (as in [2]), choose X_r to be a martingale such that $(X_1, X_2,\ldots, X_n)$ equals $(1, \alpha^{-1},\ldots, \alpha^{-r}, 0,\ldots, 0)$ with probability $\alpha^r(1-\alpha)$, $0 \le r \le n-2$, and equals $(1, \alpha^{-1},\ldots, \alpha^{-(n-1)})$ with probability α^{n-1}. For this choice, $EX_1 = 1$ and it may be checked that

$$\lim_{\alpha\to 0} E\left[\sum_1^n X_r^q\right]^{1/q} = n$$

showing that (2.6) is sharp.

REFERENCES

[1] Chow, Y.S., Robbins, H. and Siegmund, D. (1971). Great Expectations: The Theory of Optimal Stopping, Houghton-Mifflin, Boston.

[2] Hill, T.P. and Kertz, R.P. (1981). Ratio comparisons of supremum and stop rule expectations. Z. Wahr. verw. Geb. 56, 283-285.

[3] Kennedy, D.P. (1981). Optimal predictable transforms, Stochastics 5, 323-334.

[4] Kennedy, D.P. (1983). Stimulating prices in a stochastic model of resource allocation. Math. Op. Res. 8, 151-157.

[5] Kennedy, D.P. (1983). A sequential game and envelopes of stochastic processes. Stochastics (to appear).

[6] Krengel, U. and Sucheston, L. (1977). Semiamarts and finite values. Bull. Am. Math. Soc. 83, 745-747.

[7] Krengel, U. and Sucheston, L. (1978). On semiamarts, amarts and processes with finite value. in Probability on Banach Spaces (Ed. P. Ney), Marcel Dekker, New York.

[8] Pitman, J. and Speed, T.P. (1973). A note on random times. Stoch. Proc. and their Appl. 1, 369-374.

Statistical Laboratory,
University of Cambridge,
16 Mill Lane,
Cambridge CB2 1SB.
U.K.

MARTINGALE REPRESENTATION AND NONLINEAR FILTERING EQUATION FOR DISTRIBUTION-VALUED PROCESSES

H. KOREZLIOGLU

Ecole Nationale Supérieure des Télécommunications
46 Rue Barrault 75634 PARIS CEDEX 13 - FRANCE

C. MARTIAS

Centre National d'Etudes des Télécommunications
38-40 Rue du Général Leclerc, 92131 ISSY-LES-MOULINEAUX - FRANCE

The stochastic integral of operator-valued processes with respect to a distribution-valued Brownian motion is constructed and a martingale representation theorem proved. As an application the nonlinear filtering equation for distribution-valued processes is derived.

INTRODUCTION

The filtering of infinite dimensional processes started with the extension of FUJISAKI, KALLIANPUR and KUNITA's well known direct method in [3], developed for the derivation of the nonlinear filtering equation in the finite dimensional case. Complete results in this direction could only be obtained after the elaboration of the martingale representation theorem given by OUVRARD [14] and based on METIVIER and PISTONE's stochastic integral of operator-valued processes [12]. In [15] OUVRARD derived the filtering equation and the corresponding RICCATI equation for the linear model. Extending SZPIRGLAS and MAZZIOTTO's work [18] based on the reference probability method, MARTIAS studied in [10] the most general Hilbertian model with general semimartingales as state and observation processes and obtained the filtering equation for the Hilbertian version of the model considered below with some restrictions on the process h, among which, the usual boundedness condition. In order to release these restrictions KOREZLIOGLU extended in [5] the direct method of [3] to the same Hilbertian model and introduced the stochastic integration with respect to the cylindrical Brownian motion.

The filtering equation derived here concerns the following "state and observation" model

$$X_t = X_o + \int_o^t f_s \, ds + M_t$$

$$Y_t = \int_o^t h_s \, ds + W_t$$

where all the processes are distribution-valued, M is a square-integrable martingale, W is a Brownian motion and f and h satisfy adequate adaptation and integrability conditions. We give here the definition of a distribution-valued Brownian motion, contruct in an appropriate way the stochastic integral of operator-valued processes, prove the representation theorem for distribution-valued square-integrable martingales and derive by the direct method the corresponding filtering equation. The main idea of the approach lies on the reduction of the problems to a Hilbertian frame, made possible by a remark of USTUNEL's in [19] , according to which every distribution-valued square-integrable martingale can be considered as a Hilbert space-valued one. As in [5] , the stochastic integrals are expressed in terms of the cylindrical Brownian motion which enables, as shown at the end of this paper, the application of the results to the derivation of the filtering equations for two-parameter semimartingales obtained by KOREZLIOGLU, MAZZIOTTO and SZPIRGLAS [8], by means of two-parameter stochastic calculus techniques.

1. NOTATIONS AND PRELIMINARIES

$\mathcal{D}$ denotes the space of infinitely differentiable real functions with compact supports on $\mathbb{R}^n$ or an open subset of $\mathbb{R}^n$ endowed with its usual topology, [17] . $\mathcal{D}'$, the strong topological dual of $\mathcal{D}$, is the space of real distributions. $\mathcal{D}$ and $\mathcal{D}'$ are complete reflexive nuclear spaces. For a distribution d and ϕ in $\mathcal{D}$, $d(\phi)$ or (ϕ, d) will denote the value of d at ϕ and if $X = (X_t)$ is a $\mathcal{D}'$-valued process $X(\phi)$ will denote the process $(X_t(\phi))$ for $\phi \in \mathcal{D}$.

Borrowing the notations of [20] , we shall denote by F any one of the spaces $\mathcal{D}$ and $\mathcal{D}'$. If U is an absolutely convex neighborhood of 0 in F, F(U) denotes the completion of the quotient space $F/p_U^{-1}(0)$ where p_U denotes the gauge of U. k(U) denotes the canonical map of F into F(U) and for two such neighborhoods U and V such that $U \subset V$, k(V,U) is the canonical map of F(U) into F(V). If B is a closed absolutely convex subset of F then F[B] denotes the Banach space $(\bigcup_n nB, p_B)$, p_B being the gauge of B. i(B) will denote the imbedding map of F[B] into F. $\mathcal{U}_h(F)$ denotes the set of absolutely convex closed neighborhoods U of 0 in F such that F(U) is a separable Hilbert space. F being nuclear, $\mathcal{U}_h(F)$ is not empty and forms a neighborhood base of F. For all $U \in \mathcal{U}_h(F)$, F(U) is isomorphic to $F'[U^\circ]$, where F' is the dual of F and U° is the polar of U. The set $\{U^\circ, U \in \mathcal{U}_h(F)\}$ is a fundamental set of compact sets in F'. For all the properties of nuclear spaces we refer to [16].

Let H and K be real Hilbert spaces with their respective scalar products $(.,.)_H$, $(.,.)_K$ and norms $\|\cdot\|_H$, $\|\cdot\|_K$. L(H,K) is the space of continuous linear operators from H into K with the uniform norm $\|\cdot\|$, $L^1(H,K)$ the space of nuclear operators with the trace norm $\|\cdot\|_1$ and $L^2(H,K)$ the space of Hilbert-Schmidt operators with the Hilbert-Schmidt norm $\|\cdot\|_2$. $H \hat{\otimes}_1 K$ (resp. $H \hat{\otimes}_2 K$) is the projective (resp. Hilbertian) tensor product of H with K. We identify $H \hat{\otimes}_1 K$ (resp. $H \hat{\otimes}_2 K$) with $L^1(H,K)$ (resp. $L^2(H,K)$) by identifying $h \otimes k$ with the operator $(.,h)_H\, k$ for $h \in H$, $k \in K$.

For a linear operator A, from one vector space to another, Dom A, Rg A, Ker A and A^* will denote its domain, range, kernel and adjoint, respectively.

For the general theory of stochastic processes we refer to [2], for Hilbert space-valued processes to [11] and for nuclear space-valued processes to [19] (cf. also [20] in this volume). As to Hilbert space-valued martingales we consider here, we only need to introduce some notations and recall some definitions. In order not to have the usual localization problem with the Brownian motion, we shall suppose that all the processes are indexed on the finite interval [0,T]. Unless otherwise specified, all the random variables and processes are supposed to be defined on a complete probability space $(\Omega, \underline{\underline{A}}, P)$ with a filtration $\underline{\underline{F}} = (\underline{\underline{F}}_t, t \in [0,T])$ satisfying the usual right-continuity and completeness condition, [2]. We take $\underline{\underline{A}} = \underline{\underline{F}}_T$. If X is a process with values in a measurable space, the smallest filtration, denoted by $\underline{\underline{F}}^X$, to which X is adapted (i.e. $\forall$ t, X_t is $\underline{\underline{F}}_t$-measurable) is called the natural filtration of X. We denote by $\mathcal{P}(\underline{\underline{F}})$ the σ-algebra on $[0,T] \times \Omega$ of previsible sets associated with $\underline{\underline{F}}$. For a Hilbert space H, the space $L^2_H([0,T] \times \Omega, \mathcal{P}(\underline{\underline{F}}), dt \otimes dP)$, of H-valued square-integrable previsible processes will be denoted by $L^2_H(\mathcal{P}(\underline{\underline{F}}))$.

When referring to a martingale with values in a separable real Hilbert space we mean a martingale with right-continuous trajectories. For such a Hilbert space H, $\mathcal{M}(H)$ (resp. $\mathcal{M}_c(H)$) denotes the Hilbert space of right-continuous (resp. continuous) square-integrable martingales. Let, H and K be separable real Hilbert spaces and let $M \in \mathcal{M}(H)$, and $N \in \mathcal{M}(K)$. Then there exists a unique $\underline{\underline{F}}$-previsible process denoted by $<M,N>$ with values in $H \hat{\otimes}_1 K$, such that $<M,N>_o = M_o \otimes N_o$, $E\int_o^\infty \|d<M,N>_t\|_1 < \infty$ and $M \otimes N - <M,N>$ is a $H \hat{\otimes}_1 K$-valued martingale. When M = N(for H = K), $<M,M>$ is called the increasing process of M. An H-valued $(\underline{\underline{F}}, P)$ - Brownian motion $W = (W_t)$ is an H-valued continuous square-integrable martingale whose increasing process is given by tQ where Q is a symmetric non negative element of $H \hat{\otimes}_1 H$ and is called the covariance operator of W.

For the sake of completeness we reproduce here some definitions and properties of $\mathcal{D}'$-valued martingales given in [19] .

A mapping $X : [0,T] \times \Omega \longrightarrow \mathcal{D}'$ is called a weakly measurable process if for all $\phi \in \mathcal{D}$ and all $t \in [0,T]$, $X_t(\phi)$ is a real random variable.

DEFINITION 1.1

A weakly measurable $\mathcal{D}'$-valued process M is called a square-integrable martingale if for all $U \in \mathcal{U}_h(\mathcal{D}')$, $k(U)M$ (defined by $(k(U)M)_t(\omega) = k(U)M_t(\omega)$) has a modification which is a square-integrable martingale.

Such a martingale M is said to be continuous if for all $U \in \mathcal{U}_h(\mathcal{D}')$, $k(U)\, M$ has a strongly continuous modification.

At a first glance, this definition of a square-integrable martingale may not seem very natural. But the next theorem (reproducing Theorem II.4 of [19] and its extension to continuous martingales) shows that the above definition is equivalent to the weak definition. Due to its importance for the present work we reproduce here the proof.

THEOREM 1.2

A weakly measurable $\mathcal{D}'$-valued process M is a square-integrable martingale iff for all $\phi \in \mathcal{D}$, $M(\phi)$ has a modification which is a square-integrable martingale. Similarly, M is a continuous square-integrable martingale iff for all $\phi \in \mathcal{D}$, $M(\phi)$ has a modification which is a continuous square-integrable martingale.

Proof : Let M be a square-integrable martingale with values in $\mathcal{D}'$. It then holds that for all $\phi \in \mathcal{D}$, $M(\phi)$ has a modification which is a square-integrable real martingale. Similarly, if M is continuous, $M(\phi)$ has a modification which is a continuous square-integrable martingale.

Conversely, let M be such that $M(\phi)$ has a modification which is a square-integrable martingale for all ϕ, and let A be the mapping of $\mathcal{D}$ into $\mathcal{M}(\mathbb{R})$ such that for $\phi \in \mathcal{D}$, $A(\phi)$ is a modification of $M(\phi)$ belonging to $\mathcal{M}(\mathbb{R})$. A is a linear mapping. But it is also continuous and hence nuclear, because for all $V \in \mathcal{U}_h(\mathcal{D}')$, $Ai(V^\circ)$ is continuous. Therefore, A can be written as

$$A(\phi) = \sum_{i=1}^{\infty} \lambda_i F_i(\phi)\, m^i$$

where $(\lambda_i) \in l^1$, $(F_i) \subset \mathcal{D}'$ is equicontinuous and $(m^i) \subset \mathcal{M}(R)$ is bounded. Let $G \in \mathcal{U}_h(\mathcal{D})$ be such that $(F_i) \subset G^\circ$ and put

$$M'_t(\omega) = \sum_{i=1}^{\infty} \lambda_i\, m^i_t(\omega)\, F_i$$

with the series converging in $\mathcal{D}'[G^\circ]$. We have $M' \in \mathcal{M}(\mathcal{D}'[G^\circ])$ and $i(G^\circ)M'$ is a modification of M. Therefore, for all $U \in \mathcal{U}_h(\mathcal{D}')$, $k(U)i(G^\circ)M' \in \mathcal{M}(\mathcal{D}'(U))$. This shows that M is a square-integrable martingale. The proof for a continuous martingale is exactly the same with $\mathcal{M}$ replaced by $\mathcal{M}_c$. ■

Definition 1.1 does not imply the strong right-continuity (resp. continuity) of almost all trajectories of M. But the above proof shows the existence of a modification having right-continuous (resp. continuous) trajectories.

PROPOSITION 1.3

Let M be a (resp. continuous) square-integrable martingale with values in $\mathcal{D}'$. Then there is a neighborhood $G \in \mathcal{U}_h(\mathcal{D})$ and a (resp. continuous) square-integrable martingale M' with values in $\mathcal{D}'[G^\circ]$ such that $i(G^\circ)M'$ is a modification of M. Therefore, M has a right-continuous (resp. continuous) modification.

Since a (resp. continuous) square-integrable martingale has a right-continuous (resp. continuous) modification. We shall define a (resp. continuous) square-integrable martingale as a right-continuous (resp. continuous) one.

2. DISTRIBUTION-VALUED BROWNIAN MOTION

DEFINITION 2.2

Let $W = (W_t, t \in [0,T])$ be a $\mathcal{D}'$-valued continuous square-integrable martingale and let Q be a continuous symmetric semi-positive linear operator from $\mathcal{D}$ into $\mathcal{D}'$ (i.e. $\forall \phi, \Psi \in \mathcal{D}$ $(Q\phi, \Psi) = (Q\Psi, \phi)$ and $(Q\phi, \phi) \geqslant 0$). We say that W is a $\mathcal{D}'$-valued $(\underline{\underline{F}}, P)$-Brownian motion with covariance operator Q if

$$\forall \phi, \Psi \in \mathcal{D}, \quad \langle W(\phi), W(\Psi) \rangle_t = t(Q\phi, \Psi).$$

The following example will be systematically used in the sequel.

E2.1 EXAMPLE

Let $(W_{t,s} : (t,s) \in [0,T] \times [0,S])$ be a two-parameter Brownian sheet such that for all t, $W_{t,s}$ is $\underline{\underline{F}}_t$-measurable, and $\underline{\underline{F}}_t$ and $\sigma(W_{u,s} - W_{t,s} : (u,s) \in [t,T] \times [0,S])$ are independent. This defines a 1-Brownian sheet in the sense of [8]. Let W_t be the random distribution defined by $1_{[0,S]}(.)W_{t,.}$ (i.e. $\forall \phi \in \mathcal{D}$, $W_t(\phi) = \int_0^S W_{t,u}\, \phi(u)\, du$). (W_t) defines an $(\underline{\underline{F}},P)$-Brownian motion W with values in $L^2(0,S)$, (the Hilbert space of square-integrable real functions on $[0,S]$), with covariance operator Q defined as follows :

$$(2.1) \qquad \forall f \in L^2(0,S), \quad \forall s \in [0,S] \quad (Qf)(s) = \int_0^S (s \wedge u) f(u)\, du$$

where $\wedge$ stands for the infimum. As the imbedding of $L^2(0,S)$ into $\mathcal{D}'$ is continuous, the image of W by this imbedding is a $\mathcal{D}'$-valued $(\underline{\underline{F}},P)$-Brownian motion that, by an abuse of notation, we denote again by W. ■

Let (W_t) be an $(\underline{\underline{F}},P)$-Brownian motion with values in $\mathcal{D}'$ and with covariance operator Q. According to Proposition 1.3 there is a neighborhood G in $\mathcal{U}_h(\mathcal{D})$ such that W is a Brownian motion with values in $\mathcal{D}'[G°]$. Since $\mathcal{D}(G)$ and $\mathcal{D}'[G°]$ are dual Hilbert spaces, we shall identify them and denote them commonly by H. Let Q_H be the covariance operator of W as an H-valued Brownian motion. Then Q and Q_H verify the following diagram

$$\mathcal{D} \xrightarrow{\;k(G)\;} \mathcal{D}(G) \xrightarrow{\;Q_H\;} \mathcal{D}'[G°] \xrightarrow{\;i(G°)\;} \mathcal{D}' \qquad (\text{composite } \mathcal{D} \xrightarrow{\;Q\;} \mathcal{D}')$$

It is known that Q_H is a nuclear operator. As a continuous operator from $\mathcal{D}$ into $\mathcal{D}'$, Q is also nuclear.

Let $D \in L(H,H)$ be such that $Q_H = D\, D^*$. Then $D \in L^2(H,H)$, because Q_H is nuclear. Given a separable Hilbert space K, we denote by $L^2(H,K,D)$ the space of not necessarily continuous linear operators A from H into K such that Rg $D \subset$ Dom A and $AD \in L^2(H,K)$. This is a complete space under the seminorm $\|AD\|_2$. We make a Hilbert space of it by the scalar product $(A,B) = (AD, BD)_2$ and the equivalence relation : $A \sim B \Longleftrightarrow \| (A-B)D \|_2 = 0$, $((.,.)_2$ denotes the scalar product in $L^2(H,K))$. Since D is a Hilbert-Schmidt operator on H, we have $L(H,K) \subset L^2(H,K,D)$. We denote by $\tilde{L}^2(H,K,D)$ the Hilbert subspace of $L^2(H,K,D)$ generated by $L(H,K)$. We put $\tilde{H} = \tilde{L}^2(H,\mathbb{R},D)$.

H, identified with its topological dual, is dense in $\tilde{H}$ and for $h,k \in H$ we have $(h,k)_{\tilde{H}} = (D^* h, D^* k)_H$. With this definition of the scalar product on $\tilde{H}$ we see that D^* extends to an isometry I of $\tilde{H}$ into H. As shown in the following proposition, $\tilde{H}$ is, up to an isometry, independent of the chosen neighborhood G and on the factorization DD^*.

PROPOSITION 2.2

$\tilde{H}$ is isomorphic to the completion of $\mathcal{D}/\mathrm{Ker}\, Q$ endowed with the scalar product $([\phi], [\Psi]) = (Q\phi, \Psi)$ where $\phi, \Psi \in \mathcal{D}$ and $[\phi]$, $[\Psi]$ are their equivalence classes in $\mathcal{D}/\mathrm{Ker}\, Q$.

Proof : Let us put $C = i(G°)D$, where $i(G°)$ is the canonical imbedding of $H = \mathcal{D}'[G°]$ into $\mathcal{D}'$. We have $C^* = D^* k(G)$. C^* defines a bijection between $\mathcal{D}/\mathrm{Ker}\, C^*$ and $C^*(\mathcal{D})$. As $CC^* = i(G°) D D^* k(G) = Q$, we have $\mathrm{Ker}\, C^* = \mathrm{Ker}\, Q$. If $\mathcal{D}/\mathrm{Ker}\, Q$ is given the scalar product $([\phi], [\Psi]) = (Q\phi, \Psi)$ then the mapping C^* from $\mathcal{D}/\mathrm{Ker}\, Q$ into H defines an isometry between $\mathcal{D}/\mathrm{Ker}\, Q$ and $C^*(\mathcal{D})$. But since $C^* = D^* k(G)$, we have by the density of $k(G)(\mathcal{D})$ in H, $\overline{C^*(\mathcal{D})} = \overline{D^*(H)}$, with closures taken in H. ■

We shall also denote by $\tilde{H}$ the Hilbert space generated by $\mathcal{D}$ under the scalar product $(Q\phi, \Psi)$ as in the above proposition.

E2.2 EXAMPLE (CONTINUED)

Let us go back to the Brownian motion W derived from the two-parameter Brownian sheet.

The mapping $\phi \longrightarrow \| 1_{[0,S]}\phi \|_{L^2(0,S)}$ defines a continuous norm on $\mathcal{D}$. If we put $G = \{\phi \in \mathcal{D},\ \| 1_{[0,S]}\, \phi \|_{L^2(0,S)} \leq 1\}$, then $G \in \mathcal{U}_h(\mathcal{D})$ and the completion of $\mathcal{D}(G)$ coincides with $L^2(0,S)$. The Hilbert space H of the preceding setting is then $L^2(0,S)$. The covariance operator Q_H of W as an $L^2(0,S)$-valued Brownian motion is given by (2.1). It can also be written as

$$(2.2) \qquad \forall f \in L^2(0,S), \forall s \in [0,S], \quad (Q_H f)(s) = \int_0^s \left(\int_u^S f(v)\, dv\right) du$$

Let D be the linear operator on $L^2(0,S)$ defined as follows :

$$(2.3) \qquad \forall f \in L^2(0,S), \quad (Df)(s) = \int_0^s f(u)du.$$

The adjoint of D is then given by

$$(2.4) \qquad f \in L^2(0,S), \quad (D^*f)(s) = \int_s^S f(u)du.$$

We see that we have the factorization $Q_H = DD^*$. The space $\tilde{H}$ is isometric to the closure of $D^*(H)$ in $L^2(0,S)$. But $D^*(H) = \{ \int_\cdot^S f(u)du, \ f \in L^2(0,S)\}$ is dense in $H = L^2(0,S)$. Therefore, the isometry I is from $\tilde{H}$ onto $L^2(0,S)$. On the other hand, according to Proposition 2.2, $\tilde{H}$ is isomorphic to the completion of $\mathcal{D}$ under the Hilbertian norm

$$(2.5) \qquad \forall \phi \in \mathcal{D} \quad \|\phi\|^2_{\tilde{H}} = \int_0^S \left(\int_u^S \phi(v)\, dv\right)^2 du$$

The isometry I of $\tilde{H}$ onto $L^2(0,S)$ induces an isometry, that we always denote by I, from $L^2_{\tilde{H}}(\mathcal{P}(\underline{\underline{F}}))$ onto $L^2(\Omega \times [0,T] \times [0,S], \mathcal{P}(\underline{\underline{F}}) \otimes \mathcal{B}_{[0,S]})$, shortly denoted by $L^2(\mathcal{P}(\underline{\underline{F}}) \otimes \mathcal{B}_{[0,S]})$, where $\mathcal{B}_{[0,S]}$ is the Borel σ-field of $[0,S]$. Moreover, for a separable Hilbert space K, the extended isometry I still extends to an isometry from $L^2_{L^2(\tilde{H},K)}(\mathcal{P}(\underline{\underline{F}}))$ onto $L^2_K(\mathcal{P}(\underline{\underline{F}}) \otimes \mathcal{B}_{[0,S]})$. We shall also denote by I the last extended isometry. ■

3. STOCHASTIC INTEGRATION

If W is a $\mathcal{D}'$-valued Brownian motion with covariance operator Q, it coincides, in the sense of Proposition 1.3, with a Brownian motion with values in the Hilbert space $H = \mathcal{D}'[G°]$ for some $G \in \mathcal{U}_h(\mathcal{D})$ and with covariance operator Q_H as indicated in the beginning of Paragraph 2. We put $Q_H = DD^*$ for some $D \in L^2(H,H)$, and denote by I the isometry from $\tilde{H}$ onto the closure of $D^*(H)$ in H. K denotes an arbitrary real and separable Hilbert space.

A first method of stochastic integration is based on the one developed in [12] where the stochastic integral is defined for processes in $L^2_{\tilde{L}^2(H,K,D)}(\mathcal{P}(\underline{\underline{F}}))$ with $D = Q_H^{1/2}$. The method of [12] can be extended for an arbitrary factorization $Q_H = DD^*$. In this case, the stochastic integral is defined as follows : If X is an elementary process of the form :

$$X_t(\omega) = \sum_{k=o}^{n-1} A_k \, 1_{]t_k, t_{k+1}]\times F_k}(t,\omega)$$

where $0 \leq t_o < t_1 \; \ldots < t_n \leq T$, $F_k \in \underline{\underline{F}}_{t_k}$ and $A_k \in L(H,K)$, then

$$\int_o^T X_t \, dW_t = \sum_{k=o}^{n-1} 1_{F_k} A_k (W_{t_{k+1}} - W_{t_k})$$

It is easily seen that

$$E \left\| \int_o^T X_t \, dW_t \right\|_K^2 = E \int_o^T \| X_t D \|_2^2 \, dt.$$

i.e. the integral defines an isometry of the vector subspace of $L^2_{\tilde{L}^2(H,K,D)}(\mathcal{P}(\underline{\underline{F}}))$ consisting of elementary processes into a vector subspace of $L^2_K(\Omega, \underline{\underline{F}}_T, P)$. This isometry extends to the first space and the image of a process X under the extended isometry is defined as being the stochastic integral of X and is denoted by $\int_o^T X_t \, dW_t$. For a process $X \in L^2_{\tilde{L}^2(H,K,D)}(\mathcal{P}(\underline{\underline{F}}))$, the process defined by $(\int_o^T 1_{[0,t]}(s) \, X_s \, dW_s)$ is a square-integrable martingale with values in K having a continuous version which is denoted by X.W or by $(\int_o^t X_s \, dW_s)$.

We can equally develop the stochastic integral as in [5] by means of the cylindrical brownian motion corresponding to W. For an element h of $\tilde{H}$ let us put $\tilde{W}_t(h) = \int_o^t h \, dW_s$ (the integral of the constant process equal to h). $\tilde{W}_t(h)$ defines a standard cylindrical motion on $\tilde{H}$, i.e.

i) $h \in \tilde{H}$, $h \neq 0$, $\tilde{W}_t(h) / \|h\|_{\tilde{H}}$ defines a real standard brownian motion,

ii) $t \in [0,T]$, the linear mapping of $\tilde{H}$ into $L^2(\Omega, \underline{\underline{F}}, P)$ defined by $h \to \tilde{W}_t(h)$ is continuous.

The second type of stochastic integral can be developed with respect to $\tilde{W}$ as in [5] , [9] and [11]. Let $(q_n)_{n \in N}$ be the non increasing sequence of strictly positive eigenvalues of Q_H, each being counted as many times as its multiplicity and let $(e_n)_{n\in\mathbb{N}}$ be the corresponding sequence of eigenvectors. We have $Q_H = \sum_n q_n e_n \otimes e_n$ and $\sum_n q_n < \infty$. In fact, (e_n) is an orthonormal basis of $\overline{Rg(Q_H)}$. It is easily verified that $(q_n^{-1/2} e_n)_n$ is an orthonormal basis of $\tilde{H}$. Let us put

$$W_t^n = \tilde{W}_t(q_n^{-1/2} e_n) = q_n^{-1/2} (W_t, e_n)_H$$

(W^n) is a sequence of mutually independent real standard Brownian motions. Let $(k_n)_{n\in\mathbb{N}}$ be an orthonormal basis of K. Then any process $X \in L^2_{L^2(\tilde{H},K)}(\mathcal{P}(\underline{\underline{F}}))$ can be written as follows :

$$X_t = \sum_{(m,n)\in\mathbb{N}^2} X_t^{m,n} (q_n^{-1/2} e_n) \otimes k_m$$

with $X^{m,n} \in L^2_{\mathbb{R}}((\mathcal{P}(\underline{\underline{F}}))$. The stochastic integral of X with respect to $\tilde{W}$ can then be defined by

$$\int_0^T X_t d\tilde{W}_t = \sum_{(m,n)\in\mathbb{N}^2} \left(\int_0^T X_t^{m,n} dW_t^n \right) k_m$$

where the series converges in $L^2_K(\Omega, \underline{\underline{F}}_T, P)$. We see that the stochastic integral reduces to that of real processes with respect to real Brownian motions. For $X \in L^2_{L^2(\tilde{H},K)}(\mathcal{P}(\underline{\underline{F}}))$, $(\int_0^T 1_{[0,t]}(s) X_s d\tilde{W}_s)$ defines as in the preceding case, a square-integrable K-valued martingale having a continuous version denoted again by $X.\tilde{W}$ or by $(\int_0^t X_s d\tilde{W}_s)$.

The connection between the two types of stochastic integrals was shown in [5]. It can be expressed as follows : The elements of $\tilde{H}$ are (not necessarily continuous) linear functionals on H. Let $\tilde{h}(x)$ denote the value of $\tilde{h} \in \tilde{H}$, considered as a functional on H, at a point $x \in H$. Then the mapping J : $\tilde{h} \otimes k \longrightarrow \tilde{h}(.)k$ extends to an isometry from $L^2(\tilde{H},K)$ onto $\tilde{L}^2(H, K, D)$ which in turn extends to an isometry from $L^2_{L^2(\tilde{H},K)}(\mathcal{P}(\underline{\underline{F}}))$ onto $L^2_{\tilde{L}^2(H, K, D)}(\mathcal{P}(\underline{\underline{F}}))$. Let J denote again this extension. If $X \in L^2_{L^2(\tilde{H},K)}(\mathcal{P}(\underline{\underline{F}}))$ then we have $X.\tilde{W} = (JX).W$.

E.3.1 EXAMPLE (CONTINUED)

We consider again our two-parameter Brownian sheet, the corresponding

L^2-valued Brownian motion W and the cylindrical Brownian motion $\tilde{W}$. We have seen in E.2.2 that $\tilde{H}$ can be identified with the Hilbert space generated by $\mathcal{D}$ under the scalar product

$$\phi, \Psi \in \mathcal{D}, \quad (\phi, \Psi)_{\tilde{H}} = \int_0^S \left(\int_u^S \phi(v)dv \int_u^S \Psi(v)dv \right) du$$

Let us consider the stochastic integral $\int_0^T \phi \, d\tilde{W}_t$, i.e. the integral of the constant process equal to $\phi \in \mathcal{D}$.

According to two-parameter stochastic calculus rules [1], we can write

$$\tilde{W}_T(\phi) = \int_0^T \phi \, d\tilde{W}_t = \int_{R_{TS}} \phi(u) \, W(dt,u)du$$

$$= \int_{R_{TS}} \left(\int_s^S \phi(u) \, du \right) dW_{t,s}$$

where R_{TS} is the rectangle $\{(t,s) \in \mathbb{R}_+^2 : t \leq T, \; s \leq S\}$.

This correspondence between the Hilbertian stochastic integral and the two-parameter stochastic integral is easily extended to operator-valued processes. In fact, let K be a separable real Hilbert space and let $\tilde{X} \in L^2_{L^2(\tilde{H},K)}(\mathcal{P}(\underline{\underline{F}}))$ and consider its isometric image $X = I\tilde{X}$ in $L^2_K(\mathcal{P}(\underline{\underline{F}}) \otimes \mathcal{B}_{[0,S]})$, (I was described in E.2.2).

Then we have

$$\int_0^T \tilde{X}_t \, d\tilde{W}_t = \int_{R_{TS}} X_{t,s} \, dW_{t,s}. \blacksquare$$

We can deduce from what we have seen that the stochastic integration of operator-valued processes giving square-integrable martingales can be carried out without factorization of Q and independently of the space H. In fact, for a $\mathcal{D}'$-valued Brownian motion W with covariance operator Q, $\tilde{H}$ is generated by $\mathcal{D}$ under the scalar product $(Q\phi, \Psi)$ and for any $h \in \tilde{H}$, $\tilde{W}_t(h)$ is obtained as the limit, in the quadratic mean, of a sequence $(\tilde{W}_t(\phi_n))_{n \in \mathbb{N}}$ where (ϕ_n) is a sequence in $\mathcal{D}$ converging to h. As we have seen, the stochastic integration is based on $\tilde{H}$ and $\tilde{W}$.

E.3.2 ANOTHER EXAMPLE

Another interesting example of a distribution-valued Brownian motion is derived from the same two-parameter Brownian sheet considered here. We define in the usual way the derivative $\frac{\partial W_{t,s}}{\partial s}$ that we denote by $\dot{W}_t$, i.e.

$$\forall\, \phi \in \mathcal{D} \quad \dot{W}_t(\phi) = -W_t(\dot{\phi})$$

Here, $\mathcal{D}$ is the space of infinitely differentiable functions with compact support on $]0,S[$. It is obvious that $\dot{W}$ is a $\mathcal{D}'$-valued Brownian motion whose covariance operator is the canonical injection of $\mathcal{D}$ into $\mathcal{D}'$. In fact, we have

$$E[\dot{W}_t(\phi)\dot{W}_t(\Psi)] = t \int_0^S \phi(s)\Psi(s)ds$$

The scalar product on $\mathcal{D}$ associated to the covariance operator Q of $\dot{W}$ is that of $L^2(0,S)$. Therefore, we have $\tilde{H} = L^2(0,S)$. We denote by $\tilde{\dot{W}}$ the cylindrical Brownian motion corresponding to $\dot{W}$. By the very definition of $\dot{W}$ we can write (cf. E.3.1)

$$\forall\, \phi \in \mathcal{D} \quad \dot{W}_T(\phi) = \int_0^T \phi \, d\tilde{\dot{W}}_t = \int_{R_{TS}} \phi(s) dW_{t,s}$$

This relation between the Hilbertian integral and the two-parameter integral extends to operator-valued processes. Starting from the above equality we can construct an isometry from $L^2_{L^2(L^2(0,S),K)}(\mathcal{P}(\underline{\underline{F}}))$ onto $L^2_K(\mathcal{P}(\underline{\underline{F}}) \otimes \mathcal{B}_{[0,S]})$, (K is a separable real Hilbert space). If X' is a process in the first space and X is its isometric image in the second one, then we have

$$\int_0^T X'_t \, d\tilde{\dot{W}}_t = \int_{R_{TS}} X_{t,s} \, dW_{t,s}$$

This gives another representation of the stochastic integral with respect to the two-parameter Brownian sheet.■

We can extend the above construction of the stochastic integral to the integration of processes with values in the space $L(\tilde{H}, \mathcal{D}'_0)$ of continuous operators from $\tilde{H}$ into a space $\mathcal{D}'_0$ of distributions.

In the sequel, $\mathcal{D}_o$ will denote a space of infinitely differentiable real functions with compact supports and $\mathcal{D}'_o$ will denote the corresponding space of distributions.

For $U \in \mathcal{U}_h(\mathcal{D}'_o)$, since the canonical mapping $k(U)$ of $\mathcal{D}'_o$ into $\mathcal{D}'_o(U)$ is nuclear, for any $A \in L(\tilde{H}, \mathcal{D}'_o)$, the mapping $k(U)A$ of $\tilde{H}$ into $\mathcal{D}'_o(U)$ is a nuclear, hence a Hilbert-Schmidt operator. Let us consider the vector space $L^2_{L(\tilde{H}, \mathcal{D}'_o)}(\mathcal{P}(\underline{\underline{F}}))$ of all weakly predictable $L(H,\mathcal{D}'_o)$-valued processes A such that for all $U \in \mathcal{U}_h(\mathcal{D}'_o)$

$$(\Pi_U(A))^2 = \int_o^T E\,\|k(U)A_s\|_2^2\, ds < \infty$$

(A is said to be weakly predictable if for all $\phi \in \mathcal{D}_o$ and all $h \in \tilde{H}$, the process $(\phi, A\,h)$ is predictable). It is easily seen that the positive square-root $\Pi_U(A)$ of the above integrale is a seminorm on $L^2_{L(\tilde{H}, \mathcal{D}'_o)}(\mathcal{P}(\underline{\underline{F}}))$. With the vector space topology induced by the set of seminorms $(\Pi_U : U \in \mathcal{U}_h(\mathcal{D}'_o))$, this space is a locally convex space. Devided by the equivalence relation $A \sim B \Longleftrightarrow A = B$ a.e. $dt \otimes dP$, it becomes separated. We denote the quotient space by the same symbol. For a given $A \in L^2_{L(\tilde{H}, \mathcal{D}'_o)}(\mathcal{P}(\underline{\underline{F}}))$ we define the following projective system of continuous square-integrable martingales

$$M^U_t = \int_o^t k(U)A_s\, d\tilde{W}_s$$

i.e. for $U,V \in \mathcal{U}_h(\mathcal{D}'_o)$ such that $U \subset V$, we have $k(V,U)M^U = M^V$. Therefore, there is a continuous square-integrable $\mathcal{D}'_o$-valued martingale M defined by the system (M^U). Consequently, we can give the following definition.

<u>DEFINITION 3.3</u>

For $A \in L^2_{L(\tilde{H}, \mathcal{D}'_o)}(\mathcal{P}(\underline{\underline{F}}))$ we define the continuous square-integrable martingale $M = A.\tilde{W}$ as the continuous modification of the projective limit of the system $((k(U)A).\tilde{W}\,,\ U \in \mathcal{U}_h(\mathcal{D}'_o))$ and put $M_t = \int_o^t A_s\, d\tilde{W}_s$.

The following theorem gives the converse of this definition.

THEOREM 3.4

Let a $\mathcal{D}'_o$-valued continuous square-integrable martingale M be defined by the projective system

$$\{M^U = A^U.\tilde{W} : U \in \mathcal{U}_h(\mathcal{D}'_o),\ A^U \in L^2_{L^2(\tilde{H},\, \mathcal{D}'_o(U))}(\mathcal{P}(\underline{\underline{F}}))\}.$$

Then there is a process $A \in L^2_{L(\tilde{H},\, \mathcal{D}'_o)}(\mathcal{P}(\underline{\underline{F}}))$ such that $k(U).A = A^U$ and $M = A.\tilde{W}$.

Proof : Let the mapping γ of $\mathcal{D}_o$ into $L^2_{\tilde{H}}(\mathcal{P}(\underline{\underline{F}}))$ be defined by $\gamma\ \phi = (A^U)^* \phi$, where U is chosen in such a way that $\phi \in \mathcal{D}_o[U°]$. γ is well defined. In fact, if $U,V \in \mathcal{U}_h(\mathcal{D}'_o)$ with $V \subset U$ and $\phi \in \mathcal{D}_o[U°]$, as we have $k(U,V)A^V = A^U$ a.e. $dt \otimes dP$, we get $(A^V)^* (k(U,V))^* = (A^U)^*$ a.e. where $k^*(U,V)$ is the imbedding of $\mathcal{D}_o[U°]$ into $\mathcal{D}_o[V°]$. We thus have $(A^V)^* \phi = (A^U)^* \phi$ a.e..
The mapping γ which is linear is also continuous from $\mathcal{D}_o$ into $L^2_{\tilde{H}}(\mathcal{P}(\underline{\underline{F}}))$.
In fact, let (ϕ_n) converge to ϕ in $\mathcal{D}_o[U°]$ and let $(\gamma\ i(U°)\ \phi_n) = ((A^U)^* \phi_n)$ converge in $L^2_{\tilde{H}}(\mathcal{P}(\underline{\underline{F}}))$. Since $(A^U)^*$ is continuous, according to the closed graph theorem this last sequence converges to $(A^U)^* \phi = \gamma\ i(U°)\phi$. Hence, $\gamma\ i(U°)$ is continuous for all $U \in \mathcal{U}_h(\mathcal{D}'_o)$; this implies that γ is continuous. Since $\mathcal{D}_o$ is a nuclear space γ is nuclear and has the following representation

$$(\gamma\ \phi)_t(\omega) = \sum_i \lambda_i\ S_i(\phi)\ X^i_t(\omega)$$

where $(\lambda_i) \in l^1$, $(S_i) \subset \mathcal{D}'_o$ is equicontinuous and $(X^i) \subset L^2_{\tilde{H}}(\mathcal{P}(\underline{\underline{F}}))$ is bounded. The series converging in $L^2_{\tilde{H}}(\mathcal{P}(\underline{\underline{F}}))$ also converges a.e. $dt \otimes dP = d\mu$. In fact, we have

$$\mu(\sum_i |\lambda_i|\ \|X^i_t(\omega)\|_{\tilde{H}}) = \sum_i |\lambda_i|\ \ \mu(\|X^i_t(\omega)\|_{\tilde{H}})$$

$$\leq \sqrt{T}\ \ \sup_i\ \sqrt{\mu\ (\|X^i\|^2_{\tilde{H}})}\ .\sum_i\ |\lambda_i| < \infty$$

Hence, $\sum_i |\lambda_i|\ \|X^i_t(\omega)\|_{\tilde{H}} < \infty$ a.e. μ. Let N be the subset of $[0,T] \times \Omega$ on which this last relation holds. Then $\forall\ (t,\omega) \in N$ the series $\sum_i \lambda_i\ S_i(\phi)X^i_t(\omega)$ converges in $\tilde{H}$.

Let us define the operator $C_t(\omega) : \mathcal{D}_o \longrightarrow \tilde{H}$ by : $\forall \phi \;\; C_t(\omega)\phi = \sum_i \lambda_i \, S_i(\phi) \, X_t^i(\omega)$. It is defined on N, hence a.e.. For $(t,\omega) \in N$, $C_t(\omega)$ is a linear mapping. Moreover, it is continuous. In fact if (ϕ_n) converges to ϕ in $\mathcal{D}_o$, we have $\lim_n \sum_i \lambda_i \, S_i(\phi_n) \, X_t^i(\omega) = \sum_i \lim_n \lambda_i \, S_i(\phi_n) \, X_t^i(\omega)$, i.e. $\lim_n C_t(\omega) \, \phi_n = C_t(\omega)\phi$. $C_t(\omega)$ is a sequentially continuous mapping. As $\mathcal{D}_o$ is a bornological space $C_t(\omega)$ is continuous. We put $A_t(\omega) = C_t(\omega)^*$. $A_t(\omega) \in L(\tilde{H}, \mathcal{D}'_o)$ and for $\phi \in \mathcal{D}_o$, $h \in \tilde{H}$, $(\phi , A_t(\omega)h) = (C_t(\omega)\phi , h) = ((\gamma\phi)_t(\omega),h)$ a.e.. Therefore, A defined a.e. $dt \otimes dP$, has values in $L(\tilde{H}, \mathcal{D}'_o)$ and is weakly measurable.

Finally, for $U \in \mathcal{U}_h(\mathcal{D}'_o)$, $\phi \in \mathcal{D}_o[U°]$, $h \in \tilde{H}$ we have $(k(U)A_t(\omega)h,\phi) = (h, A_t^*(\omega)k^*(U)\phi) = (h, A_t^*(\omega)\phi) = (h, C_t(\omega)\phi) = (h, (A^U)^*\phi)$. Hence $k(U) \, A_t(\omega) = A_t^U(\omega)$ a.e. $dt \otimes dP$. We then conclude that $M = A.\tilde{W}$. ■

REMARK 3.5

As in the finite dimensional case, we do not need to define the stochastic integral $\int_o^t A_s \, d\tilde{W}_s$ only for previsible processes A. It can also be defined for optional processes giving exactly the same set of square-integrable martingales, [13]. In fact, one can show that there is a previsible process A' such that $A'.\tilde{W} = A.\tilde{W}$.

Spaces $L^2(\mathcal{O}(\underline{\underline{F}}))$ of optional processes are defined as those of previsible processes.

4. MARTINGALE REPRESENTATION

The following representation theorem given in [9] will be used in the representation of distribution-valued processes. W, H, $\tilde{H}$, D and I are as defined in the preceding paragraphs.

THEOREM 4.1

Let M be a square-integrable martingale with values in a separable real Hilbert space K. Then there exists a process $X \in L^2_{L^2(\tilde{H},K)}(\mathcal{P}(\underline{\underline{F}}))$ such that

$$(4.1) \qquad M_t = M_o + \int_o^t X_s d\tilde{W}_s + M_t^{\perp}$$

where $M^{\perp}$ is a square integrable martingale orthogonal to W.

W considered as an H-valued Brownian motion has then the representation given below in terms of $\tilde{W}$.

PROPOSITION 4.2

$$(4.2) \qquad W_t = \int_o^t DI \; d\tilde{W}.$$

Proof : Let $(q_n)_{n \in \mathbb{N}}$ be the decreasing sequence of eigenvalues of the covariance operator Q_H of W, (e_n) the orthonormal system of corresponding eigenvectors. The sequence $(q_n^{-1/2} e_n)$ is a complete orthonormal basis of $\tilde{H}$ and we can write

$$\forall\, h \in \tilde{H}, \qquad h = \sum_n (h, q_n^{-1/2} e_n)_{\tilde{H}}\, q_n^{-1/2} e_n$$

Then

$$DI(h) = \sum_n (h, q_n^{-1/2} e_n)_{\tilde{H}}\, DI(q_h^{-1/2} e_n)$$

$$= \sum_n (h, q_n^{-1/2} e_n)_{\tilde{H}}\, q_n^{1/2} e_n$$

We thus have

$$DI = \sum_n q_n^{1/2} (q_n^{-1/2} e_n \otimes e_n)$$

as an operator from $\tilde{H}$ into H. According to the definition of the stochastic integral with respect to $\tilde{W}$, we have

$$\int_o^t DI \; d\tilde{W}_s = \sum_n (W_t, e_n)_H\, e_n$$

where the series converges a.s. and also in the quadratic mean. But for each $t \in [0,T]$, $W_t \in \overline{R_g Q_H}$ a.s. In fact, $\forall h \in H$, $E((W_t,h)_H^2) = t(Q_H h,h)_H$ and if $h \in \text{Ker } Q_H$ then $(W_t,h)_H = 0$ a.s. because, by the separability of Ker Q_H, we have a.s. $(W_t,h)_H = 0 \ \forall h \in \text{Ker } Q_H$, i.e. $W_t \in (\text{Ker } Q_H)^{\perp} = \overline{R_g Q_H}$.

Since (e_n) is an orthonormal basis of $\overline{R_g Q_H}$, we have

$$W_t = \sum_n (W_t, e_n)e_n = \int_0^t DI \, d\tilde{W}_s \quad \text{a.s.} \ \blacksquare$$

Representation Theorem 4.1 implies an orthogonal decomposition of distribution - valued square-integrable martingales.

PROPOSITION 4.3

Let $\mathcal{D}_o'$ be a space of distributions for a space $\mathcal{D}_o$ of infinitely differentiable real functions with compact supports. Let M be a square-integrable $\mathcal{D}_o'$-valued martingale. Then

$$(4.3) \qquad M = N + M^{\perp}$$

where N is a continuous square-integrable $(\underline{\underline{F}}^W, P)$-martingale and $M^{\perp}$ is orthogonal to W, in the sense that $\forall \phi \in \mathcal{D}_o$, $\forall \Psi \in \mathcal{D}$, $M(\phi)$ and $W(\Psi)$ are orthogonal martingales.

Proof : Let $G \in \mathcal{U}_h(\mathcal{D}_o)$ be such that M induces a square-integrable martingale M' with values in $\mathcal{D}_o'[G°]$ such that M is a modification of $i(G°)\,M'$, (cf. Proposition 1.3). According to the preceeding theorem M' has the orthogonal decomposition : $M' = N' + M'^{\perp}$ where N' is a continuous square-integrable $(\underline{\underline{F}}^W,P)$ - martingale and $M'^{\perp}$ is a square-integrable martingale orthogonal to W. We obtain the decomposition (4.3) from :

$$i(G°)M_t' = i(G°)N_t' + i(G°)M_t'^{\perp} \quad \text{a.s.} \ \blacksquare$$

Now we give the representation theorem for $\mathcal{D}_o'$-valued square-integrable martingales.

THEOREM 4.4

Let $\mathcal{D}_o$ and $\mathcal{D}'_o$ be as in Proposition 4.3 and let M be a square-integrable $(\underline{\underline{F}}^W, P)$ - martingale with values in $\mathcal{D}'_o$ such that $M_o = 0$. Then there is a process $A \in L^2_{L(\tilde{H}, \mathcal{D}'_o)}(\mathcal{P}(F^W))$ such that $M = A.\tilde{W}$.

Proof : According to Theorem 4.1, for each $U \in \mathcal{U}_h(\mathcal{D}'_o)$ there is an element $A^U \in L^2_{L^2(\tilde{H}, \mathcal{D}'_o(U))}(\mathcal{P}(\underline{\underline{F}}^W))$ such that $k(U)M = A^U.\tilde{W}$. But Theorem 3.4 says that there is a process $A \in L^2_{L(\tilde{H}, \mathcal{D}'_o)}(\mathcal{P}(\underline{\underline{F}}^W))$ such that the continuous martingale $A.\tilde{W}$ is a modification of M. Since M is right-continuous, we have $M = A.\tilde{W}$.■

5. THE GIRSANOV THEOREM

We use the notations of the preceding paragraphs.

THEOREM 5.1

Let W be a $\mathcal{D}'$-valued Brownian motion with covariance operator Q and let $G \in \mathcal{U}_h(\mathcal{D})$ be such that W is considered as a Brownian motion with trajectories in $H = \mathcal{D}'[G°]$, (cf. Paragraph 2). Let $\ell \in L^2_{\tilde{H}}(\mathcal{P}(\underline{\underline{F}}))$ and $h = i(G°)\, D\, I\, \ell$ and put

$$(5.1) \quad Y_t = \int_o^t h_s\, ds + W_t,$$

$$(5.2) \quad Z_t = \exp\left(-\int_o^t \ell_s\, d\tilde{W}_s - \frac{1}{2}\int_o^t \|\ell_s\|^2_{\tilde{H}}\, ds\right).$$

If $E(Z_T) = 1$, then $\tilde{P} = Z_T P$ is a probability on $\underline{\underline{F}}_T$, equivalent to P and Y is a $\mathcal{D}'$-valued $(\underline{\underline{F}}, \tilde{P})$-Brownian motion with covariance operator Q.

Proof : This theorem is an immediate consequence of the Girsanov Theorem corresponding to the case where Equation (5.1) is written for H-valued processes, [5] . We may hence suppose that $h = D\, I\, \ell$ and that processes in (5.1) are H-valued. Let us first remark that according to Proposition 4.2, we have

$$< \int_o \ell_s\, d\tilde{W}_s\, , W >_t \; = \; < \int_o \ell_s\, d\tilde{W}_s, \int_o DI\, d\tilde{W}_s >_t$$

$$= \int_o^t DI\, \ell_s\, ds \; = \; \int_o^t h_s\, ds$$

By using the classical Girsanov Theorem, we see that for $\phi \in \mathcal{D}$, the process $Y(\phi) = W(\phi) + \int_0^{\cdot} h_s(\phi)ds$ is a continuous $(\underline{\underline{F}},\tilde{P})$ - local martingale with increasing process $t\|\phi\|^2_{\tilde{H}}$. From this we deduce that for $\phi, \Psi \in \mathcal{D}$, we have $< Y(\phi), Y(\Psi)>_t = t(Q\phi,\Psi)$. ■

REMARK 5.2

The process h in (5.1) does not depend on the particular factorization $Q = DD^*$ we have been considering here. It actually depends only on the martingale $M = \ell.\tilde{W}$, since $\int_0^t h_s\, ds = <M,W>_t$. But, once M is given, it can also be obtained by means of another factorization of Q, for the set of martingales $\{\ell.\tilde{W},\ \ell \in L^2_{\tilde{H}}\ (\mathcal{P}(F))\}$ is the same for all the factorizations of Q.

E.5.3 EXAMPLE (CONTINUED)

As a consequence of the above theorem we have the following Girsanov theorem which is a particular case of the one given in [8].

Let $(W_{t,s})$ be the Brownian sheet considered here, let $H \in L^2(\mathcal{P}(\underline{\underline{F}}) \otimes \mathcal{B}_{[0,S]})$ and put

$$(5.3)\quad Y_{t,s} = \int_0^t \left(\int_0^s H_{u,v}\, dv\right) du + W_{t,s}$$

$$(5.4)\quad Z_t = \exp\left(- \int_0^t \int_0^S H_{u,v}\, dW_{u,v} - \frac{1}{2} \int_0^t \left(\int_0^S H^2_{u,v}\, dv\right)du\right)$$

If $E(Z_T) = 1$, then $\tilde{P} = Z_T\, P$ is a probability measure on $\underline{\underline{F}}_T$, equivalent to P and, under $\tilde{P}$, Y is a Brownian sheet such that, $\forall\, t$, $\underline{\underline{F}}_t$ is independent of $\sigma\,(Y_{u,s} - Y_{t,s}\ ;\ u \in [t, T],\ s \in [0,S])$.

In fact, if $\ell \in \tilde{H}$ and the process H is its isometric image under the isometry considered in E 3.1 (H can also be considered as an $L^2(0,S)$ - valued process), then the process $(\int_0^{\cdot} H_{t,v}\, dv)_{t \in [0,T]}$ is defined by $D\, H_{t,\cdot}$. ■

6. FILTERING EQUATION

The "state and observation" model presented below extends the Hilbertian model of [5].

$\mathcal{D}$, $\mathcal{D}'$, $\mathcal{D}_o$, $\mathcal{D}'_o$ and W will be as described in the preceding paragraphs. We shall use the same notations.

The state process X is a $\mathcal{D}'_o$-valued process defined by the equation

$$(6.1) \qquad X_t = X_o + \int_o^t f_s \, ds + M_t$$

where X_o is a weakly measurable random variable on $(\Omega , \underline{\underline{F}}_o, P)$ such that, $\forall\, U \varepsilon\ \mathcal{U}_h(\mathcal{D}'_o)$, $E p_U^4(X_o) < \infty$, f is a weakly progressive process (i.e. $\forall\, \phi \varepsilon\ \mathcal{D}_o$ $f(\phi)$ is progressive) such that $\forall\, U \varepsilon\ \mathcal{U}_h(\mathcal{D}'_o)$, $\int_o^T E\, p_U^4(f_t) dt < \infty$ and M is a $(\underline{\underline{F}},P)$ - martingale such that $M_o = 0$ and $\forall\, U \varepsilon\ \mathcal{U}_h(\mathcal{D}'_o)$, $E\,[\, p_U^4(M_t)] < \infty$. X_o and M are supposed to be independent.

The observation process is the process Y defined by (5.1), with $\ell \varepsilon\ L^4_{\tilde{H}}(\mathcal{P}(\underline{\underline{F}}))$.

In what follows $\underline{\underline{G}} = (\underline{\underline{G}}_t)_{t \varepsilon [0,T]}$ denotes the natural filtration of Y and $\mathcal{O}(\underline{\underline{G}})$ the σ- algebra of optional sets in $\Omega \times [0,T]$, associated with $\underline{\underline{G}}$.

We note that according to Theorems 4.3 and 4.4, the martingale M has the following representation

$$(6.2) \qquad M = C.\tilde{W} + M^{\perp}$$

where $C \varepsilon\ L^2_{L(\tilde{H},\ \mathcal{D}'_o)}(\mathcal{P}(\underline{\underline{F}}))$.

In order to express the filtering equation, we shall need the following lemma for the definition of projections.

LEMMA 6.1

Let A be a weakly measurable process with values in $L(\tilde{H}, \mathcal{D}'_o)$ such that, $\forall\, U \varepsilon\ \mathcal{U}_h(\mathcal{D}'_o)$, $\int_o^T E\, \| k(U) A_t \|_2^2\, dt < \infty$. Then there is a process

$\hat{A} \in L^2_{L(\tilde{H},\ \mathcal{D}'_o)}(\mathcal{O}(\underline{\underline{G}}))$ such that $k(U)\hat{A}$ is the projection of $k(U)\,A$ onto $L^2_{L^2(\tilde{H},\ \mathcal{D}'_o(U))}(\mathcal{O}(\underline{\underline{G}}))$. $\hat{A}$ is then the conditional expectation of A given $\mathcal{O}(\underline{\underline{G}})$ under the measure $dt \otimes dP$ and, for almost all t, $\hat{A}_t$ is a version of $E(A_t/\underline{\underline{G}}_t)$.

The construction of $\hat{A}$ is the same as that of A in Theorem 3.4.

In case A is a measurable square-integrable process (with respect to $dt \otimes dP$) with values in a separable Hilbert space, $\hat{A}$ will again denote its conditional expectation given $\mathcal{O}(\underline{\underline{G}})$ under $dt \otimes dP$.

THEOREM 6.2

(i) The innovation process ν defined by

$$(6.3)\qquad \nu_t = Y_t - \int_o^t \hat{h}_s\, ds$$

is a $\mathcal{D}'$-valued $(\underline{\underline{G}},P)$ - Brownian motion with the same covariance operator Q as W.

We denote by $\tilde{\nu}$ the corresponding cylindrical Brownian motion on $\tilde{H}$.

(ii) $\hat{X}$ has a continuous version given by the following equation.

$$(6.4)\quad \hat{X}_t = E(X_o) + \int_o^t \hat{f}_s\, ds + \int_o^t [\widehat{(\ell \otimes X)}_s - (\hat{\ell} \otimes \hat{X})_s + \hat{C}_s]\, d\tilde{\nu}_s$$

where $\ell \otimes X$ is the $L(\tilde{H},\ \mathcal{D}'_o)$-valued process defined by $(\ell \otimes X)k = (\ell,k)_{\tilde{H}}\, X$ for all $k \in \tilde{H}$. The definition of $\hat{\ell} \otimes \hat{X}$ is the same.

Proof : The proof is the same as in the Hilbertian case of [5] and [15]. We do not reproduce it entirely.

For any nonrandom measurable bounded function f and any $k \in \tilde{H}$ we have, by an application of the Ito formula,

$$E(\exp i \int_o^t f_s\, d\tilde{\nu}_s(k) \,/\, \underline{\underline{G}}_u) = \exp(-\tfrac{1}{2} \|k\|^2_{\tilde{H}} \int_u^t f_s^2\, ds).$$

where $d\tilde{\nu}_s(k) = (h_s - \hat{h}_s)(k)ds + d\tilde{W}_s(k)$, $((h_s(\omega) - \hat{h}_s(\omega))(k)$ is the value of the functional $h_s(\omega) - \hat{h}_s(\omega)$ at $k \in H)$. From this (i) is concluded.

Let N be defined by :

$$N_t = \hat{X}_t - E(X_o) - \int_o^t \hat{f}_s \, ds$$

It is proved as in [3] that N is a square-integrable (G,P) - martingale with values in $\mathcal{D}'_o$. Then again as in the finite dimensional case of [3], N is represented as a stochastic integral with respect to $\tilde{\nu}$. In fact, the following lemma holds.

LEMMA 6.3

Let N be any square-integrable ($\underline{\underline{G}}$,P)-martingale with values in $\mathcal{D}'_o$. Suppose $N_o = 0$. Then there is an element $U \in L^2_{L(\tilde{H}, \mathcal{D}'_o)}(\mathcal{P}(\underline{\underline{G}}))$ such that $N = U.\tilde{\nu}$. (According to Remark 3.5, U can be chosen to be optional).

The proof of this lemma is similar to that of [15] in the linear case. But the exponential process used for the change of probability is given here by (5.2).

Following the lines of [15], we first find the filtering equation for $k(U)X$, $U \in \mathcal{U}_h(\mathcal{D}'_o)$, by using the fact that processes in the observation equation (5.1) can be considered as Hilbert space-valued processes, and then obtain equation (6.4) by passing to projective limits. ■

REMARK 6.4

When X is defined as a Hilbert space-valued process the filtering equation would have exactly the same expression as the one given in [5].

7. FILTERING EQUATIONS FOR TWO-PARAMETER SEMINARTINGALES

Filtering equations for two-parameter semi-martingales were obtained in [8] by means of two-parameter stochastic calculus techniques. The same types of filtering equations for two-parameter diffusion processes were deduced in [4] and in [7] in a rather formal way from the filtering equation for Hilbert space-valued processes. We want to show here that these equations can be deduced rigorously from Equation (6.4).

Let B be a Brownian sheet and M be defined by

$$M_{t,s} = \int_{R_{ts}} G_{u,v} dB_{u,v}$$

where G is a non-random continuous function. Let X be a continuous process satisfying the following system of equations.

$$(S_1)\ X_{t,s} = x(s) + \int_o^t f(u,s,X_{u,s})du + \int_o^t g(u,s,X_{u,s})\ M(du,s)$$

$$(S_2)\ X_{t,s} = \tilde{x}(t) + \int_o^s \tilde{f}(t,v,X_{t,v})dv + \int_o^s \tilde{g}(t,v,X_{t,v})M(t,dv)$$

with the boundary conditions :

$$x(s) = x_o + \int_o^s \tilde{f}(o,v,x(v))dv$$

$$\tilde{x}(t) = x_o + \int_o^t f(u,o,\tilde{x}(u))du$$

where x_o is a real number and x, $\tilde{x}$, f, $\tilde{f}$, g, $\tilde{g}$ are nonrandom continuous real functions. The existence of Markovian solutions of such a system is studied in [6].

We shall only consider equation (S_1) as the state equation. We suppose that the observation equation is given by (5.3) considered in E 5.3 with W independent of B. We shall systematically use the one-to-one correspondance between the Hilbertian stochastic integral and the two-parameter stochastic integral and the isometry I considered in E 3.1. Rather than supposing that $(X_{t,s})$ can be considered as a distribution valued process, we shall take it as a real process depending on the parameter s. We suppose that all the 4-integrability conditions on the model of Paragraph 6 are also satisfied here. For a process $(Z_{t,u,v,\ldots})_t$ depending on parameters u,v, ... the process $(\hat{Z}_{t,u,v,\ldots})_t$ is a measurable version of the process $(E(Z_{t,u,v,\ldots}/\underline{\underline{G}}_t))_t$. Such versions exist for all the processes that we consider here for the derivation of the filtering equation.

THEOREM 7.1

(i) The innovation process ν defined by

$$\nu_{t,s} = Y_{t,s} - \int_o^t \left(\int_o^s \hat{H}_{u,v}\, dv\right) du$$

is a Brownian sheet such that, for all s, $(\nu_{\cdot,s})$ is adapted to $\underline{\underline{G}}$ and $\forall t$, $\underline{\underline{G}}_t$ and $\sigma\{\nu_{u,s}-\nu_{t,s} : (u,s) \in [t,T] \times [0,S]\}$, are independent.

(ii) $\hat{X}$ satisfies the following equation :

$$(7.2)\quad \hat{X}_{t,s} = x(s) + \int_0^t \widehat{f(u,s,X_{u,s})}\,du + \int_0^t \int_0^S (\widehat{X_{u,s}\,H_{u,v}} - \hat{X}_{u,s}\,\hat{H}_{u,v})\,d\nu_{u,v}.$$

Proof : By taking $h_t = \int_0^{\cdot} H_{t,v}\,dv$, we see that Equation (5.3) can be considered as an equation between $\mathcal{D}'$-valued processes. We then immediately deduce the assersion about the innovation process.

Equation (6.4) becomes here :

$$(7.3)\quad \hat{X}_{t,s} = x(s) + \int_0^t \widehat{f(u,s,X_{u,s})}\,du + \int_0^t (\widehat{X_{u,s}\,\ell_u} - \hat{X}_{u,s}\,\hat{\ell}_u)\,d\tilde{\nu}_u.$$

According to E 5.3. H is the image of ℓ under the isometry I constructed in E 3.1 and we have

$$\int_0^t (\widehat{X_{u,s}\,\ell_u} - \hat{X}_{u,s}\,\hat{\ell}_u)\,d\tilde{\nu}_u = \int_0^t \int_0^S I(\widehat{X_{u,s}\,\ell_u} - \hat{X}_{u,s}\,\hat{\ell}_u)_v\,d\nu_{u,v}$$

$$= \int_0^t \int_0^S (\widehat{X_{u,s}\,I\ell_u} - \hat{X}_{u,s}\,I\hat{\ell}_u)_v\,d\nu_{u,v}$$

$$= \int_0^t \int_0^S (\widehat{X_{u,s}\,H_{u,v}} - \hat{X}_{u,s}\,\hat{H}_{u,v})\,d\nu_{u,v}$$

From this and (7.3), equation (7.2) is deduced. ■

Equation (7.2) was called in [8] the horizontal filtering equation. The vertical filtering equation is obtained by interchanging the roles of the two parameters. The filtering equation where the two parameters play a symmetric role (called in [8] the diagonal filtering equation) can be derived from the horizontal and vertical ones as in [8].

REFERENCES

[1] R. CAIROLI and J.B. WALSH, Stochastic integrals in the plane, Acta Math. 134 (1975) 111-183.

[2] C. DELLACHERIE and P.A. MEYER, Probabilité et Potentiel, Hermann, Paris, Tome 1 (1975), Tome 2 (1980).

[3] M. FUJISAKI, G. KALLIANPUR and H. KUNITA, Stochastic differential equation for the nonlinear filtering problem, Osaka J. Math. 9 (1972) 19-40.

[4] H. KOREZLIOGLU, Two-parameter Gaussian Markov processes and their recursive filtering, Ann. Scientifiques de l'Université de Clermont 67 (1979) 69-93.

[5] H. KOREZLIOGLU, Nonlinear filtering equation for Hilbert space valued processes, in Nonlinear Stochastic Problems, Ed. R.S. Bucy and J.M.F. Moura, Reidel Publ. Co. Dordrecht 1983, 279-289.

[6] H. KOREZLIOGLU, P. LEFORT and G. MAZZIOTTO, Une propriété markovienne et diffusions associées, in Lecture Notes in Math. Nr. 863, Springer-Verlag (1981), 245-274.

[7] H. KOREZLIOGLU, G. MAZZIOTTO and J. SZPIRGLAS, 2-D filtering via infinite dimensional filtering, Note technique NT/PAA/ATR/MTI/394, CNET, Paris (1981).

[8] H. KOREZLIOGLU, G. MAZZIOTTO and J. SZPIRGLAS, Nonlinear filtering equations for two-parameter semimartingales, Stochastic Processes and their Applications, 15 (1983) 239-269.

[9] D. LEPINGLE and J.-Y. OUVRARD, Martingales Browniennes hilbertiennes, C.R. Acad. Sc. Paris, t.276 (1973) , Série A 1225-1228.

[10] C. MARTIAS, Filtrage non-lineare dans des espaces de Hilbert réels et séparables, Ann. Scientifiques de l'Université de Clermont-Ferrand II, 69 (1981) 87-113.

[11] M. METIVIER and J. PELLAUMAIL, Stochastic Integration, Academic Press, 1980.

[12] M. METIVIER and G. PISTONE, Une formule d'isométrie pour l'intégrale stochastique et équation d'évolution linéaires stochastiques, Z. Wahrsch. Verw. Gebiete, 33 (1975) 1-18.

[13] P.A. MEYER, Théorie des intégrales stochastiques, Lecture Notes in Math. 511, Springer-Verlag (1976) 246-400.

[14] J.Y. OUVRARD, Représentation de martingales vectorielles de carré intégrable à valeurs dans les espaces de Hilbert réels séparables, Z. Wahrsch. Verw. Gebiete, 33 (1975) 195-208.

[15] J.Y. OUVRARD, Martingale projection and linear filtering in Hilbert spaces, SIAM J. Control and Optimization, 16, 6 (1978) 912-937.

[16] H.H. SCHAEFER, Topological Vector Spaces, Macmillan, New-York (1966).

[17] L. SCHWARTZ, Théorie des Distributions, Hermann, Paris (1966).

[18] J. SZPIRGLAS and G. MAZZIOTTO, Modèle général du filtrage non-lineare et équations différentielles stochastiques associées, Ann. Inst. H. Poincaré, Série B 15 (2) (1979) 147-173.

[19] S. USTUNEL, Stochastic integration on nuclear spaces and its applications, Ann. Inst. H. Poincaré, Série B, 18 (2) (1982) 165-200.

[20] S. USTUNEL, Distributions-valued semimartingales and applications to control and filtering, in this volume.

JEU DE DYNKIN
AVEC COUT DEPENDANT D'UNE STRATEGIE CONTINUE

par

J.P.LEPELTIER

M.A.MAINGUENEAU (*)

—

INTRODUCTION

L'objet de ce travail est l'étude d'un jeu de somme nulle de critère :

$$E\,(X_S\, 1\!\!1_{(S\leq T)} - X'_T\, 1\!\!1_{(T<S)} + (C^u - C'^v)_{S\wedge T})$$

où le premier (resp.second) joueur choisit à la fois une stratégie continue et une stratégie d'arrêt, soit (u,S) (resp.(v,T)) et cherche à maximiser (resp.minimiser) le critère.

Lorsque C^u et C'^v sont identiquement nuls pour tout u et v, on retrouve le jeu de Dynkin classique d'où le titre de notre article, qui se présente donc comme une généralisation des travaux de J.M.Bismut [3] et M.Alario, J.P.Lepeltier, B.Marchal [1].

En fait, grâce à des hypothèses de Mokobodski généralisées nous montrons que ce problème se ramène à un problème de contrôle mixte. On donne alors de "bonnes' conditions suffisantes d'existence d'une valeur ou d'un point-selle. Notre travail se décompose donc en quatre parties :

Dans un premier paragraphe nous introduisons avec précision le modèle étudié, puis dans un second nous étudions le problème de contrôle mixte associé à notre jeu. Le troisième paragraphe est alors consacré à l'application des résultats du contrôle mixte en vue de la réalisation de notre jeu, alors qu'enfin le quatrième paragraphe énonce les résultats d'existence.

(*) Département de mathématiques de l'Université du Maine, Route de Laval 72017 - LE MANS Cedex.

§ 1 - LE MODELE

Le modèle étudié ci-dessous consiste en la donnée :

- d'un espace de probabilité filtré $(\Omega, \underline{\underline{F}}, \underline{\underline{F}}_t, P)$, la filtration $(\underline{\underline{F}}_t)_{t\geq 0}$ satisfaisant aux "conditions habituelles" [5],

- d'un espace métrique compact U qui jouera le rôle d'espace où les contrôles continus prennent leurs valeurs,

- d'un ensemble de stratégies identique pour chacun des deux joueurs, soit $\mathcal{U}$ ensemble des couples (u,S) où :

i) S est un $\underline{\underline{F}}_t$ temps d'arrêt,

ii) u est une application de $R_+ \times \Omega$ dans U, prévisible,

- d'un critère de la forme :

$$J\,(u,S),(v,T)) = E\,(X_S\, 1\!\!1_{(S\leq T)} - X'_T\, 1\!\!1_{(T<S)} + (C^u - C'^v)_{S\wedge T})$$

où pour tout u et v, C^u et C'^v sont des processus à variation finie vérifiant pour tout $\underline{\underline{F}}_t$ temps d'arrêt S :

i) si u et u' (resp.v et v') coïncident jusqu'à S, alors :

$$C^u_{t\wedge S} = C^v_{t\wedge S} \qquad (\text{resp. } C'^v_{t\wedge S} = C'^{v'}_{t\wedge S})$$

ii) si u et u' (resp.v et v') coïncident après S, alors :

$$C^u_{t\vee S} - C^u_S = C^{u'}_{t\vee S} - C^{u'}_S \quad (\text{resp.} C'^v_{t\vee S} - C'^{v'}_S = C'^{v'}_{t\vee S} - C'^{v'}_S)$$

et X, X' sont des processus optionnels de la classe (D).

La règle du jeu est alors la suivante : le premier joueur choisit (u,S) de manière à maximiser le critère, alors que le second joueur cherche à choisir (v,T) de manière à minimiser ce même critère.

On rappelle les notions suivantes :

DEFINITION 1.1.

On appelle valeur supérieure (resp.inférieure) du jeu l'expression :

$$V^+ = \inf_{(v,t)\in\mathcal{U}} \ \sup_{(u,S)\in\mathcal{U}} J((u,S),(v,T)) \quad (\text{resp.} V^- = \sup_{(u,S)\in\mathcal{U}} \ \inf_{(v,T)\in\mathcal{U}} J((u,S),(v,T))$$

On dira que le jeu possède une valeur V lorsque :

$$V^+ = V^- = V$$

DEFINITION 1.2.

Un couple $((u^*,S^*),(v^*,T^*))$ *est appelé point-selle du jeu s'il est tel que :*

$$J((u,S),(v^*,T^*)) \leqslant J((u^*,S^*),(v^*,T^*)) \leqslant J((u^*,S^*),(v,T))$$

pour tous $(u,S),(v,T)$ *appartenant à* $\mathcal{U}$.

Il est bien connu [1] que s'il existe un point-selle, il existe une valeur réalisée en particulier par ce point-selle. Le résultat suivant permet toutefois d'obtenir l'existence d'une valeur sans nécessairement celle d'un point-selle.

PROPOSITION 1.3.

Supposons que pour tout $\varepsilon > 0$, *on puisse trouver* $(u^\varepsilon,S^\varepsilon),(v^\varepsilon,T^\varepsilon)$ *dans* $\mathcal{U}$, *qui soient tels que :*

$$J((u,S),(v^\varepsilon,T^\varepsilon)) - \varepsilon \leqslant J((u^\varepsilon,S^\varepsilon),(v,T)) + \varepsilon$$

pour tous $(u,S),(v,T)$ *appartenant à* $\mathcal{U}$. *Alors le jeu possède une valeur.*

De manière analogue au jeu de Dynkin où sous de bonnes hypothèses ([1],[3]) la résolution repose essentiellement sur les résultats de l'arrêt optimal, la résolution de ce problème passera par les résultats d'une certaine forme simplifiée de contrôle mixte [7]. L'outil essentiel sera l'opérateur "plus petite surmartingale compatible qui majore" [7], qui jouera un rôle analogue à celui de l'enveloppe de Snell dans l'arrêt optimal. Ses propriétés sont rappelées dans le paragraphe suivant :

§ 2 - UN PROBLEME DE CONTRÔLE MIXTE

A. Réduite compatible d'une surmartingale compatible sur un ensemble optionnel.

DEFINITION 2.1.

Une famille de v.a.r. $X(S,u)$ *indexée par les éléments de* $\mathcal{U}$, *est appelée un* $\mathcal{U}$ *-système si :*

i) sur $(S = T)$ $X(S,u) = X(T,u)$ P ps. $\forall u$

ii) les v.a.r. $X(S,u)$ *sont* $\underline{\underline{F}}_S$*-mesurables*

iii) si u *et* v *coïncident jusqu'à* S, *et* $u = v$ *sur* $A \in \underline{\underline{F}}_S$ *alors*

$X(S,u) = X(S,v)$ *sur* A, P ps.

Un $\mathcal{U}$ *-surmartingal-système est un* $\mathcal{U}$*-système tel que :*

iv) pour tout $(S,u) \in \mathcal{U}$, $X(S,u)$ *est* P*-intégrable*

v) si S *et* T *sont deux* $\underline{\underline{F}}_t$*-temps d'arrêt tels que* $S \leqslant T$:

$E(X(T,u)/\underline{\underline{F}}_S) \leqslant X(S,u)$ P ps. $\forall u$

vi) si u *et* v *coïncident jusqu'à un* $\underline{\underline{F}}_t$*-temps d'arrêt* S

$X(S,u) = X(S,v)$ P ps.

La proposition suivante nous donne pour tout $\mathcal{U}$-système l'existence et la caractérisation du "plus petit surmartingal-système" qui majore cet $\mathcal{U}$ -système.

THEOREME 2.2.

Soit $(X(S,u))_{(S,u) \in \mathcal{U}}$ un $\mathcal{U}$ *-système. Le gain maximal conditionnel défini par :*

$$J(S,u) = \underset{v^S = u^S, T \geqslant S}{P\text{-ess sup}} \; E(X(T,v) / \underline{\underline{F}}_S)$$

est un $\mathcal{U}$ *-surmartingal-système qui pour tout* $u \in \mathcal{U}$ *s'agrège en une* $\underline{\underline{F}}_t$*-surmartingale* J^u. J^u_S *est le plus petit surmartingal-système qui majore* $X(S,u)$, *c'est-à-dire*

$$J^u_S \geqslant X(S,u) \text{ P ps. } \forall (S,u) \in \mathcal{U}.$$

Preuve :

On utilise d'abord le fait que $\{ \hat{\Gamma}(S,T,v) \; ; \; v^S = u^S \; , \; T \geqslant S \}$ est filtrant croissant si $\hat{\Gamma}(S,T,v) = E(X(T,v) / \underline{\underline{F}}_S)$; d'où l'on déduit le caractère surmartingal-système de $J(S,u)$.

L'agrégation de $J(S,u)$ en J^u_S provient alors du fait que pour tout u, $J(S,u)$ est un Σ-surmartingal-système [6].

Enfin, la construction de $J(S,u)$ entraîne que tout surmartingal-système qui majore $X(S,u)$ majore $J(S,u)$ et donc J^u_s.

Dans toute la suite nous appellerons surmartingal-systeme compatible toute famille J^u de $\underline{\underline{F}}_t$-surmartingales, telle que $(J^u_S)_{(S,u)\in\mathcal{U}}$ soit un $\mathcal{U}$-surmartingal-systeme.

DEFINITION 2.3.

On appelle réduite sur un ensemble optionnel A, *de la surmartingale compatible* X^u, *le plus petit surmartingal-système compatible qui majore le -système* $X^u_S \, 1\!\!1_A(S)$; *cette réduite sera notée* $[\mathcal{R}^A(X)]^u$.

La propriété fondamentale qui va suivre va donner une décomposition importante de $\mathcal{R}^A(X)$, d'où l'on pourra déduire des résultats de régularité.

THEOREME 2.4.

Soit X^u *un surmartingal-système compatible, et* S *un* $\underline{\underline{F}}_t$*-temps d'arrêt. Soit* A *optionnel, et* D^A_S *le début de l'ensemble* $A \cap [\![S,+\infty[\![$. *Alors :*

$$[\mathcal{R}^A(X)]^u_S = \underset{v^S=u^S}{\text{P-ess sup}}\; E\,(X^v_{D^A_S}\, 1\!\!1_A(D^A_S) + (X^{v,+}_{D^A_S}\, 1\!\!1_{A^c}(D^A_S) \,/\, \underline{\underline{F}}_S)$$

où $X^{v,+}$ *désigne le processus régularisé à droite de la surmartingale* X^v.

Preuve :

On remarque déjà les inégalités évidentes : $X^v \geq [\mathcal{R}^A(X)]^v \geq X^v\, 1\!\!1_A$; en particulier $[\mathcal{R}^A(X)]^v$ et X^v coïncident sur A. On en déduit que $\mathcal{R}^A(X) = (\mathcal{R}^A(X)\; 1\!\!1_A)$ P.ps pour tout v, et on peut donc écrire :

$$[\mathcal{R}^A(X)]^u_S = \underset{v^S=u^S,\,T\geq S}{\text{P-ess sup}}\; E\left[(\mathcal{R}^A(X))^v_T\; 1\!\!1_A(T) \,/\, \underline{\underline{F}}_S\right]$$

$$= \underset{v^S=u^S,\,T\geq D^A_S}{\text{P-ess sup}}\; E\left[(\mathcal{R}^A(X))^v_T\; 1\!\!1_A(T) \,/\, \underline{\underline{F}}_S\right]$$

$$\leq \underset{v^S=u^S}{\text{P-ess sup}} \Big[\underset{T\geq D^A_S}{\text{P-ess sup}}\; E\,((\mathcal{R}^A(X))^v_T\; 1\!\!1_A(D^A_S) \,/\, \underline{\underline{F}}_S)$$

$$+ \underset{T\geq D^A_S}{\text{P-ess sup}}\; E\,((\mathcal{R}^A(X))^v_T\; 1\!\!1_{A^c}(D^A_S) \,/\, \underline{\underline{F}}_S)\Big]$$

$\mathcal{R}^A(X)$ est un surmartingal-système compatible. Ceci nous permet de majorer le membre de droite de l'inégalité et d'obtenir :

$$\left[\mathcal{R}^A(X)\right]^u_S \leq \underset{v^S = u^S}{\text{P-ess sup}}\ E\left((\mathcal{R}^A(X))^v_{D^A_S}\ 1\!\!1_A(D^A_S) + (\mathcal{R}^A(X))^{v,+}_{D^A_S}\ 1\!\!1_{A^c}(D^A_S)\ /\ \underline{\underline{F}}_S\right)$$

D'autre part, l'inégalité inverse est évidente, d'où l'égalité :

$$\left[\mathcal{R}^A(X)\right]^u_S = \underset{v^S = u^S}{\text{P-ess sup}}\ E\left((\mathcal{R}^A(X))^v_{D^A_S}\ 1\!\!1_A(D^A_S) + (\mathcal{R}^A(X))^{v,+}_{D^A_S}\ 1\!\!1_{A^c}(D^A_S)\ /\ \underline{\underline{F}}_S\right)$$

Enfin il suffit de remarquer que :

- d'une part puisque X^v et $(\mathcal{R}^A(X))^v$ coïncident sur A, on a :

$$(\mathcal{R}^A(X))^v_{D^A_S}\ 1\!\!1_A(D^A_S) = X^v_{D^A_S}\ 1\!\!1_A(D^A_S)$$

- d'autre part si $D^A_S(\omega) \notin A$, par définition du début d'un ensemble il existe une suite t_n dépendant de ω, appartenant à la coupe suivant ω de A qui converge vers $D^A_S(\omega)$, et alors :

$$\left[R^A(X)\right]^{v,+}_{D^A_S(\omega)} = \lim_{n\to\infty} (R^A(X))^v_{t_n} = \lim_{n\to\infty} X^v_{t_n} = X^{v,+}_{D^A_S(\omega)} ,$$

pour avoir le résultat cherché.

B. Un problème de contrôle mixte.

Le contrôleur doit faire choix à la fois d'une stratégie d'arrêt, et d'une stratégie continue en vue de maximiser un critère de la forme : $\hat{\Gamma}(T, u) = E(C^u_T + Y_T)$. Les stratégies sont donc l'ensemble $\mathcal{U}$, C^u représente le gain d'évolution, et Y le gain d'arrêt.

On supposera dans toute la suite que Y est optionnel borné de même que C^u qui est pour tout u un processus à variation finie borné uniformément en u. De plus C^u vérifie les hypothèses i) et ii) des processus à variation finie du critère du jeu du premier paragraphe.

Ce modèle n'est alors qu'un cas particulier du contrôle mixte de [7] ; nous avons donc les résultats suivants :

PROPOSITION 2.5. [7]

Posons pour tout (S,u) *appartenant à* $\mathcal{U}$:

$$\hat{J}(S,u) = \underset{v^S = u^S, T \geq S}{\text{P-ess sup}} E(C_T^v + Y_T / \underline{\underline{F}}_S)$$

Alors :

1) $\hat{J}(S,u)$ *est un surmartingal-système qui s'agrège pour tout* u *en un surmartingal-système compatible* $\hat{J}^u$, *qui est le plus petit surmartingal-système compatible qui majore* $C^u + Y$.

2) Pour tout u, $\hat{J}^u$ *admet la décomposition* $\hat{J}^u = C^u + \hat{W}$ *où* $\hat{W}$ *est un processus optionnel ne dépendant pas de* u.

3) Un contrôle (T^*,u^*) *est optimal si et seulement si* :

i) $\hat{W}_{T^*} = Y_{T^*}$ P ps.

ii) $\hat{J}^{u^*}_{t \wedge T^*}$ *est une* $\underline{\underline{F}}_t$*-martingale.*

La propriété suivante va permettre de préciser les discontinuités de $\hat{J}^u$ et sera à la base des théorèmes d'existence.

PROPOSITION 2.6.

Pour tout (S,u) *appartenant à* $\mathcal{U}$, *pour tout* $\varepsilon > 0$:

$$\hat{J}^u_S = \underset{v^S = u^S}{\text{P-ess sup}} E(\hat{J}^v_{D_S^\varepsilon} 1\!\!1_{A^\varepsilon}(D_S^\varepsilon) + \hat{J}^{v,+}_{D_S^\varepsilon} 1\!\!1_{(A^\varepsilon)^c}(D_S^\varepsilon) / \underline{\underline{F}}_S)$$

où D_S^ε *désigne le début après* S *de* $A^\varepsilon = \{\hat{W} < Y + \varepsilon\}$.

Preuve :

Pour établir cette propriété on procède comme dans le cadre de l'arrêt optimal, en considérant $\hat{J}^{\lambda,u}$, le plus petit surmartingal-système compatible qui majore $\hat{J}^u 1\!\!1_{\{\lambda \hat{J}^u \leq C^u + Y\}}$, l'ensemble $(\lambda \hat{J}^u \leq C^u + Y)$ étant non vide par définition de $\hat{J}^u$. Il suffit alors de remarquer que $(1-\lambda)\hat{J}^{\lambda,u} + \lambda \hat{J}^u$ est un surmartingal-système compatible qui majore $C^u + Y$, et qui est naturellement majoré par $\hat{J}^u$ pour établir que $\hat{J}^u$ et $\hat{J}^{\lambda,u}$ sont indistinguables pour tout u, et pour tout $\lambda \in [0,1[$.

Puisque avec les hypothèses faites sur C^u et Y, $\hat{J}^u$ est borné uniformément en u :

$$\{\lambda\ \hat{J}^u \leqslant C^u + Y\} = \{\hat{W} \leqslant (1-\lambda)\ \hat{J}^u + Y\} \subset \{\hat{W} \leqslant Y + \varepsilon\}$$

si ε est bien choisi ($\varepsilon = (1-\lambda)$ K, K borne de $\hat{J}^u$).

Par suite puisque pour tout $0 \leqslant \lambda < 1$, $\hat{J}^{\lambda,u}$ et $\hat{J}^u$ sont indistingables pour tout u, on a également, pour tout $\varepsilon > 0$, $\hat{J}^{\varepsilon,u}$ et $\hat{J}^u$ indistinguables pour tout u. Le résultat souhaité est alors immédiat en utilisant le théorème 2.4. avec $A^\varepsilon = \{\hat{W} \leqslant Y + \varepsilon\}$. ∎

Pour tout u, $\hat{J}^u$ est une surmartingale bornée donc en particulier de classe (D) ; elle admet donc une décomposition de la forme $J^u = M^u - A^{u,-} - B^u$, où A^u est un processus croissant continu à droite, intégrable, B^u un processus croissant continu à droite, purement discontinu, prévisible, et M^u une martingale de classe (D). On se propose ici d'étudier les sauts des processus croissants A^u et B^u. La preuve est identique à celle de [7], proposition 2.34.

<u>*PROPOSITION 2.7.*</u> ([7])

Pour tout $\underline{\underline{F}}_t$*-temps d'arrêt* S, *et tout* $\varepsilon > 0$:

$$B^u_{D^\varepsilon_S} = B^u_S$$

et sur $\{\hat{W}_{D^\varepsilon_S} \leqslant Y_{D^\varepsilon_S} + \varepsilon\}$, $A^{u-}_S = A^{u-}_{D^\varepsilon_S}$ P p.s.

sur $\{\hat{W}_{D^\varepsilon_S} > Y_{D^\varepsilon_S} + \varepsilon\}$, $A^{u-}_S = A^u_{D^\varepsilon_S}$.

En particulier les temps de saut de A^u *sont inclus dans* $\{\hat{W} = Y\}$ *et ceux de* B^u *dans* $\{\hat{W}^- < {}^pY\}$ *où* ${}^pY$ *désigne la projection prévisible de* Y.

On va à présent pouvoir énoncer des conditions suffisantes, pour obtenir un système de contrôle ε-optimal de la forme $(u^\varepsilon, D^\varepsilon_o)$ puis un système de contrôle optimal. On notera pour simplifier l'écriture $D^\varepsilon_o = D^\varepsilon$.

PROPOSITION 2.8.

Si Y *et pour tout* u, C^u *sont scs à droite sur les trajectoires* c.a.d.

$\limsup_{s \searrow t} C^u_s = \overline{C}^{u+}_t \leqslant C^u_t$, $\limsup_{s \searrow t} Y_s = \overline{Y}^+_t \leqslant Y_t$, *alors* :

1°) $D^\varepsilon \in A^\varepsilon$ *et donc* :

$$E\,(\hat{J}^u_{D^\varepsilon}) < (C^u_{D^\varepsilon} + Y_\varepsilon) + \varepsilon \quad \textit{pour tout } u$$

2°) *Il existe* u^ε *tel que* $(D^\varepsilon, u^\varepsilon)$ *appartient à* $\mathcal{U}$ *avec* :

$$\sup_{\mathcal{U}} E\,(C^u_T + Y_T) \leqslant E\,[C^{u^\varepsilon}_{D^\varepsilon} + Y_{D^\varepsilon}] + 2\,\varepsilon$$

Preuve :

1°) On a toujours l'inégalité :

$$\overline{\hat{W}}^+_{D^\varepsilon} \leqslant \overline{Y}^+_{D^\varepsilon} + \varepsilon \qquad \text{ou encore :}$$

$$\hat{J}^{u,+}_{D^\varepsilon} \leqslant \overline{C}^{u,+}_{D^\varepsilon} + \overline{Y}^+_{D^\varepsilon} + \varepsilon \quad P \text{ ps.}$$

* Si D_ε n'est pas un temps de saut de A, ce n'est pas un temps de saut à droite de $\hat{J}^u$ et l'inégalité s'écrit :

$$\hat{J}^u_{D^\varepsilon} \leqslant \overline{C}^{u+}_{D^\varepsilon} + \overline{Y}^+_{D^\varepsilon} + \varepsilon \leqslant C^u_{D^\varepsilon} + Y_{D^\varepsilon} + \varepsilon$$

en vertu des hypothèses, soit encore :

$$\hat{W}_{D^\varepsilon} \leqslant Y_{D^\varepsilon} + \varepsilon \quad P \text{ ps.}$$

* Si D^ε est un temps de saut de A, d'après la proposition 2.7, l'inégalité $\hat{W}_{D^\varepsilon} \leqslant Y_{D^\varepsilon} + \varepsilon$ P ps. est automatiquement vérifiée.

Puisque $D^\varepsilon \in A^\varepsilon$, on a donc $\hat{J}^u_{D^\varepsilon} \leqslant C^u_{D^\varepsilon} + Y_{D^\varepsilon} + \varepsilon$ d'où le résultat cherché en passant à l'espérance.

2°) D'après la proposition 2.6., et puisque $D^\varepsilon \in A^\varepsilon$ on a en particulier :

$$\hat{J}_o = \sup_v E\left(\hat{J}^v_{D^\varepsilon}\right) \leqslant \sup_v E\,(C^v_{D^\varepsilon} + Y_{D^\varepsilon}) + \varepsilon$$

or $\hat{J}_o = \sup_{\mathcal{U}} E\,(C^u_T + Y_T)$ et par définition du sup on peut donc trouver u^ε tel que :

$$\hat{J}_o \leqslant E\,(C^{u^\varepsilon}_{D^\varepsilon} + Y_{D^\varepsilon}) + 2\varepsilon \qquad \text{CQFD.}$$

On se propose alors d'établir sous de bonnes hypothèses l'existence de stratégies optimales pour un tel problème. On pourrait pour cela procéder comme dans [7] en faisant tendre ε vers 0, mais ici la loi du système n'étant pas contrôlée on a beaucoup plus simple. Les résultats obtenus précédemment nous permettrons toutefois d'obtenir l'existence d'une valeur sous des hypothèses très générales dans les jeux définis au premier paragraphe.

Nous ferons désormais l'hypothèse :

(H_1) $\forall u$ C^u_t est de la forme $\int_o^t e^{-\alpha s}\, c\,(s,\omega,u(s,\omega))\, ds$ où c est une fonction bornée, continue en u, $\mathcal{P} \otimes \underline{\underline{U}}$ mesurable ($\mathcal{P}$ désignant la tribu prévisible).

En utilisant le lemme de Benes ([2]) on peut trouver $u^*(t,\omega)$ prévisible, tel que : $c\,(t,\omega,u^*(t,\omega)) = \sup\limits_{u \in U} c(t,\omega,u)$ pour tout (t,ω). Il est alors clair que u^* est optimal pour le problème de contrôle associé à :

$$E\,(\int_o^T e^{-\alpha s}\, c\,(s,u_s)\, ds + Y_T)$$

et ceci pour tout $\underline{\underline{F}}_t$-temps d'arrêt T. En fait ce résultat est un cas "trivial" du cas général où un contrôle optimal est trouvé en maximisant l'hamiltonien qui ici se trouve être simplement $c(t,\omega,u)$. On a donc :

$$\sup_{\mathcal{U}} E\,(\int_o^T e^{-\alpha s} c(s,u_s)ds + Y_T) = \sup_T E \int_o^T e^{-\alpha s} c(s,u^*_s)\, ds + Y_T$$

On est ainsi ramené à un problème d'arrêt optimal, où la réduite est égale à :

$$\underset{S \geqslant T}{\text{P-ess sup}}\; E\,(\int_o^S e^{-\alpha s}\, c(s,u^*_s)ds + Y_S/\underline{\underline{F}}_T) = \underset{S \geqslant T}{\text{P-ess sup}}\; \underset{v^s = u^{*s}}{\text{P-ess sup}}\; E\,(\int_o^S e^{-\alpha s} c(s,v_s)\, ds + Y_S/\underline{\underline{F}}_T)$$

$$= \hat{J}^{u^*}_T = C^{u^*}_T + W_T$$

en utilisant le fait que si u^* est optimal il est encore conditionnellement optimal d'après le critère de Bellman [7].

En utilisant les résultats de l'arrêt optimal [7], on obtient :

THEOREME 2.9.

Si Y *est scs sur les trajectoires, et sous l'hypothèse* (H_1) *le couple* (u^*,D) *est une stratégie optimale pour le problème posé où :*

- u^* est tq. $c(t,\omega,u^*(t,\omega)) = \sup_U c(t,\omega,u) \quad \forall (t,\omega)$

- D *est le début de l'ensemble* $(\hat{W} = Y)$

§ 3. APPLICATION DES RESULTATS DU CONTROLE MIXTE AU JEU

La proposition suivante donne une conditon suffisante de point-selle et justifie l'étude de la "double équation" qui va suivre. Dans toute la suite on supposera C^u et C'^v nuls en zéro pour tout u et pour tout v.

PROPOSITION 3.1.

Si (S^*,u^*) *et* (T^*,v^*) *sont des éléments de* $\mathcal{U}$ *tels qu'il existe un processus optionnel* $\tilde{W}$ *avec :*

a) $X \leq \tilde{W} \leq - X'$

b) $\tilde{W}_{S^*} = X_{S^*} \quad \tilde{W}_{T^*} = - X'_{T^*} \quad$ P ps.

c) $\forall v \quad (\tilde{W} + C^{u^*} - C'^{v})_{s \wedge S^*}$ *est une* $(\underline{\underline{F}}_t, P)$ *sous-martingale*

$\forall u \quad (\tilde{W} + C^{u} - C'^{v^*})_{s \wedge T^*}$ *est une* $(\underline{\underline{F}}_t, P)$ *surmartingale.*

Alors $((S^*,u^*),(T^*,v^*))$ *est un point selle pour le jeu.*

Preuve :

Sous les hypothèses énoncées le processus $(\tilde{W} + C^{u^*} - C'^{v^*})_{s \wedge S^* \wedge T^*}$ est une martingale, et donc en particulier :

$$\tilde{W}_o = E\,(\tilde{W}_{S^* \wedge T^*} + (C^{u^*} - C'^{v^*})_{S^* \wedge T^*})$$

$$= E\,(X_{S^*}\, 1\!\!1_{(S^* \leq T^*)} - X'_{T^*}\, 1\!\!1_{(T^* < S^*)} + (C^{u^*} - C'^{v^*})_{S^* \wedge T^*})$$

Par ailleurs, pour tout (S,u) appartenant à $\mathcal{U}$:

$$\tilde{W}_o \geqslant E((\tilde{W} + C^u - C'^{v*})_{S \wedge T*})$$

$$\geqslant E(\tilde{W}_S \, 1\!\!1_{(S \leqslant T*)} + \tilde{W}_{T*} \, 1\!\!1_{(T*<S)} + (C^u - C'^{v*})_{S \wedge T*})$$

$$\geqslant E(X_S \, 1\!\!1_{(S \leqslant T*)} - X'_{T*} \cdot 1\!\!1_{(T*<S)} + (C^u - C'^{v*})_{S \wedge T*})$$

De manière symétrique on montrerait que pour tout (T,v) appartenant à $\mathcal{U}$:

$$\tilde{W}_o \leqslant E(X_{S*} \, 1\!\!1_{(S* \leqslant T)} - X'_T \, 1\!\!1_{(T<S*)} + (C^{u*} - C'^{v})_{S* \wedge T})$$

ce qui montre à la fois le caractère de point-selle de (((S*,u*),(T*,v*)), et la caractérisation de $\tilde{W}_o$ comme valeur du jeu.

Supposons alors que (W,W') soit solution de la double équation :

$$(*) \quad \begin{aligned} W + C^u &= \mathcal{R}(C^u + X + W') \\ W' + C'^v &= \mathcal{R}(C'^v + X' + W) \end{aligned}$$

où $\mathcal{R}$ désigne l'opérateur "plus petit surmartingal-système compatible". Il est alors facile de remarquer que si (S*,u*) (resp(T*,v*))est solution du problème de contrôle mixte de critère $E(C^u_T + X_T + W'_T)$ (resp.$E(C'^v_T + X'_T + W_T)$) et que si l'on pose $\tilde{W} = W - W'$, les hypothèses de la proposition précédente sont vérifiées.

En effet, les résultats du contrôle mixte (2.5) permettent alors d'écrire que l'on a :

$$W_{S*} = (X + W')_{S*} \quad , \quad (W + C^{u*})_{s \wedge S*} \text{ est une martingale}$$

de même :

$$W'_{T*} = (X' + W)_{T*} \quad , \quad (W' + C'^{v*})_{t \wedge T*} \text{ est une martingale.}$$

Il est donc tout a fait naturel d'étudier (*). Nous avons le résultat suivant :

<u>*PROPOSITION 3.2.*</u>

Sous les hypothèses :

. $X_\infty = -X'_\infty = 0$

. *Il existe deux surmartingales-systèmes compatibles* $\tilde{Z}^u$, $\mathcal{Z}'^u$ *positifs, bornés, tels que :*

$$C^u + X \leqslant \tilde{Z}^u - \tilde{Z}'^u \leqslant -C'^u - X'$$

La double équation (*) *admet une solution* (W,W').

Preuve :

On construit par itération deux suites croissantes $(Z^{n,u})$ et $(Z'^{n,u})$ de surmartingales positives majorées respectivement par $\tilde{Z}^u$ et $\tilde{Z}'^u$. Posons pour tout u :

$$Z^{1,u} = \mathcal{R}(C^u + X) \quad , \quad W^1 = Z^{1,u} - C^u$$

$$Z'^{1,u} = \mathcal{R}(C'^u + X') \quad , \quad W'^1 = Z'^{1,u} - C'^u$$

et pour tout $n \geqslant 1$:

$$Z^{n+1,u} = \mathcal{R}(C^u + X + W'^n)$$

$$Z'^{n+1,u} = \mathcal{R}(C'^u + X' + W^n)$$

Pour tout n, $Z^{n,u} - C^u$, $Z'^{n,u} - C'^u$ ne dépendent pas de u, on construit donc par récurrence :

$$W^n = Z^{n,u} - C^u \quad , \quad W'^n = Z'^{n,u} - C'^u$$

Les suites ainsi définies sont croissantes, en effet, sous les hypothèses faites :

$$\begin{aligned} W'^1_T + C'^u_T = Z'^{1,u}_T &= (C'^u + X')_T \\ &= \underset{S \geqslant T}{\text{P-ess sup}}\ E\left[C'^u_S + X'_S / \underline{\underline{F}}_T\right] \\ &\geqslant C'^u_T + \underset{S \geqslant T}{\text{P-ess sup}}\ E(X'_S / \underline{\underline{F}}_T) \\ &\geqslant C'^u_T + X'_\infty \\ &\geqslant C'^u_T \end{aligned}$$

Donc W'^1_T est positif pour tout temps d'arrêt T.

Il suffit alors de remarquer que $Z^{2,u}$ est supérieur à $C^u + X + W'^1$ donc à $C^u + X$ pour conclure.

Pour tout temps d'arrêt T :

$$Z^{2,u}_T = C^u_T + W^2_T \geqslant \mathcal{R}(C^u + X)_T = C^u_T + W^1_T$$

c'est-à-dire : $W_T^2 \geqslant W_T^1$; on montrerait de même l'inégalité pour tout T : $W_T'^2 > W_T'^1$.

La croissante des suites (W^n) et donc celle des suites $(Z^{n,u})$,$(Z'^{n,u})$ s'en déduit alors facilement par récurence sur n.

On obtient ainsi deux suites croissantes de surmartingales majorées respectivement par $\mathcal{Z}^u$ et $\mathcal{Z}'^u$.

Posons alors :

$$Z^u = \lim_n Z^{n,u} , \quad W = Z^u - C^u$$

$$Z'^u = \lim_n Z'^{n,u} , \quad W' = Z'^u - C'^u$$

Pour montrer que (W,W') est une solution à la double équation (*), il suffit d'établir que la limite croissante de surmartingales compatibles est une surmartingale compatible. On en déduira alors (*) par passage à la limite dans les deux égalités.

Soit donc $(Z^{n,u})$ une suite croissante de surmartingales compatibles de limite Z^u. Z^u est clairement un $\mathcal{U}$ système.

De plus si u et v coïncident jusqu'à un $\underline{\underline{F}}_t$ temps d'arrêt S :

$$Z_S^u = Z_S^v \quad P \text{ p.s.}$$

Soient enfin S et T deux $\underline{\underline{F}}_t$-temps d'arrêt tels que $S \leqslant T$. Pour tout n on a l'inégalité :

$$E\left(Z_T^{n,u} / \underline{\underline{F}}_S\right) \leqslant Z_S^{n,u}$$

Les suites étant croissantes on obtient par passage à la limite :

$$E\left(Z_T^u / \underline{\underline{F}}_S\right) \leqslant Z_S^u$$

C.Q.F.D.

On se propose à présent d'énoncer des conditions suffisantes pour l'existence d'une valeur au jeu, en étudiant l'existence de solutions ε-optimales aux problèmes de contrôle mixte définis par la double équation (*).

Pour un ε positif fixé on considère les ensembles optionnels suivants :

$$A^{\varepsilon} = \{W \leq X + W' + \varepsilon\}$$

$$A'^{\varepsilon} = \{W' \leq X' + W + \varepsilon\}$$

On peut alors énoncer :

PROPOSITION 3.3.

Si les débuts des ensembles A^{ε} et A'^{ε} notés respectivement D^{ε} et D'^{ε} leur appartiennent, alors il existe une valeur pour le jeu.

Preuve :

On considère le problème de contrôle mixte associé au critère :

$$E\left(C_T^u + X_T + W'_T\right)$$

Les notations seront celles utilisées au paragraphe 2. Y représente ici le processus X + W', $\hat{J}^u$ est le coût maximal conditionnel et se décompose pour tout temps d'arrêt T sous la forme :

$$\hat{J}_T^u = C_T^u + W_T$$

Sous les hypothèses faites on sait qu'il existe un contrôle u^{ε} tel que $(D^{\varepsilon}, u^{\varepsilon})$ soit une solution ε-optimale, c'est-à-dire :

$$\hat{J}_o \leq E\left(C^{u^{\varepsilon}}_{D^{\varepsilon}} + W_{D^{\varepsilon}}\right) + \varepsilon$$

$\hat{J}^u$ étant une surmartingale on en déduit pour tout temps d'arrêt T, l'inégalité suivante :

$$\hat{J}_o \leq E\left(C^{u^{\varepsilon}}_{D^{\varepsilon}\wedge T} + W_{D^{\varepsilon}\wedge T}\right) + \varepsilon$$

En appelant $\hat{J}'^u$ le coût maximal conditionnel associé au critère du deuxième contrôle mixte, on peut toujours écrire :

$$\hat{J}'_o \geq E\left(\hat{J}'^v_{D^{\varepsilon}\wedge T}\right) = E\left(C'^v_{D^{\varepsilon}\wedge T} + W'_{D^{\varepsilon}\wedge T}\right)$$

où v est un contrôle admissible quelconque.

De ces deux inégalités on déduit :

$$\hat{J}_o - \hat{J}'_o < E\left((W - W')_{D^{\varepsilon}\wedge T} + (C^{u^{\varepsilon}} - C'^v)_{D^{\varepsilon}\wedge T}\right) + \varepsilon$$

Il suffit alors de remarquer que :

- sur l'ensemble $\{D^\varepsilon \leqslant T\}$ $T \wedge D^\varepsilon = D^\varepsilon$ et $(W - W')_{D^\varepsilon} \leqslant X_{D^\varepsilon} + \varepsilon$

puisque nous avons supposé que D^ε appartient à A^ε ,

- sur l'ensemble $\{D^\varepsilon > T\}$ on a toujours $(W - W')_T < - X'_T$

On obtient la majoration suivante pour tout contrôle v :

$$\hat{J}_o - \hat{J}'_o \leqslant E\left(X_{D^\varepsilon}\, 1_{\{D^\varepsilon \leqslant T\}} - X'_T\, 1_{\{D^\varepsilon > T\}} + (C^{u^\varepsilon} - C'^{v})_{T\wedge D^\varepsilon}\right) + 2\,\varepsilon$$

On montrerait de manière analogue qu'il existe un contrôle v^ε tel que $(D'^\varepsilon, v^\varepsilon)$ soit élément de $\mathcal{U}$ vérifiant l'inégalité pour tout temps d'arrêt S et tout contrôle u :

$$\hat{J}_o - \hat{J}'_o \geqslant E\left(X_S\, 1_{\{D'^\varepsilon > S\}} - X'_{D'^\varepsilon}\, 1_{\{D'^\varepsilon \leqslant S\}} + (C^u - C'^{v^\varepsilon})_{S\wedge D'^\varepsilon}\right) - 2\,\varepsilon$$

Le lemme de Stettner (Proposition 1.3) permet alors de conclure à l'existence d'une valeur du jeu égale à $W_o - W'_o$.

C.Q.F.D.

§ 4 - RESULTATS D'EXISTENCE

Le paragraphe précédent a permis sous certaines conditions d'établir des résultats d'existence d'une valeur et d'un point-selle pour le jeu de Dynkin étudié, à partir de l'existence de solutions ε-optimales ou optimales pour les problèmes de contrôle mixte associés aux équations (*). On se propose ici d'en déduire des conditions suffisantes d'existence pour le jeu en utilisant les propositions 2.8 et 2.9.

Les hypothèses faites ici sont celles énoncées dans le modèle du §1 ; on supposera de plus que les processus C^u (resp.C'^v) sont nuls en zéro, à variation finie, bornés uniformément en u (resp.en v).

Des propositions 2.8 et 3.2 on déduit immédiatement un critère suffisant d'existence d'une valeur.

PROPOSITION 4.1.

Si pour tout contrôle u admissible les processus C^u, C'^u, X + W', X' + W sont s.c.s. à droite sur les trajectoires, alors il existe une valeur au jeu égale à $W_o - W'_o$.

Ces hypothèses où interviennent W et W' seront en particulier vérifiées dans le cas suivant :

PROPOSITION 4.2.

Si les hypothèses suivantes sont vérifiées :

a) X et X' s.c.s. à droite sur les trajectoires,

b) Pour tout u, C^u et C'^u scs à droite sur les trajectoires,

c) Il existe (u_o, v_o) deux contrôles tels que C^{u_o} et C'^{v_o} soient c.a.d.

Alors il existe une valeur pour le jeu de Dynkin associé.

Preuve :

Il suffit de montrer que sous ces hypothèses W et W' sont s.c.s. à droite sur les trajectoires, ce qui est une conséquence triviale de c) et du fait que $\hat{J}^{u_o}$ et $\hat{J}'^{v_o}$ étant des surmartingales sont s.c.s. à droite sur les trajectoires.

Pour l'existence d'un point-selle, on est amené à supposer de plus :

- l'ensemble {X = -X'} evanescent,
- que les coûts C^u (C'^u) sont pour tout u de la forme :

$$\int_o^t e^{-\alpha s} c(s, a, u(s,\omega))\, ds$$

(hypothèse H_1 du paragraphe 2). Ils sont en particulier pour tout u continus sur les trajectoires.

On peut dans ce cadre énoncer le système de conditions suffisantes d'existence d'un point-selle, qui se déduit directement des propositions 2.9 et 3.1.

PROPOSITION 4.3.

Si les processus X et X' sont s.c.s. sur les trajectoires (et sous l'hypothèse H_1) alors il existe un point-selle pour le jeu : ((D,u),(D',v*))*

- u^* *est tel que* : $\forall\ (t,\omega),\ c\ (t,\ \omega,\ u^*(t,\omega)) = \sup_{u\in\mathcal{U}} c\ (t,\omega,u)$
- v^* *est tel que* : $\forall\ (t,\omega),\ c'(t,\ \omega,\ v^*(t,\omega)) = \sup_{u\in\mathcal{U}} c'(t,\omega,u)$
- D (*resp.*D') *est le début de* $\{W = X + W'\}$ (*resp. de* $\{W' = X' + W\}$).

Preuve :

Il suffit en effet de montrer que W et W' n'ont pas de temps de saut prévisibles à gauche. On remarque que si T est un temps de saut prévisible à gauche pour W, il l'est aussi pour $\hat{J}^u$. Son graphe est donc inclus dans l'ensemble :

$$\{W^- = X^P + W'^-\}$$

De même ceux de W' sont inclus dans :

$$\{W'^- = X'^P + W^-\}$$

L'ensemble $\{X = -X'\}$ étant évanescent, les temps de saut prévisibles de W et W' ont des graphes presque sûrement disjoints. Soit donc T temps de saut prévisible pour W. On peut écrire pour tout u :

$$E\left(C_T^u + W_T^-\right) = E\left(C_T^u + X_T^P + W_T'^-\right) \leqslant E\left(C_T^u + X_T + W_T'\right)$$

car X est s.c.s. à gauche et T n'est pas un temps de saut de W'.

Donc :

$$E\left(C_T^u + W_T^-\right) \leqslant E\left(C_T^u + W_T\right)$$

$\hat{J}^u$ étant une surmartingale, on peut donc conclure :

$$E\left(W_{T^-}\right) = E\left(W_T\right)$$

On procèderait de même pour W'.

C.Q.F.D.

BIBLIOGRAPHIE

1 M.ALARIO, J.P.LEPELTIER, B.MARCHAL, Jeux de Dynkin, 2e Bad Honnef Workshop on stochastic processes. Lecture Notes in Control and Information Sciences. Springer Verlag (1982).

2 V.E.BENES, Existence of optimal strategies bases on specified information, for a class of stochastic decision processes. SIAM J.of control 8, 179-188 (1970).

3 J.M.BISMUT, Contrôle de processus alternants et applications. Z.f.Wahr.V.Geb.47, 241-288 (1979).

4 J.M.BISMUT, temps d'arrêt optimal, quasi-temps d'arrêt et retournement du temps. Ann. of Proba. 7, 933-964 (1979).

5 C.DELLACHERIE, Capacités et processus stochastiques. Springer n°67 (1972).

6 C.DELLACHERIE-E.LENGLART, Sur des problèmes de régularisation, de recollement et d'interpolation en théorie des martingales. Séminaire de probabilités XV 1979-1980. Lecture Notes in mathematics. Springer Verlag n°850.

7 N.EL KAROUI, Cours sur le contrôle stochastique. Ecole d'Eté de probabilités de St Flour IX 1979. Lecture Notes in maths. n°876. Springer Verlag.

OPTIMAL CONTROL OF REFLECTED DIFFUSION PROCESSES

P.L. LIONS
Ceremade
Université Paris IX-Dauphine
Place de Lattre de Tassigny- 75775 Paris Cedex 16

I- Introduction

We consider stochastic systems $\underline{\underline{S}}$ given by the collection of
i) a complete probability space $(\Omega, \underline{\underline{F}}, \underline{\underline{F}}_t, \mathbb{P})$ with some m dimensional Brownian motion B_t,
ii) a progressively measurable process - the control process - denoted by α_t taking its values in a given separable metric space $\underline{\underline{A}}$.
The state of the system, denoted by X_t, is given by the solution of the following stochastic differential equation with reflection:

(1) $dX_t = \sigma(X_t, \alpha_t)\, dB_t + b(X_t, \alpha_t)\, dt - \gamma(X_t)\, dK_t \ , \ X_0 = x$

where X_t, K_t are continuous $\underline{\underline{F}}_t$-adapted processes satisfying:

(2) $X_t \in \overline{\underline{\underline{O}}}$, for all $t \geq 0$

(3) K_t is nondecreasing, $K_t = \int_0^t 1\!\!1_{(X_s \in \partial \underline{\underline{O}})}\, dK_s$,

where $\underline{\underline{O}}$ is a given smooth open set in $\mathbb{R}^N$, $x \in \overline{\underline{\underline{O}}}$, $\sigma(x,\alpha)$, $b(x,\alpha)$ are coefficients satisfying conditions stated below and γ is a smooth vector field on $\mathbb{R}^N$ (say C_b^2) satisfying:

(4) $\exists \nu > 0 \ , \ (\gamma(x), n(x)) \geq \nu > 0 \ , \ \forall x \in \partial \underline{\underline{O}}$

and n is the unit outward normal to $\partial\underline{\underline{O}}$ at x.

Given $\underline{\underline{S}}$ and some assumptons on σ,b, it is well known that problem (1)-(2)-(3) yields a unique solution (X_t,K_t): we refer, for example, to N. Ikeda and S. Watanabe (3), P.L. Lions and A.S. Sznitman (9). If $f(x,\alpha)$ $c(x,\alpha)$ are given real valued functions (satisfying conditions stated below), we introduce, for any <u>admissible system</u> $\underline{\underline{S}}$ and any initial condition x, a <u>cost function</u>:

(5) $J(x,\underline{\underline{S}}) = E\int_0^\infty f(X_t,\alpha_t)\exp(-\int_0^t c(X_s,\alpha_s)\,ds)\,dt$.

We will always assume, at least,

(6) $\inf(\, c(x,\alpha) \,/\, x \in \overline{\underline{\underline{O}}},\ \alpha \in \underline{\underline{A}}\,) = \lambda > 0$.

Then, of course, the optimal stochastic control problem we consider is to minimize the cost function J over all possible systems $\underline{\underline{S}}$:

(7) $u(x) = \inf_{\underline{\underline{S}}} J(x,\underline{\underline{S}})\ ,\quad \forall x \in \overline{\underline{\underline{O}}}$;

This function u is called the <u>value function</u> of the control problem.

The results we present here concern

i) the determination of u via the solution of the associated Hamilton-Jacobi-Bellman equation,

ii) the existence of optimal Markovian controls.

They are basically taken from P.L. Lions (5), (6), P.L. Lions and N. Trudinger (10); and many of the arguments used in their proofs are adapted from the solution of the analogous problem where the state process is stopped at the first exit time from $\overline{\underline{\underline{O}}}$, and we refer to N.V. Krylov (4), P.L. Lions (7), (8), (5) and the bibliography therein.

To simplify the presentation, we will always assume:

(8) $\sup_{\alpha} (\|\phi(.,\alpha)\|_\infty + \|D_x\phi(.,\alpha)\|_\infty + \|D_x^2\phi(.,\alpha)\|_\infty) < \infty$

$\phi(x,.)$ is continuous on it, for all $x \in \bar{\mathcal{O}}$

where $\phi = \sigma_{ij}$ $(1 \leq i \leq N,\ 1 \leq j \leq m)$, b_i $(1 \leq i \leq N)$, f, c ;

and we will denote by A_α the operator:

$$(9) \quad A_\alpha = -a_{ij}\partial_{ij} + b_i\partial_i + c$$

where we use the convention on repeated indices and $a(x,\alpha) = \frac{1}{2}\sigma\sigma^T$

II- Auxiliary results

Theorem 1: The value function $u \in C_b^{0,\theta}(\bar{\mathcal{O}})$ where $\theta = \lambda/\lambda_\circ$ if $\lambda < \lambda_\circ$, θ is arbitrary in $]0,1[$ if $\lambda = \lambda_\circ$, $\theta = 1$ if $\lambda > \lambda_\circ$ and $\lambda_\circ$ depends only on σ, b, γ and $\mathcal{O}$.

Remark: If $\mathcal{O}$ is convex, then we may take:

$$\lambda_\circ = \sup\Big(\frac{1}{2}\,\mathrm{Tr}(\sigma(x,\alpha) - \sigma(x',\alpha)).(\sigma^T(x,\alpha) - \sigma^T(x',\alpha)).|x-x'|^{-2} + (b(x,\alpha) - b(x',\alpha), x-x').|x-x'|^{-2} \;/\; x,x' \in \bar{\mathcal{O}},\ x \neq x',\ \alpha \in \mathcal{A}\Big)$$

Theorem 2: For all system S, let τ be a stopping time. We have:

$$(10) \quad u(x) = \inf_S E\Big(\int_0^\tau f(X_t,\alpha_t)\exp\big(-\int_0^t c(X_s,\alpha_s)\,ds\big)\,dt + u(X_\tau)\exp\big(-\int_0^\tau c(X_t,\alpha_t)\,dt\big)\Big)$$

This identity is usually called the optimality principle of the programming principle and may easily reformulated in terms of sub and supermartingales.

Theorem 3: The value function u is a viscosity solution of the Hamilton-Jacobi-Bellman equation:

(11) $$\sup_{\alpha \in \underline{\underline{A}}} (A_\alpha u(x) - f(x,\alpha)) = 0 \quad \text{in } \underline{\underline{O}}$$

Remarks:

i) These results are proved in P.L. Lions (5).

ii) It is also possible to show that without changing the value function, we may restrict the infimum in (7) to systems S such that $(\Omega, \underline{\underline{F}}, \underline{\underline{F}}_t, P, B_t)$ is fixed, and where the control process belongs to some "dense" class of processes with values in $\underline{\underline{A}}$ (like step-processes)

iii) For the definition of viscosity solutions and main properties, we refer to P.L. Lions (5)- recall that this notion was first introduced for first-order equations by M.G. Crandall and P.L. Lions (1).

III- The nondegenerate case

In this section we assume:

(12) $\exists\, \nu > 0,\ \forall (x,\alpha) \in \overline{\underline{\underline{O}}} \times \underline{\underline{A}},\ \forall \xi \in \mathbb{R}^N,\ a_{ij}(x,\alpha)\xi_i\xi_j \geqq \nu|\xi|^2$.

We then have the following result due to P.L. Lions and N. Trudinger (10).

Theorem 4: The value function u belongs to $C^{2,\theta}(\underline{\underline{O}}) \cap C^{1,1}(\overline{\underline{\underline{O}}})$ for some $\theta \in]0,1[$ depending only on bounds on $a(x,\alpha)$, ν, $\underline{\underline{O}}$; and u is the unique solution of (11) satisfying:

(13) $$u \in C^2(\overline{\underline{\underline{O}}}),\quad \frac{\partial u}{\partial \gamma} = 0 \text{ on } \partial \underline{\underline{O}} .$$

Remark: This result is proved in (10) by PDE techniques; if we compare the method of proof with the one in the case of Dirichlet boundary conditions i.e. stopped diffusion processes, the main point lies in the obtention of a priori estimates on $\partial\underline{\underline{O}}$.

It is then straightforward to deduce the

Corollary 5: If $\underline{\underline{A}}$ is compact, then there exists an optimal markovian control.

IV- The degenerate case

We will assume that $\partial\underline{\underline{O}} = \Gamma_+ \cup \Gamma_-$, where Γ_+ , Γ_- are disjoint closed and possibly empty and that we have:

(14) $\exists\ \nu > 0,\ \forall\ (x,\alpha) \in \Gamma_+ x\underline{\underline{A}}\ ,\ \forall\ \xi \in \mathbb{R}^N\ ,\ a_{ij}(x,\alpha)\xi_i\xi_j \geq \nu|\xi|^2$,

(15) $\forall\ (x,\alpha) \in \Gamma_- x\underline{\underline{A}}\ ,\ b_i(x,\alpha)n_i(x) - a_{ij}(x,\alpha)\partial_{ij}d(x) \leq 0$,

$$a_{ij}(x,\alpha)n_i(x)n_j(x) = 0$$

where $d(x) = \text{dist}(x,\partial\underline{\underline{O}})$.

Theorem 6: There exists λ_1 – depending only on bounds on the first order derivatives of σ, b – such that if $\lambda > \lambda_1$, we have

i) $u \in W^{1,\infty}(\underline{\underline{O}})$, u is of class $C^{2,\theta}_{loc} \cap C^{1,1}$ in a neighborhood of Γ_+, and $\frac{\partial u}{\partial \gamma} = 0$ on Γ_+ ;

ii) $A_\alpha u \in L^\infty(\underline{\underline{O}})$ for all $\alpha \in \underline{\underline{A}}$ and $\sup_{\alpha\in\underline{\underline{A}}} \|A_\alpha u\|_\infty < \infty$;

iii) u is semi-concave i.e.: $\exists c > 0,\ \forall|\xi| = 1\ ,\ \partial^2_\xi u \leq c$ in $\underline{\underline{D}}'(\underline{\underline{O}})$

iv) The HJB equation holds a.e. :

(11') $\sup_{\alpha\in\underline{\underline{A}}} (\ A_\alpha u(x) - f(x,\alpha)\ ) = 0$ a.e. in $\underline{\underline{O}}$.

Remarks: i) It is possible to give an explicit formula for λ_1(cf. (5) and (4)): recall that if σ, b do not depend on x, then $\lambda_1 = 0$. Similar results hold for time-dependent problems: in that case the assumption on λ is no more necessary.

ii) By the proof of Theorem 4 (see (10)) one can show that u is of class $C^{2,\theta}$ near Γ_+, the result is then deduced from the general regularity results of P.L. Lions (5).

We conclude with some uniqueness result:

Theorem 7: Let $w \in C_b(\bar{\underline{\underline{O}}}) \cap W^{1,N}_{loc}(\underline{\underline{O}})$ be of class $W^{2,N}$ in a neighborhood of Γ_+ and assume that w satisfies:

i) $\frac{\partial w}{\partial \gamma} = 0$ on Γ_+, in the sense of traces;

ii) $g \in L^N_{loc}(\underline{\underline{O}})$, $\Delta w \leq g$ in $\underline{\underline{D}}'(\underline{\underline{O}})$;

iii) $\forall \alpha \in \underline{\underline{A}}$, $A_\alpha u \leq f(.,\alpha)$ in $\underline{\underline{D}}'(\underline{\underline{O}})$;

iv) $\sup_{\alpha \in \underline{\underline{A}}} (A_\alpha u - f(.,\alpha)) = 0$ in the sense of measures on $\underline{\underline{O}}$.

Then: $w = u$ in $\bar{\underline{\underline{O}}}$.

Remark: In view of the celebrated Schwartz theorem on positive distributions, iii) implies that $A_\alpha u - f(.,\alpha)$ is a non-positive measure, and iv) makes sense.

Bibliography:

(1) M.G. Crandall and P.L. Lions: Viscosity solutions of Hamilton-Jacobi equations. Trans. Amer. Math. Soc., 277(1983),1-42. Announced in C.R.Acad.Sci.Paris,292(1981),183-186.

(2) M.G. Crandall, L.C. Evans and P.L. Lions: Some properties of viscosity solutions of Hamilton-Jacobi equations. Trans.Amer. Math.Soc. (1983).

(3) N. Ikeda and S. Watanabe: Stochastic differential equations and diffusion processes, North-Holland,1981,Amsterdam.

(4) N.V. Krylov: Controlled diffusion processes, Springer,1980,Berlin

(5) P.L. Lions: Optimal control of diffusion processes and Hamilton-Jacobi-Bellman equations. Part 1 and 2, Comm.P.D.E. (1983); Part 3, in Nonlinear PDE and their applications, College de France Seminar, Vol V, Pitman,1983, London. Announced in C.R.Acad.Sci. Paris,289(1979)329-382 and 295(1982)567-570.

(6) P.L. Lions: Some recent results in the optimal control of diffusion processes. In Stochastic analysis, Proc. of the Taniguchi Intern. Symp. on Stochastic Analysis, Katata and Kyoto, 1982; Kinokunya, 1983, Tokyo.

(7) P.L. Lions: Résolution analytique des problèmes de Bellman-Dirichlet. Acta Math. 146 (1981),151-166. Announced in C.R. Acad. Sc. Paris, 287 (1978), 747-750.

(8) P.L. Lions: Control of diffusion processes in $\mathbb{R}^N$. Comm. Pure Appl. Math. 34 (1981), 121-147. Announced in C.R. Acad. Sc. Paris 288 (1979), 339-342.

(9) P.L. Lions and A.S. Sznitman: Stochastic differential equations with reflecting boundary conditions. To appear in Comm. Pure Appl. Math. (1984).

(10) P.L. Lions and N. Trudinger: work in preparation.

ON A FORMULA RELATING THE SHANNON INFORMATION TO THE FISHER INFORMATION FOR THE FILTERING PROBLEM

Eddy Mayer-Wolf and Moshe Zakai

1. Introduction

Consider the following standard filtering problem: Let $(\Omega,\underline{\underline{F}},P)$ be a probability space and let $\{\underline{\underline{F}}_t\}_{t\geq o}$ be an increasing right continuous family of sub-sigma fields of $\underline{\underline{F}}$. Let W_t, $t\geq 0$ and B_t, $t\geq 0$ be $\underline{\underline{F}}_t$ adapted, independent standard Brownian motions on $\mathbb{R}^\ell$ and $\mathbb{R}^m$, respectively. The signal process $x_t\in\mathbb{R}^n$ and the observation process $y_t\in\mathbb{R}^m$ satisfy

$$dx_t = m(x_t)dt + \sigma(x_t)dW_t \tag{1}$$

$$dy_t = h(x_t)dt + dB_t \tag{2}$$

where $\sigma(x)\in\mathbb{R}^{n\times\ell}$, $m(x)\in\mathbb{R}^n$, $h(x)\in\mathbb{R}^m$. The functions $\sigma(\cdot)$ and $m(\cdot)$ are assumed to satisfy sufficient conditions for the existence of a solution to (1) (c.f., e.g., Section 4.4 of [1]). The path $\{y_s,\ 0\leq s\leq t\}$ will be denoted by y_o^t. Let $\pi_t(\beta)$, $\beta\in\mathbb{R}^n$ denote the conditional density of x_t conditioned on y_o^t (cf. the next section), $\rho_t(\beta)$ will denote the corresponding unnormalized conditional density [2], and $p_t(\beta)$ will denote the unconditioned density of x_t. Let $I(x_t,y_o^t)$ denote the mutual information between the vector valued random variable x_t and the path y_o^t ([3],[4]):

$$I(x_t,y_o^t) = E\log\frac{\pi_t(x_t)}{p_t(x_t)}\ , \tag{3}$$

where the expectation is over the state x_t and the observation y_o^t. Define similarly,

$$J(x_t,y_o^t) = E\log\frac{\rho_t(x_t)}{p_t(x_t)}\ . \tag{4}$$

Assume, for a moment, that x_t is scalar-valued $(n=1)$ and consider

$$F_t^\pi = E\left(\sigma(\beta)\left.\frac{\partial\log\pi_t(\beta)}{\partial\beta}\right|_{\beta=x_t}\right)^2 = E\left(\sigma(\beta)\left.\frac{\partial\log\rho_t(\beta)}{\partial\beta}\right|_{\beta=x_t}\right)^2 , \tag{5}$$

for $\sigma(\beta)\equiv 1$, this is the Fisher information for x_t based on the measurements y_o^t ([5],[6]). We will refer to (5) as the a-posteriori Fisher information. Similarly, $F_t^p = E((\sigma(\beta)\partial\log p_t(\beta)/\partial\beta)_{\beta=x_t})^2$ will be denoted by the a-priori Fisher information.

In this note we consider the following identities:

$$\frac{d}{dt} I(x_t, y_o^t) = \frac{1}{2}\left\{E(h(x_t) - \hat{h}_t)^2 - [F_t^\pi - F_t^p]\right\} , \tag{6}$$

$$\frac{d}{dt} J(x_t, y_o^t) = \frac{1}{2}\left\{E(h(x_t))^2 - [F_t^\pi - F_t^p]\right\} , \tag{6a}$$

where $\hat{h}_t = E(h(x_t)|y_o^t)$. Equation (6) is a relation between a filtering error, the mutual information between x_t and y_o^t and the difference between the Fisher a-posteriori and a-priori information quantities. Directly related to (6) is a "symmetric" form, (16b), relating Shannon information quantities with Fisher information quantities. This result was first derived by Lipster in [7] for the one-dimensional case ($m(x) = m\cdot x$ and $\sigma(x) = \sigma$) and a similar result was obtained by Bucy for the time-discrete, vector nonlinear case [8]. It was pointed out in [9] that, at least for the one-dimensional case, the time continuous version of Bucy's result can be derived along the same lines as Liptser's proof. The purpose of this note is to carry out this extension and to further extend the result to a more general case. The notation and assumptions are given in Section 2. Preliminary results are derived in Section 3; in particular, it is shown that if K_t is the normalization constant

$$K_t = \int_{\mathbf{R}^n} \rho_t(\beta)\,d\beta \quad \text{then} \quad E\,\ell n\,K_t = \frac{1}{2}\int E|\hat{h}_s|^2 ds .$$

The proof of the vector-valued case of (6) and (6a) is given in Section 4. Finally, Section 5 deals with the case where the observation equation (2) is replaced by a more general model (cf. equation (17)).

2. Notation and Assumptions

Standard vector notation will be used. If $a \in \mathbf{R}^k$, $|a|^2 = \sum_1^k a_i^2$, ∇ will denote the symbolic $k \times 1$ matrix $\left(\frac{\partial}{\partial x_i}\right)$ and ∇^2 will denote the Hessian $k \times k$ matrix $\left(\frac{\partial^2}{\partial x_i \partial x_j}\right)$. For any matrix B, B^T will denote its transpose, if B is square then $\mathrm{tr}B = \Sigma b_{ii}$. Recall that Kolmogorov's forward operator $\mathcal{L}^*$ associated with (1) is

$$\mathcal{L}^*\phi = -\nabla^T(m\phi) + \frac{1}{2}\,\mathrm{tr}(\nabla^2(A\phi)) = -\sum_1^n \frac{\partial}{\partial x_i}(m_i\phi) + \frac{1}{2}\sum_{i,j=1}^n \frac{\partial^2}{\partial x_i \partial x_j}(a_{ij}\phi) , \tag{7}$$

where $A = (a_{ij}) = \sigma\sigma^T$, m is assumed differentiable and a_{ij} and ϕ are assumed twice differentiable. For later reference we rewrite (7) in the following form:

$$\mathcal{L}^*\phi = -(\nabla^T m)\phi - (\nabla^T\phi)m + (\nabla^T A)(\nabla\phi) + \frac{1}{2}\,\mathrm{tr}[(\nabla^2 A)\phi] + \frac{1}{2}\,\mathrm{tr}[(\nabla^2\phi)A] . \tag{7a}$$

In addition to the restrictions made on the coefficients of (1) in order to assure the existence of a solution x_t, we further assume:

(A) For every $t > 0$, the probability distribution of x_t possesses a twice continuously differentiable density $p_t(\beta)$, $\beta \in \mathbb{R}^n$ satisfying

$$\frac{\partial}{\partial t} p_t(\beta) = (\mathcal{L}^* p_t)(\beta) \quad . \tag{8}$$

(B) The filtering problem (1) and (2) possesses a twice continuously differentiable unnormalized conditional density $\rho_t(\beta)$ satisfying (cf. [2])

$$d\rho_t(\beta) = (\mathcal{L}^* \rho_t)(\beta)dt + \rho_t(\beta)h^T(\beta)dy_t \quad , \tag{9}$$

and a normalized conditional density satisfying (cf. [1])

$$d\pi_t(\beta) = (\mathcal{L}^* \pi_t)(\beta)dt + \pi_t(\beta)(h(\beta) - \hat{h}_t)^T d\nu_t \quad , \tag{10}$$

where $\nu_t = y_t - \int_o^t \hat{h}_s ds$ is the associated innovation process.

(C) The Fisher a-priori and a-posteriori information quantities were defined in the introduction for $n = 1$. In the vector case

$$F_t^\pi = E\left|\sigma(x_t)^T \frac{\nabla\pi_t(x_t)}{\pi\ (x_t)}\right|^2 = E\left|\sigma(x_t)^T \nabla\, \ell n\, \pi_t(x_t)\right|^2 = E\left|\sigma(x_t)^T \frac{\nabla\rho_t(x_t)}{\rho_t(x_t)}\right|^2 \quad , \tag{11}$$

and

$$F_t^p = E\left|\sigma(x_t)^T \frac{\nabla p_t(x_t)}{p_t(x_t)}\right|^2 = E\left|\sigma(x_t)^T \nabla\, \ell n\, p_t(x_t)\right|^2 \quad . \tag{12}$$

We assume that for every $t \geq 0$, F_t^p and F_t^π are finite, and

$$\int_o^t F_\theta^p d\theta < \infty \text{ and } \int_o^t F_\theta^\pi d\theta < \infty \quad .$$

(D) $E\int_o^T |h(x_t)|^2 dt < \infty$ for all finite T.

(E) Let $\mathcal{D}$ be any one of the operators $\frac{\partial}{\partial n_i}$, $\frac{\partial^2}{\partial n_i \partial n_j}$, $1 \leq \nu, j \leq n$ then $E\,\mathcal{D}\pi_t(\beta) = \mathcal{D}p_t(\beta)$ for all t and β. The requirement $E \sup_{\beta-\varepsilon\leq\theta\leq\beta+\varepsilon} |\mathcal{D}\pi_t(\theta)| < \infty$, for some $\varepsilon > 0$, suffices to guarantee the above condition.

<u>Note:</u> The finiteness of $dI(x_t,y_o^t)/dt$ and $dJ(x_t,y_o^t)/dt$ is not assumed here as it follows from Theorem 1.

3. Preliminary Results

From (1), (9), (10) it follows that equations (3), (4) and (11) are in fact expectations of compositions of Ito processes. In order to evaluate these compositions we need the following extension of Ito's formula.

Lemma 1 ([10]),[11],[12]): Let $(\Omega,\underline{\underline{F}},[\underline{\underline{F}}_t]_{t>o},P)$ be as above, and $(\eta_t,\underline{\underline{F}}_t)$ a k-dimensional Wiener process. Consider the family of processes, indexed by $\beta\in\mathbb{R}^q$,

$$G_t(\beta) = G_o(\beta) + \int_o^t f_s(\beta)\,ds + \int_o^t g_s(\beta)^T\,d\eta_s \tag{13}$$

with $f_s(\beta)\in\mathbb{R}$ and $g_s(\beta)\in\mathbb{R}^k$ adapted processes for each β. Assume further that

(i) $G_o(\beta)$ is continuous in β and independent of η_t.

(ii) f, g and G are continuous in (t,β).

(iii) For each fixed t almost surely $f_t(\cdot)$, $g_t^{(i)}(\cdot)(1\le i\le k)\in C^1(\mathbb{R}^q)$ and $G_t(\cdot)\in C^2(\mathbb{R}^q)$.

Let Z_t be a continuous adapted q-dimensional semimartingale, then

$$G_t(Z_t) = G_o(Z_o) + \int_o^t f_s(Z_s)d_s + \int_o^t g_s(Z_s)^T d\eta_s + \int_o^t (\nabla G_s(Z)^T dZ_s + \frac{1}{2}\int_o^t tr(\nabla^2 G_s(Z_s)d\langle Z,Z\rangle_s)$$
$$+ \int_o^t tr(\nabla(g(Z_s))^T\,d\langle\eta,Z\rangle_s)\ , \tag{14}$$

where $\langle\eta,Z\rangle_t$ denotes the cross-quadratic variation matrix between the processes η_t and Z_t.

Returning to the filtering problem, the normalization constant K_t $(K_t = \int_{\mathbb{R}^n}\rho_t(\beta)d\beta)$ satisfies $E\,\ell n\,K_t = J(x_t,y_o^t) - I(x_t,y_o^t)$ and is related to the filtering problem by:

Lemma 2:

$$\frac{d}{dt}\,I(x_t,y_o^t) = \frac{d}{dt}\,J(x_t,y_o^t) - \frac{1}{2}E|\hat{h}_t|^2\ .$$

Proof: It is well known [2], [13], that K_t is given by

$$K_t = \exp\left\{\int_o^t \hat{h}_s^T dy_s - \frac{1}{2}\int_o^t |\hat{h}_s|^2 ds\right\}\ ,$$

so that

$$E\,\ell n\,K_t = E\int_o^t (\hat{h}_s^T(h(x_s)ds + dB_s)) - \frac{1}{2}E\int_o^t|\hat{h}_s|^2 ds = \frac{1}{2}\int_o^t E|\hat{h}_s|^2 ds + E\int_o^t \hat{h}_s^T\,dB_s\ .$$

It follows from condition (D) that the second integral vanishes. Therefore $E\,\ell n\,K_t = \frac{1}{2}\int_o^t E|\hat{h}_s|^2 ds$ and the result of the lemma follows.

4. The Evolution of the Mutual Information

Let $G_t(\beta) = \ell n(\pi_t(\beta)/p_t(\beta))$ and $H_t(\beta) = \ell n(\rho_t(\beta)/p_t(\beta))$. Then, by (3) and (4) $I(x_t,y_o^t) = EG_t(x_t)$ and $J(x_t,y_o^t) = EH_t(x_t)$. Applying Ito's formula to G and H, it follows from (8), (9) and (10) that:

$$dG_t(u) = \left[\frac{(\mathcal{L}^*\pi_t)(u)}{\pi_t(u)} - \frac{(\mathcal{L}^*p_t)(u)}{p_t(u)} - \frac{1}{2}|h(u) - \hat{h}_t|^2\right]dt + (h(u) - \hat{h}_t)^T d\nu_t \,, \tag{15}$$

$$dH_t(u) = \left[\frac{(\mathcal{L}^*\rho_t)(u)}{\rho_t(u)} - \frac{(\mathcal{L}^*p_t)(u)}{p_t(u)} - \frac{1}{2}|h(u)|^2\right]dt + h(u)^T dy_t \,. \tag{15a}$$

Theorem 1: Under conditions (A) to (E), the following hold:

$$\frac{d}{dt} I(x_t,y_o^t) = \frac{1}{2}\left\{E|h(x_t) - \hat{h}_t|^2 - (F_t^\pi - F_t^p)\right\} \tag{16}$$

$$\frac{d}{dt} J(x_t,y_o^t) = \frac{1}{2}\left\{E|h(x_t)|^2 - (F_t^\pi - F_t^p)\right\} \,. \tag{16a}$$

Remark: Note that $\frac{1}{2}E|h(x_t) - \hat{h}_t|^2$ is equal to $d(I(x_o^t,y_o^t))/dt$ (cf. [1]) Theorem 16.3, or [4]) therefore (16) can be rewritten as:

$$\frac{d}{dt}\left[I(x_t,y_o^t) - I(x_o^t,y_o^t)\right] = \frac{1}{2}(F_t^\pi - F_t^p) \,. \tag{16b}$$

Proof: We shall prove (16a) in detail from which (16) follows immediately. Comparing (15a) and (13), we obtain from Lemma 1 a representation of $\ell n\frac{\rho_t(x_t)}{p_t(x_t)}$ in the form of (14), namely,

$$\ell n\left(\frac{\rho_t(x_t)}{p_t(x_t)}\right) = \int_o^t \left[\frac{(\mathcal{L}^*\rho_s)(x_s)}{\rho_s(x_s)} - \frac{(\mathcal{L}^*p_s)(x_s)}{p_s(x_s)} - \frac{1}{2}|h(x_s)|^2\right] ds + \int_o^t h^T(x_s)dy_s$$
$$+ \int_o^t \left(\frac{\nabla^T\rho_s(x_s)}{\rho_s(x_s)} - \frac{\nabla^T p_s(x_s)}{p_s(x_s)}\right) dx_s + \frac{1}{2}\int_o^t \left\{ tr\left[\left(\frac{\nabla^2\rho_s(x_s)}{\rho_s(x_s)} - \frac{\nabla^2 p_s(x_s)}{p_s(x_s)}\right) A(x_s)\right] - \left(\left|\frac{\sigma(x_s)^T\nabla\rho_s(x_s)}{\rho_s(x_s)}\right|^2 - \left|\frac{\sigma(x_s)^T\nabla p_s(x_s)}{p_s(x_s)}\right|^2\right)\right\} ds \,.$$

Note that the last term in (14) vanishes since the cross-quadratic variation between the signal and observation processes is zero: also $d\langle x,x\rangle_t = A(x_t)dt =: \sigma(x_t)\sigma^T(x_t)dt$. We now substitute the appropriate expressions for $\mathcal{L}^*$ (from (7a)), dx_s (from (1)), and dy_s (from (2)) to obtain;

$$\ln \frac{\rho_t(x_t)}{p_t(x_t)} = -\int_0^t \left(\frac{\nabla^T\rho_s(x_s)}{\rho_s(x_s)} - \frac{\nabla^T p_s(x_s)}{p_s(x_s)}\right) m(x_s)\,ds + \int_0^t \nabla^T A(x_s)\left(\frac{\nabla\rho_s(x_s)}{\rho_s(x_s)} - \frac{\nabla p_s(x_s)}{p_s(x_s)}\right) ds$$

$$+ \frac{1}{2}\int_0^t \mathrm{tr}\left[\left(\frac{\nabla^2\rho_s(x_s)}{\rho_s(x_s)} - \frac{\nabla^2 p_s(x_s)}{p_s(x_s)}\right) A(x_s)\right] ds + \frac{1}{2}\int_0^t |h(x_s)|^2 ds + \int_0^t h^T(x_s)\,dB_s$$

$$+ \int_0^t \left(\frac{\nabla^T\rho_s(x_s)}{\rho_s(x_s)} - \frac{\nabla^T p_s(x_s)}{p_s(x_s)}\right) m(x_s)\,ds + \int_0^t \left(\frac{\nabla^T\rho_s(x_s)}{\rho_s(x_s)} - \frac{\nabla^T p_s(x_s)}{p_s(x_s)}\right) \sigma(x_s)\,dW_s$$

$$+ \frac{1}{2}\int_0^t \mathrm{tr}\left[\left(\frac{\nabla^2\rho_s(x_s)}{\rho_s(x_s)} - \frac{\nabla^2 p_s(x_s)}{p_s(x_s)}\right) A(x_s)\right] ds$$

$$- \frac{1}{2}\int_0^t \left\{\left|\frac{\sigma(x_s)^T\nabla\rho_s(x_s)}{\rho_s(x_s)}\right|^2 - \left|\frac{\sigma(x_s)^T\nabla p_s(x_s)}{p_s(x_s)}\right|^2\right\} ds \quad .$$

Reordering and taking expectations yields:

$$J(x_t, y_o^t) = \frac{1}{2}\int_0^t \left[E|h(x_s)|^2 - \left(E\left|\frac{\sigma(x_s)^T\nabla\rho_s(x_s)}{\rho_s(x_s)}\right|^2 - E\left|\frac{\sigma(x_s)^T\nabla p_s(x_s)}{p_s(x_s)}\right|^2\right)\right] ds$$

$$+ \int_0^t \left\{E\left[\nabla^T A(x_s)\left(\frac{\nabla\rho_s(x_s)}{\rho_s(x)} - \frac{\nabla p_s(x_s)}{p_s(x_s)}\right)\right] + E\,\mathrm{tr}\left[\left(\frac{\nabla^2\rho_s(x_s)}{\rho_s(x_s)} - \frac{\nabla^2 p_s(x_s)}{p_s(x_s)}\right) A(x_s)\right]\right\} ds$$

$$+ E\int_0^t h^T(x_s)\,dB_s + E\int_0^t \left(\frac{\nabla^T\rho_s(x_s)}{\rho_s(x_s)} - \frac{\nabla^T p_s(x_s)}{p_s(x_s)}\right)\sigma(x_s)\,dW_s \quad .$$

In view of conditions (C) and (D), the last two integrals vanish; thus (16a) will follow if the second integral vanishes as well. Indeed, using condition (E):

$$E\left[\nabla^T A(x_s)\,\frac{\nabla\rho_s(x_s)}{\rho_s(x)}\right] = E\left[\nabla^T A(x_s)\,\frac{\nabla\pi_s(x_s)}{\pi_s(x_s)}\right] = \int_{\mathbf{R}^n} \nabla^T A(\beta) E\nabla\pi_s(\beta)\,d\beta = \int_{\mathbf{R}^n} \nabla^T A(\beta)\nabla p_s(\beta)\,d\beta =$$

$$= E\left[\nabla^T A(x_s)\,\frac{\nabla p_s(x_s)}{p_s(x_s)}\right];$$

$$E\left[\frac{\nabla^2\rho_s(x_s)}{\rho_s(x_s)}\,A(x_s)\right] = E\left[\frac{\nabla^2\pi_s(x_s)}{\pi_s(x_s)}\,A(x_s)\right] = \int_{\mathbf{R}^n} E\nabla^2\pi_s(\beta)A(\beta)\,d\beta = \int_{\mathbf{R}^n} \nabla^2 p_s(\beta)A(\beta)\,d\beta =$$

$$= E\left[\frac{\nabla^2 p_s(x_s)}{p_s(x_s)}\,A(x_s)\right]$$

from which (16a) follows. Finally (16a) implies (16) by Lemma 2 and the obvious identity $E|h(x_s)|^2 - E|\hat{h}_s|^2 = E|h(x_s) - \hat{h}_s|^2$.

5. An Extension of the Filtering Model

In this section we consider an extension of the filtering model. Let x_t be as before and let y_t satisfy

$$dy_t = h(x_t,y_t)dt + b(y_t)dW_t \quad , \tag{17}$$

where $b(\cdot) \in R^{m\times \ell}$. We assume that bb' is nonsingular, which implies that $\ell \geq m$. The generalization consists in (a) the dynamic noise N_t^1 $(= \int \sigma(x_t)\,dW_t)$ and the observation noise N_t^2 $(= \int b(y_t)dW_t)$ may be correlated:

$$<N^1,N^2>_t = <\int \sigma(x)dW, \int b(y)dW>_t = \int_0^t Q_s ds \quad ,$$

where

$$Q_t = \sigma(x_t)b'(y_t) \quad ,$$

and (b) feedback of the observation variables is allowed in the observation equation (17).

The extension of (16) to (1) and (17) is as follows:

Theorem 2: Under the above assumptions

$$\frac{dI(x_t,y_o^t)}{dt} = \frac{1}{2}\left\{E\left|h(x_t) - \hat{h}_t\right|^2_{c^{-1}} - (F_t^\pi - F_t^p) - E\left|\nabla^T(\pi_t(x_t)Q_t\right|^2_{c^{-1}}\right\}$$
$$+ E\left\{tr\left[Q_t c^{-1}(y_t)\left\{\nabla\left[(h(x_t)-\hat{h}_t)^T - \frac{\nabla^T(\pi_t(x_t)Q_t)}{\pi_t(x_t)}\right]\right\}^T\right]\right\} \tag{18}$$

where $c(y_t) = b(y_t)b'(y_t)$ and for any $u \in \mathbb{R}^k$, $A \in \mathbb{R}^{k\times k}$, positive definite $|u|_A^2 = u'Au$. Note that if $Q = 0$, (18) simplifies to:

$$\frac{dI(x_t,y_o^t)}{dt} = \frac{1}{2}\left\{E\left|h(x_t,y_t) - h_t\right|^2_{c^{-1}} - (F_t - F_t^p)\right\} \quad . \tag{19}$$

Regarding the proof of (18), the evolution of the normalized conditional density is given by [cf. theorem 11.2.1 of [14]]:

$$d\pi_t(\beta) = (\mathcal{L}^*\pi_t)(\beta)dt + \left\{\pi_t(\beta)\left[c(y_t)^{-1/2}(h(\beta,y_t) - \hat{h}_t)\right]^T - \nabla^T(\pi_t(\beta)Q_t)c(y_t)^{-1/2}\right\}d\tilde{\nu}_t \quad , \tag{20}$$

where

$$\tilde{\nu}_t = \int_0^t [c(y_s)^{-1/2}(h(x_s,y_s) - \hat{h}_s)]ds + \int_0^t c_s(y_s)^{-1/2}b(y_s)dW_s \quad .$$

From this point on, the proof of Theorem 2 follows along the same lines as that of the previous results and therefore the details are omitted.

References

[1] R.S. Liptser and A.N. Shiryayev: Statistics of Random Processes, Vols. I and II, Springer-Verlag, New York (1977).

[2] M. Zakai: "On the optimal filtering of diffusion processes", Z. Wahr. Verw. Geb., 11 (1969), pp. 230-243.

[3] M. Pinsker: Information and Information Stability of Random Variables and Processes, translated (from Russian) and edited by A. Feinstein, San Francisco, Holden Day (1964).

[4] T.T. Kadota, M. Zakai and J. Ziv: "Mutual information of the white Gaussian channel with and without feedback", IEEE Trans. Infor. Theory, Vol. IT-17, 4 (1971), pp. 368-371.

[5] H.L. Van Trees: Detection, Estimation and Modulation Theory, Part 1, New York, Wiley (1968).

[6] B.Z. Bobrovski and M. Zakai: "A lower bound on the estimation of certain diffusion processes", IEEE Trans. Infor. Theory, Vol. IT-22, 1 (1976), pp. 45-52.

[7] R.S. Liptser: "Optimal coding and decoding for transmission of a Gaussian Markov signal in a noiseless feedback channel", Probl. Peredachi Inform., 10, No. 4 (1974), pp. 3-15.

[8] R.S. Bucy: "Information and Filtering", Information Sciences, 18 (1979), pp. 179-187.

[9] B.Z. Bobrovski and M. Zakai: "Asymptotic a-priori estimates for the error in the nonlinear filtering problem", IEEE Trans. Infor. Theory, Vol. IT-28, 2 (1982), pp. 371-376.

[10] B.L. Rozovskii, "On the Ito-Wentzel formula", Vestnik Moskov. Univ., No. 1 (1973), pp. 26-32.

[11] H. Kunita: "Some extensions of Ito's formula", Séminaire de Probabilités XV, Lect. Notes in Math., 850, Springer-Verlag, Berlin-Heidelberg-New York, (1982), pp. 118-141.

[12] J.M. Bismut: "A generalized formula of Ito and some other properties of stochastic flows", Z. Wahr. Verw. Geb., 55 (1981), pp. 331-350.

[13] M.H.A. Davis and S.I. Marcus: "An introduction to nonlinear filtering", in Stochastic Systems: The Mathematics of Filtering and Identification, M. Hazewinkel and J.C. Willems (eds.), D. Reidel, Dortrecht (1981).

[14] G. Kallianpur: Stochastic Filtering Theory, Springer-Verlag, New-York-Heidelberg-Berlin (1980).

E. Mayer-Wolf
Applied Mathematics

M. Zakai
Dept. of Electrical Eng.

Technion - Israel Institute of Technology
Haifa 32000, Israel.

OPTIMAL STOPPING OF BI-MARKOV PROCESSES

G. MAZZIOTTO

PAA/TIM/MTI
Centre National d'Etudes des Télécommunications
38-40, rue du Général Leclerc
92 131 - ISSY LES MOULINEAUX - FRANCE

In this paper, we solve the optimal stopping problem for a particular class of two-parameter processes, that is the bi-Markov processes defined by a set of stochastic differential equations.

The first chapter deals with bi-Markov processes, and the corresponding bi-potential theory. The class of bi-Markov processes considered here is defined and explicitly constructed in the first paragraph. In the second paragraph, we introduce some definitions concerning two-variable functions, which are similar to those of the classical potential theory. We then study, in the third paragraph, the two-parameter processes associated to these functions and to the bi-Markov process previously defined. Various decomposition results are obtained for two-parameter supermartingales. In the fourth paragraph, we present the definition of weak harmonicity which will be useful in optimal stopping.

The second chapter is devoted to the optimal stopping problem of a bi-markov process. In the first paragraph, various results on two-parameter optimal stopping are recalled in a presentation adapted to the markovian situation. Existence results are obtained. In the second paragraph we give a different approach to the problem, leading to a characterization of the Snell envelop. The problem of optimal stopping for a bi-Markov process is treated in full. The notion of Snell reduite is introduced, and the problem is solved under mild assumptions. In the fourth paragraph we study links between weak harmonicity and Snell reduite. A characterization of it as a solution of a system of variational inequations is given.

PRELIMINARIES:

The processes we consider in this paper are indexed on $\mathbb{R}_+^2$, and extended to its one-point compactification, $\overline{\mathbb{R}}_+^2 = \mathbb{R}_+^2 \cup \{\infty\}$, as being null at infinity. The partial order is defined by

$\forall\, s=(s_1,s_2),\ t=(t_1,t_2) :\ s \leq t \iff s_1 \leq t_1$ and $s_2 \leq t_2$; with $t \leq \infty\ \ \forall\, t \in \mathbb{R}_+^2$.

Defined on a complete probability space $(\Omega, \underline{\underline{A}}, \mathbb{P})$, a filtration is a family $\underline{\underline{F}} = (\underline{\underline{F}}_t;\ t\in \mathbb{R}_+^2)$ of sub-σ-fields of $\underline{\underline{A}}$, such that ([10], [29], [36]) : $\underline{\underline{F}}_0$ contains all the $\mathbb{P}$-negligible sets of $\underline{\underline{A}}$ (Axiom F1), family $\underline{\underline{F}}$ is increasing with respect to the partial order on $\mathbb{R}_+^2$ (Axiom F2), and $\underline{\underline{F}}$ is right-continuous (Axiom F3). In addition, we also assume that filtration $\underline{\underline{F}}$ satisfies the following conditional independence property (Axiom F4)

| $\forall\, t=(t_1,t_2)$: σ-fields $\underline{\underline{F}}^1_{t_1} = \bigvee_u \underline{\underline{F}}_{(t_1,u)}$ and $\underline{\underline{F}}^2_{t_2} = \bigvee_u \underline{\underline{F}}_{(u,t_2)}$
| are conditionaly independent given $\underline{\underline{F}}_t$.

The optional σ-field on $\Omega \times \mathbb{R}_+^2$, and the optional projection of a bounded process X, say ${}^oX$, are defined in ([1]).

A stopping point (s.p.) is a random variable (r.v.) T, taking its values in $\overline{\mathbb{R}}_+^2$, such that $\{T \leq t\} \in \underline{\underline{F}}_t$, $\forall\, t \in \mathbb{R}_+^2$. The set of all s.p. is denoted by $\underline{\underline{T}}$. To any s.p. T, we associate a σ-field $\underline{\underline{F}}_T$, which is the σ-field of all events A such that $A \cap \{T \leq t\} \in \underline{\underline{F}}_t$, $\forall\, t$. All the classical properties of stopping times ([11]) do not extend to stopping points (see ([35])). The graph of a s.p. T, denoted by $[\![T]\!]$, is the optional set defined by : $[\![T]\!] = \{(\omega,t) : T(\omega) = t\ ,\ t \in \mathbb{R}_+^2\}$.

Given a random set H in $\Omega \times \mathbb{R}_+^2$, we denote by $[\![H,\infty[\![$ the random set $[\![H,\infty[\![= \{(\omega,t) : \exists\, s \leq t$ such that $(\omega,s) \in H\}$. The début of H, denoted by L_H, is the lower boundary of the set $[\![H,\infty[\![$, with the convention that $L_H = \infty$ if the section is empty. A stopping line (s.l.) is the début of an optional random set ([27]). We denote by $\underline{\underline{L}}$ the set of all stopping lines. $\underline{\underline{T}}$ can be taken as a subset of $\underline{\underline{L}}$, by identifying any s.p. T with the s.l. which is the début of the set $[\![T]\!]$.

The partial order is extended to $\underline{\underline{T}}$ by:

$$\forall\, T, T' \in \underline{\underline{T}} : T \leq T' \iff T \leq T' \text{ a.s. },$$

as well as to $\underline{\underline{L}}$ by:

$$\forall\, L, L' \in \underline{\underline{L}} : L \leq L' \iff [\![L',\infty[\![\subset [\![L,\infty[\![\text{ a.s. }.$$

1- BI-MARKOV PROCESSES

Bi-Markov processes are particular two-parameter processes, analogous to the well-known bi-Brownian motion ([6,35]). In this chapter various bi-Markov processes are constructed, and notions of an associated bi-potential theory are presented. In connection with future optimal stopping problems, different properties of supermartingales are obtained, and a notion of weak harmonicity on an open set is proposed.

1-1- Construction of bi-Markov processes:

Roughly speaking, bi-Markov processes are defined as the tensor product of two classical — one-parameter — Markov processes. In this paper we work on a special class of such Markov processes, those which can be obtained as strong solutions of independent stochastic differential equations.

In the sequel, super- or sub-script i will take values 1 and 2. Let $B^i = (B^i_u\,;\, u \in \mathbb{R}_+)$ be a Brownian motion, defined on the probability space $(\Omega^i, \underline{\underline{M}}^i, \mathbb{P}^i)$ with respect to the filtration $(\underline{\underline{M}}^i_u\,;\, u \in \mathbb{R}_+)$. Consider the stochastic differential equation on $\mathbb{R}^{d^i} = E^i$:

$$dX^i_u = b^i(X^i_u)\, du + \sigma^i(X^i_u)\, dB^i_u \quad ,$$

where b^i and σ^i are matrices of appropriate dimension such that, for any $y \in E^i$, there exists a unique strong solution $X^{iy} = (X^{iy}_u\,;\, u \in \mathbb{R}_+)$ with initial value $X^{iy}_0 = y$. Processes $(X^{iy}\,;\, y \in E^i)$ form a Markov family to which one can associate a canonical Markov process X^i, ([13]). Let $P^i = (P^i_u\,;\, u \in \mathbb{R}_+)$ be the semi-group of this Markov process, and denote by $U^i = (U^i_p\,;\, p \in \mathbb{R}_+)$ its resolvent family. The set of bounded Borel (resp. bounded uniformly continuous) functions on E^i is denoted by $b(E^i)$ (resp. $C(E^i)$). The generator of X^i is the second-order differential operator $\underline{\underline{L}}^i$, defined on the domain $\underline{\underline{D}}(\underline{\underline{L}}^i) \subset C(E^i)$, by :

$$\forall\, f \in \underline{\underline{D}}(\underline{\underline{L}}^i) : \underline{\underline{L}}^i f = \sum_{k=1}^{d^i} b^i_k \frac{\partial f}{\partial x_k} + \frac{1}{2} \sum_{k,j=1}^{d^i} \sigma^i_{jk} \frac{\partial^2 f}{\partial x_j \partial x_k}$$

We suppose that b^i and σ^i are such that X^i is a strong Feller process ([13]).

For $p \in \mathbb{R}_+$, let $\underline{\underline{L}}_p^i$ be the operator $\underline{\underline{L}}^i$ - p(Identity). Then,

$$\forall\, f \in \underline{\underline{D}}(\underline{\underline{L}}^i) : f = U_p^i\, g \quad <=> \quad g = -\underline{\underline{L}}_p^i\, f \quad .$$

The family of bi-Markov processes $X = (X^x;\ x=(x^1,x^2)\quad E=E^1 \times E^2)$ is defined by the following:

$$\forall\, x = (x^1,x^2),\ \forall\, t = (t_1,t_2) \in \mathbb{R}_+^2 : \quad X_t^x = (X_{t_1}^{1x^1}, X_{t_2}^{2x^2})$$

on the product probability space $(\Omega = \Omega^1 \times \Omega^2, \underline{\underline{A}} = \underline{\underline{M}}^1 \otimes \underline{\underline{M}}^2, \mathbb{P} = \mathbb{P}^1 \otimes \mathbb{P}^2)$, endowed with the two-parameter filtration $\underline{\underline{F}}^\circ = (\underline{\underline{F}}_t^\circ = \underline{\underline{M}}_{t_1}^1 \otimes \underline{\underline{M}}_{t_2}^2\ ;\ t=(t_1,t_2))$. It may be noticed that these processes enter in the class of two-parameter Markov processes, as defined in ([17], [32]) for example. Denote by $\underline{\underline{F}} = (\underline{\underline{F}}_t\ ;\ t \in \mathbb{R}_+^2)$ the smallest filtration which contains $\underline{\underline{F}}^\circ$ and which is right-continuous, such that all the $\mathbb{P}$-negligible sets belong to $\underline{\underline{F}}_0$. In addition, $\underline{\underline{F}}$ satisfies the conditional independence property of Axiom F4.

Let b(E) (resp. C(E)) denote the set of bounded Borel functions (resp. bounded uniformly continuous functions) on the product space $E = E^1 \times E^2$. We define a two-parameter semi-group on b(E) by setting:

$$\forall\, t = (t_1,t_2) : P_t = P_{t_1}^1 \otimes P_{t_2}^2 \quad ,$$

the associated resolvent is the two-parameter family of operators on b(E) defined by:

$$\forall\, p = (p_1,p_2) : U_p = U_{p_1}^1 \otimes U_{p_2}^2 \quad .$$

Operators $P_{t_1}^1$, $U_{p_1}^1$, $P_{t_2}^2$, $U_{p_2}^2$ will be considered as operating on b(E) as well as operating on spaces $b(E^1)$ or $b(E^2)$ with no risk of ambiguity. Similarly, generators $\underline{\underline{L}}^1$ and $\underline{\underline{L}}^2$ will be considered on the domain $\underline{\underline{D}}(\underline{\underline{L}}^1,\underline{\underline{L}}^2)$ of functions $f \in C(E)$ such that functions $\underline{\underline{L}}^1 f$ and $\underline{\underline{L}}^2 f$ are well defined and belong to C(E).

The Markov property of processes $X = (X^x\ ;\ x \in E)$ can be resumed by the following:

> $\forall\, f \in b(E)$, $\forall\, s \in \mathbb{R}_+^2$, and for any $x \in E$, the optional projection of process $f(X_{s+.}^x) = (f(X_{s+t}^x)\ ;\ t \in \mathbb{R}_+^2)$ is such that:
>
> $$^{o}(f(X_{s+.}^x))_t = P_s f(X_t^x) \quad , \quad \forall\, t \in \mathbb{R}_+^2 \quad .$$

Let us denote by $\underline{\underline{T}}$ the set of all stopping points with respect to the filtration $\underline{\underline{F}}$. Processes X satisfy a strong Markov property with respect to stopping points; namely,

$$\forall\, T \in \underline{\underline{T}} \;,\; \forall \text{ a } \underline{\underline{F}}_T\text{-measurable r.v. } S,\; \forall\, f \in b(E),\; \forall\, x \in E :$$

$$E(f(X^x_{T+S}) \,/\, \underline{\underline{F}}_T) = P_S f(X_T) \quad .$$

This result can be proved as in ([28]) by using the optional sampling theorem of ([19]).

In the sequel, $E_x(f(X))$ will represent the expectation $E(f(X^x))$.

1-2- <u>Notions of a bi-potential theory</u>:

Given two classical semi-groups on two spaces E^1 and E^2, different classes of functions on the product space $E = E^1 \times E^2$ can be defined separately, according to their properties on each space E^1 or E^2. Such a study has been done in ([7],[8]) , dealing with the one-parameter semi-group constructed as the tensor product of two classical semi-groups. In this paragraph we recall definitions and results of ([7]) with the preceding two-parameter semi-group.

<u>Definition 1-2-1</u>: Let f be a positive function on the product space $E = E^1 \times E^2$, and let $p \in \mathbb{R}_+$. For i=1,2, function f is called p-i-supermedian (resp. p-i-excessive, p-i-harmonic) on E, if the function on E^i, defined by: $x^i \rightarrow f(x^1,x^2)$, $\forall\, x^i \in E^i$, is p-supermedian (resp. p-excessive, p-harmonic) when the other variable is fixed.

Let f be a positive function on E, and let $p=(p_1,p_2) \in \mathbb{R}^2_+$. Function f is called p-bisupermedian (resp. p-biexcessive, p-biharmonic) iff f is both p_1-1-supermedian (resp. p_1-1-excessive, p_1-1-harmonic) and p_2-2-supermedian (resp. p_2-2-excessive, p_2-2-harmonic).

We refer to ([28]) for the definitions of the classical potential theory. It is proved in ([7]) that any positive function on E which is both p_1-1-excessive and p_2-2-excessive, is measurable on E, and is lower semi-continuous when processes X^1 and X^2 are strongly Fellerian .

For $p=(p_1,p_2)$ and $t=(t_1,t_2) \in \mathbb{R}_+^2$, denote by p.t the scalar product $p_1t_1 + p_2t_2$. It should be noted that if function f is p-bisuper-median then:

$$\forall\, t \in \mathbb{R}_+^2 : e^{-p.t}\, P_t f \leq f \quad ,$$

and if moreover, function f is p-biexcessive, then:

$$\lim_n e^{-p.t(n)}\, P_{t(n)} f = f \quad , \text{ for any sequence } (t(n); n \in \mathbb{N})$$

decreasing to zero.

If function f is p-biharmonic, then it is p-biexcessive, and

$$\forall\, t \in \mathbb{R}_+^2 : e^{-p.t}\, P_t f = f \; .$$

For any function $g \in b(E)$ (not necessarely positive), the function $f = U_p g$ is p-biexcessive (hence positive) iff functions $U^1_{p_1} g$ and $U^2_{p_2} g$ are positive. Such a function is called p-potential in the sequel. More generally, a function $f \in \underline{\underline{D}}(\underline{\underline{L}}^1, \underline{\underline{L}}^2)$ such that $\underline{\underline{L}}^1_{p_1} f \leq 0$ and $\underline{\underline{L}}^2_{p_2} f \leq 0$ is p-biexcessive.

The following result proves that any p-biexcessive function can be approximated by p-potentials.

<u>Proposition 1-2-1</u>: For $p=(p_1,p_2)$ such that $p_1 > 0$ and $p_2 > 0$, any p-biexcessive function f is the limit of a non-decreasing sequence of p-potentials $(U_p g^n ; n \in \mathbb{N})$.

The proof is similar to that of the one-parameter case ([13]). The situation of p belonging to the coordinate axis could be handled with the same technical caution as in the classical potential theory.

We note that the p-biexcessive functions f which could be assimilated to the excessive functions of the classical theory, are those corresponding to a sequence of positive functions $(g^n ; n \in \mathbb{N})$ of Proposition 2-1-1. This class of functions satisfies:

$$\forall\, t \in \mathbb{R}_+^2 : f + e^{-p.t}\, P_t f - e^{-p_1t_1}\, P^1_{t_1} f - e^{-p_2t_2}\, P^2_{t_2} f \geq 0 \; .$$

Some of them had been introduced in ([8]), where a Riesz type decomposition was obtained.

1-3- Two-parameter supermartingales:

In this paragraph we study the two-parameter processes associated to the functions defined before. Dynkin formulas w hich generalize the classical one and those of ([20], [34]), are obtained for various classes of supermartingales. Trajectorial regularity properties of general supermartingales are given.

Given a complete probability space $(\Omega,\underline{\underline{A}},\mathbb{P})$ endowed with a filtration $\underline{\underline{F}} = (\underline{\underline{F}}_t \,;\, t \in \mathbb{R}^2_+)$ satisfying Axioms F1, F2, F3 and F4, we recall that a supermartingale (resp. strong supermartingale) is a process $J = (J_t \,;\, t \in \mathbb{R}^2_+)$ adapted to the filtration $\underline{\underline{F}}$ (resp. optional), integrable (resp. of class (D) i.e., $\{J_T \,;\, T \in \underline{\underline{T}}\}$ is uniformly integrable) such that:

$$\forall\, s, t \in \mathbb{R}^2_+ \;:\; s \leq t \;\Rightarrow\; E(J_t / \underline{\underline{F}}_s) \leq J_s \text{ a.s. }, \text{ (resp.:}$$

$$\forall\, S, T \in \underline{\underline{T}} \;:\; S \leq T \;\Rightarrow\; E(J_T / \underline{\underline{F}}_S) \leq J_S \text{ a.s.) .}$$

Coming back to our bi-Markov process X, let us define the two-parameter processes which correspond to the functions of section 1-2. To any function f on E and p $\mathbb{R}^2_+$ we associate process J^x for $x \in E$, as follows:

$$\forall\, t \in \mathbb{R}^2_+ \;:\; J^x_t = e^{-p.t} f(X^x_t) \quad .$$

In the sequel, in order to simplify notations and computations we only consider point $p = (p_1,p_2)$ such that $p_1 = p_2$ (say p). For $t = (t_1,t_2)$, we put $|t| = t_1 + t_2$, and $|\infty| = \infty$. Then, for $p \in \mathbb{R}_+$, p.t will stand for $p|t|$.

It is easy to verify that if f is p-supermedian, then the associated processes J^x are non-negative supermartingales, $\forall\, x \in E$. We begin by studying the case of p-potentials. It is clear that, for any $g \in b(E)$, $U_p g$ is continuous (X^1 and X^2 are strong Feller process). It follows that the associated processes J^x are continuous bounded supermartingales; hence strong supermartingales ([35]). If f is any bounded p-biexcessive function, then by Proposition 1-2-1, f is the limit of an increasing sequence of p-potentials. This proves that the associated process J^x is the limit of an increasing sequence of strong supermartingales and, by a result of ([35]), a strong supermartingale.

The following result concerns the processes associated to a p-potential.

Proposition 1-3-1: For p > 0, let f be a p-potential: $f = U_p g$ where $g \in b(E)$. Then $\forall\, x \in E$, the associated process J^x is undistinguishable from the optional projection of the process C^x defined by:

$$\forall\, t \in \mathbb{R}_+^2 : C_t^x = \int_{t_1}^{\infty} \int_{t_2}^{\infty} e^{-p.s}\, g(X_s^x)\, ds_1\, ds_2 \quad .$$

Proof: Let $x \in E$ be fixed. Coming back to the definition of an optional projection ([1]), we verify directly that:

$$\forall\, t \in \mathbb{R}_+^2 : ({}^{o}C^x)_t = \int_{t_1}^{\infty} \int_{t_2}^{\infty} e^{-p.s}\ {}^{o}(g(X_s^x))_t\, ds_1\, ds_2 \quad .$$

Then using the Markov property of section 1-1, we get:

$$\forall\, s \text{ fixed}, \forall\, t \leq s : {}^{o}(g(X_s^x))_t = P_{s-t}\, g(X_t^x) \quad .$$

This proves the proposition.

The first consequence of this result is a Dynkin-type formula. Let T be any stopping point. Using properties of an optional projection ([1]), we obtain

$$E_x(e^{-p.T}\, f(X_T)) = E_x\left(\int_{T_1}^{\infty} \int_{T_2}^{\infty} e^{-p.s}\, g(X_s)\, ds_1\, ds_2 \right) \quad .$$

The second consequence is a decomposition of process J^x which is analogous, in some sense, to the Doob-Meyer decomposition of supermartingales ([11]) :

$$\forall\, t \in \mathbb{R}_+^2 : J_t^x = m_t + \int_0^{t_1} \int_0^{t_2} e^{-p.s}\, g(X_s^x)\, ds_1\, ds_2 \quad ,$$

where m is a weak martingale (see [29] for this definition).

It can be noticed that enters in these results the fact that $f = U_p g$, but not the fact that f was a p-potential (i.e. $U_p^1 g \geq 0$ and $U_p^2 g \geq 0$).

To conclude this paragraph, we study the restrictions of two-parameter supermartingales to optional increasing paths. A second type of Dynkin formula will be obtained. Some preliminaries are necessary.

The notion of an optional increasing path has been introduced [35], as a generalization of the discrete tactics of [18] and [22]. An optional increasing path (o.i.p.) is a one-parameter family $(Z_u ; u \in \mathbb{R}_+)$ of stopping points, such that the mapping $u \to Z_u$ is increasing and continuous a.s.. Moreover, any o.i.p. can be parametrized "canonically" by taking

$$\forall u \in \mathbb{R}_+ : Z_u = (Z_u^1, Z_u^2) \quad \text{with} \quad u = Z_u^1 + Z_u^2 = |Z_u| .$$

For $m \in \mathbb{N}$, let $\mathbb{D}_m$ denote the set of dyadic numbers of order m in $\mathbb{R}_+^2$: $\mathbb{D}_m = \{ t = (j2^{-m}, k2^{-m}) ; j,k \in \mathbb{N} \}$. Then, a tactic in $\mathbb{D}_m$ is an increasing sequence of stopping points $(T_n ; n \in \mathbb{N})$ such that $\forall n: T_n \in \mathbb{D}_m$ a.s. and $T_{n+1} = T_n + (2^{-m},0)$ or $T_n + (0,2^{-m})$, and T_{n+1} is a $\underline{\underline{F}}_{T_n}$-measurable random variable, [18,22,35]. Such a tactic is said of order m.
By interpolating between each s.p. T_n, we associate to any tactic in $\mathbb{D}_m$ $(T_n ; n \in \mathbb{N})$, an o.i.p. $Z = (Z_u ; u \in \mathbb{R}_+)$ whose trajectories are increasing step functions, with corners in $\mathbb{D}_m$ (a corner being a point where Z changes direction). Moreover, using the definition of a tactic, it can be verified that the corners form a sequence of stopping points.
It is proved in [35] that any o.i.p. can be approximated by a sequence of tactics of increasing order. We denote by $\underline{\underline{Z}}$ the set of all o.i.p., and by $\underline{\underline{Z}}_m^d$ the set of all tactics of order m.
Given an o.i.p. $Z = (Z_u ; u \in \mathbb{R}_+)$, $\underline{\underline{F}}^Z$ is the one-parameter filtration defined by $\underline{\underline{F}}^Z = (\underline{\underline{F}}_u^Z = \underline{\underline{F}}_{Z_u} ; u \in \mathbb{R}_+)$, and $\underline{\underline{T}}^Z$ is the set of all $\underline{\underline{F}}^Z$-stopping times.

Let J be a two-parameter process. For any optional increasing path $Z=(Z_u\,;\,u\in\mathbb{R}_+)$, the restriction of J to the o.i.p. Z is the one-parameter process J^Z, defined by $J^Z=(J^Z_u=J_{Z_u}\,;u\in\mathbb{R}_+)$. This process is $\underline{\underline{F}}^Z$-optional if J is itself $\underline{\underline{F}}$-optional, and it is a strong supermartingale with respect to $\underline{\underline{F}}^Z$ if J is a strong supermartingale.

In case the strong supermartingale J is associated to a p-biexcessive function f of $\underline{\underline{D}}(\underline{\underline{L}}^1,\underline{\underline{L}}^2)$, we obtain the following Dynkin formula.

<u>Proposition 1-3-2</u>: Let $Z=(Z_u\,;\,u\in\mathbb{R}_+)$ be a given optional increasing path and let f be a p-biexcessive function of $\underline{\underline{D}}(\underline{\underline{L}}^1,\underline{\underline{L}}^2)$ for $p>0$. There exist two one-parameter $\underline{\underline{F}}^Z$-adapted processes, λ^{1Z} and λ^{2Z}, non-vanishing simultaneously and taking their values in $[0,1]$, such that, for every pair of ordered $\underline{\underline{F}}^Z$-stopping times, $\sigma\le\tau$, one has $\forall\, x\in E$:

$$E_x\left(e^{-p\tau}\, f(X^Z_\tau) - e^{-p\sigma}\, f(X^Z_\sigma)\;\Big/\;\underline{\underline{F}}^Z_\sigma\right) =$$

$$= E_x\left(\int_\sigma^\tau (\underline{\underline{L}}^1_p f(X^Z_u)\,\lambda^{1Z}_u + \underline{\underline{L}}^2_p f(X^Z_u)\,\lambda^{2Z}_u)\, e^{-pu}\, du\;\Big/\;\underline{\underline{F}}^Z_\sigma\right)$$

<u>Proof</u>: If the o.i.p. $Z=(Z_u\,;\,u\in\mathbb{R}_+)$ is a tactic of order m, say $(T_n\,;\,n\in\mathbb{N})$, the formula can be computed step by step, i.e. between each pair of successive points T_n, T_{n+1} by means of the classical Dynkin formula. We obtain the following

$$E_x(e^{-p(n+1)2^{-m}}\, f(X^Z_{(n+1)2^{-m}}) - e^{-pn2^{-m}}\, f(X^Z_{n2^{-m}})\;/\;\underline{\underline{F}}^Z_{n2^{-m}}) =$$

$$= E_x\Big(\int_{n2^{-m}}^{(n+1)2^{-m}} (\underline{\underline{L}}^1_p f(X^Z_u)\; 1\!\!1_{\{Z^2_{(n+1)2^{-m}} = Z^2_{n2^{-m}}\}} +$$

$$\underline{\underline{L}}^2_p f(X^Z_u)\; 1\!\!1_{\{Z^1_{(n+1)2^{-m}} = Z^1_{n2^{-m}}\}})\, e^{-pu}\, du\;\Big/\;\underline{\underline{F}}^Z_{n2^{-m}}\Big)$$

Therefore, we define processes λ^{1Z} and λ^{2Z} by the following formula:

$$\lambda_u^{1Z} = 1 = 1 - \lambda_u^{2Z} \quad \text{on } \{Z^2_{(k+1)2^{-n}} = Z^2_{k2^{-n}}\} \text{ , and}$$
$$\lambda_u^{2Z} = 1 = 1 - \lambda_u^{1Z} \quad \text{on } \{Z^1_{(k+1)2^{-n}} = Z^1_{k2^{-n}}\} \text{ ,}$$

for $k2^{-n} \leq u \leq (k+1)2^{-n}$.

Then the formula is extended to any stopping times, as stated in the proposition.

Now let us consider a general o.i.p. Z. It can be approximated by a sequence of tactics of increasing orders $(Z^n \, ; n \in \mathbb{N})$, such that the paths $((Z^n_u \, ; u \in \mathbb{R}_+) \, ; n \in \mathbb{N})$ converge a.s. uniformly on any finite interval to the path $(Z_u \, ; u \in \mathbb{R}_+)$. For each tactic Z^n we can write the preceding Dynkin formula with processes λ^{1Z^n} and λ^{2Z^n}. By continuity, processes $(\underline{\underline{L}}^1_p(X^{Z^n}) \, ; n \in \mathbb{N})$ and $(\underline{\underline{L}}^2_p(X^{Z^n}) \, ; n \in \mathbb{N})$ converge to processes $\underline{\underline{L}}^1_p(X^Z)$ and $\underline{\underline{L}}^2_p(X^Z)$ respectively. It remains to verify that the sequence of processes $(\lambda^{1Z^n} \, ; n \in \mathbb{N})$ and $(\lambda^{2Z^n} \, ; n \in \mathbb{N})$ converge. For that purpose we modify a method developed in ([10]) to define stochastic integration on increasing paths. Namely, we remark that processes λ^{1Z^n} and λ^{2Z^n} can be associated to Radon-Nikodym derivatives of measures on $\mathbb{R}_+$, with respect to the Lebesgue measure. For arbitrary $u, v \in \mathbb{R}_+$ such that $u \leq v$ we define the quantity $\Lambda^{1Z^n}([u,v])$ (resp. $\Lambda^{2Z^n}([u,v])$) to be the Lebesgue measure on $\mathbb{R}^2_+$ of the domain determined by Z^n, the vertical lines of abcissas u and v, and the horizontal line of ordinate -1 (resp. determined by Z^n, the horizontal lines of ordinates u and v, and the vertical line of abcissa -1). It is clear that Λ^{1Z^n} and Λ^{2Z^n} are random measures on $\mathbb{R}_+$ absolutely continuous with respect to the Lebesgue measure. Let $\tilde{\lambda}^{1Z^n}$ and $\tilde{\lambda}^{2Z^n}$ be their Radon-Nikodym derivatives. It is a matter of verification to see that the processes λ^{1Z^n} and λ^{2Z^n} previously defined are exactly the processes $(\tilde{\lambda}^{1Z^n}_u / 1+Z^{n2}_u \, ; u \in \mathbb{R}_+)$ and $(\tilde{\lambda}^{2Z^n}_u / 1+Z^{n1}_u \, ; u \in \mathbb{R}_+)$. Moreover the convergence of $(Z^n \, ; n \in \mathbb{N})$ to Z implies that the sequences of measures $(\Lambda^{1Z^n} \, ; n \in \mathbb{N})$ and $(\Lambda^{2Z^n} \, ; n \in \mathbb{N})$ converge weakly to measures Λ^{1Z} and Λ^{2Z} similarly constructed. This permits to define λ^{1Z} and λ^{2Z}, and the Dynkin formula for the o.i.p. Z follows by arguments of weak convergence. That achieves the proof.

1-4- Weak harmonic functions:

Bi-harmonic functions are well known; their connections with one-parameter or two-parameter processes have been studied in [6,35] particularly. Another notion has been introduced in [34]. The definition we propose here is different. It is motivated by the optimal stopping problem; it is analogous to the notion considered in [22] dealing with the optimal stopping of several Markov chains. In this paragraph we only give basic definitions and properties. Additional results will be given in the next chapter.

But before, we need a new definition of a début. Let H be a random set in $\Omega \times \mathbb{R}_+^2$. For any optional increasing path $Z = (Z_u\,;\, u \in \mathbb{R}_+)$, denote by D_H^Z the random variable defined by

$D_H^Z = Z_\tau$ with $\tau = \inf\{u : Z_u \in H\}$ and $D_H^Z = \infty$ if the set is empty.

This variable belonging to $Z \cup \{\infty\}$ is called "the début of H along Z".

<u>Lemma 1-4-1</u>: If H is an optional set, then for any optional increasing path Z, D_H^Z is a stopping point.

<u>Proof</u>: The graph of Z : $\{(\omega,t) \in \Omega \times \mathbb{R}_+^2 : t \in Z(\omega)\}$ is optional, and consequently, so is $[\![Z]\!] \cap H$. Then its début is a stopping line [27]. This stopping line has only one minimal element, which is D_H^Z ; this implies that D_H^Z is a stopping point.

Given a subset $A \subset E$ and an o.i.p. Z, the début along Z of the optional random set $\{(\omega,t) : X_t^x(\omega) \in A\}$, is called the entrance point of X^x in A along Z and denoted by $D_A^{Z,x}$. The exit point of X^x in A along Z is defined by $S_A^{Z,x} = D_{A^c}^{Z,x}$, where A^c is the complement of A.

First, we define a harmonic operator similar to those of the classical theory. For $A \subset E$ and $p \in \mathbb{R}_+$, let H_A^p be the operator defined from b(E) in the set of all bounded functions on E, by

$$\forall f \in b(E)\ ,\ \forall x \in E : H_A^p f(x) = \sup_{Z \in \underline{\underline{Z}}} E_x\left(e^{-p.D_A^{Z,x}}\, f(X^x_{D_A^{Z,x}})\right) .$$

It may be noticed that $H_A^p f$ has no reason to be measurable, but it is always analytical ([12]). Although H_A^p is non-linear, it verifies several properties of classical harmonic operators.

<u>Proposition 1-4-1</u>: Operator H_A^p satisfies the following;

i) If A is closed, then $H_A^p(1\!\!1_A f) = H_A^p(f)$

ii) If $f \geq g$, then $H_A^p f \geq H_A^p g$

iii) $\forall x \in A : H_A^p f(x) = f(x)$

iv) If f is p-biexcessive and A is closed, then

$$\forall x \in E : H_A^p f(x) = \sup_{T \in \underline{\underline{T}}} E_x(e^{-p.T} 1\!\!1_{\{X_T \in A\}} f(X_T))$$

<u>Proof</u>: If A is closed, then $X^x_{D_A^{Z,x}} \in A$, $\forall Z \in \underline{\underline{Z}}$: this implies i). ii) is obvious. If $x \in A$, then $D_A^{Z,x} = 0$. This proves iii). Let Z be an o.i.p., denote by τ the $\underline{\underline{F}}^Z$-stopping time such that $Z_\tau = D_A^{Z,x}$, and let σ be any $\underline{\underline{F}}^Z$-stopping time. Then the following inequalities hold:

$$E_x(e^{-p.Z_\sigma} (1\!\!1_A f)(X^Z_\sigma)) \leq E_x(e^{-p.Z_{\sigma\vee\tau}} (1\!\!1_A f)(X^Z_{\sigma\vee\tau}))$$
$$\leq E_x(e^{-p.Z_{\sigma\vee\tau}} f(X^Z_{\sigma\vee\tau})) \leq E_x(e^{-p.Z_\tau} f(X^Z_\tau)) .$$

We deduce from this :

$$\sup_{Z \in \underline{\underline{Z}}} \sup_{\sigma \in \underline{\underline{T}}^Z} E_x(e^{-p.Z_\sigma} (1\!\!1_A f)(X_{Z_\sigma})) = H_A^p f(x) \quad , \quad \forall x \in E .$$

Using the fact that for any stopping T, there exists $Z \in \underline{\underline{Z}}$ and $\sigma \in \underline{\underline{T}}^Z$ such that $T = Z_\sigma$ a.s. ([18,22,35]), we deduce iv).

The last assertion of Proposition 1-4-1 suggests that harmonic operators are connected with optimal stopping of functions of bi-Markov processes. This will be made clear in the next chapter. To conclude, we we only give here the definition of weak harmonicity, which extends those of ([22,23]).

<u>Definition 1-4-1</u>: A function f on E will be said to be p-weakly harmonic on a given subset $A \subset E$, if and only if

$$\forall\, x \in A : f(x) = H^{p}_{A^{c}} f(x) \quad \text{i.e.} \quad f(x) = \sup_{Z \in \underline{\underline{Z}}} E_x\left(e^{-p.S^{Z}_{A}} f(X_{S^{Z}_{A}})\right)$$

where S^{Z}_{A} denotes the exit point of X out of A along Z.

If a p-biexcessive function f is p-weakly harmonic on A, then f is p-weakly harmonic on any subset B contained in A. Moreover

$$\forall\, B \subset A \ : H^{p}_{A}\, H^{p}_{B}\, f = H^{p}_{B}\, H^{p}_{A}\, f = H^{p}_{B}\, f.$$

The set of p-weakly harmonic functions on a given subset contains the set of p-biharmonic functions ([23]).

2- OPTIMAL STOPPING OF A BI-MARKOV PROCESS

In the first paragraph, various results about two-parameter optimal stopping are recalled. They are not given in their strongest form, but our presentation will be adapted to the markovian situation considered here. In the second paragraph, we give a different approach to the problem. It leads to a characterization of the Snell envelop in terms of a réduite. The third paragraph is devoted to the optimal stopping of a pay-off process which is a given function of a bi-Markov process. In the fourth paragraph, we study relations between réduite , weak harmonicity and variational inequations.

2-1- Optimal stopping problem for two-parameter processes:

The optimal stopping problem for a process indexed on the set $\mathbb{N}^2$ is well known:recent contributions are ([18],[22],[26],[30]). For processes with $\mathbb{R}_+^2$ as parameter set, an existence result is obtained in ([30]), by developing arguments of convexity and compactness similar to those of ([14]) in the classical theory. In ([25]) we gave another existence result by using a method analogous to that of ([26]) in the discrete case. In this paragraph, we recall the approach of this last work with some slight modifications adapted to the frame of this paper.

On a complete probability space $(\Omega, A, \mathbb{P})$ endowed with a two-parameter filtration $\underline{\underline{F}} = (\underline{\underline{F}}_t \; ; t \in \mathbb{R}_+^2)$ verifying the axioms F1, F2, F3 and F4, let us consider an optional non-negative process of class D $Y = (Y_t \; ; t \in \mathbb{R}_+^2)$. It is called the pay-off process. The optimal stopping problem consists in finding a stopping point T* such that

$E(Y_{T^*}) = \sup_{T \in \underline{\underline{T}}} E(Y_T)$. Such a s.p. will be said optimal.

We first study the particular situation of a pay-off process Y which differs from zero only on a given stopping line L , that is to say : $\forall\, t \in \mathbb{R}_+^2 : Y_t = Y_t \, \mathbb{1}_{\{t \in L\}}$. This problem is intrinsically one-dimensional. It is solved in ([25]) in a general framework, and the method will be briefly recalled here under stronger hypotheses. Then we show how the general optimal stopping problem reduces to this particular case.

To prove the following result in a simpler way we do assume that σ-field $\underline{\underline{A}}$ is countably generated.

Proposition 2-1-1: Let Y be an optional non-negative process of class D and let a stopping line L be given. If the process Y is continuous on $\mathbb{R}_+^2 \cup \{\infty\}$, then the optimal stopping problem associated to the pay-off $Y\, 1\!\!1_{[\![L]\!]}$ admits a solution (which belongs to L a.s.).

Proof: Let $\underline{\underline{T}}(L)$ be the set of all stopping points T such that $[\![T]\!] \subset [\![L]\!]$, i.e. T belongs a.s. to $L \cup \{\infty\}$. The method is the same as in ([14]) and ([30]) for one- and two-parameter situations, respectively. The first step consists in introducing a set containing $\underline{\underline{T}}(L)$, which is convex compact, and such that the set of its extremal elements is exactly $\underline{\underline{T}}(L)$. In the second step we prove the existence of an optimal element in the larger set, and by a classical argument ([14,30]) in $\underline{\underline{T}}(L)$.

Let $\underline{\underline{M}}°$ be the set of all random probabilities on $\mathbb{R}_+^2 \cup \{\infty\}$ endowed with the Baxter-Chacon topology ,i.e. the coarsest topology such that, for every continuous bounded process X, the application $\forall\, \mu \in \underline{\underline{M}}°: \mu \to E(\int X\, d\mu)$ is continuous. Let $\underline{\underline{M}}$ be the subset of $\underline{\underline{M}}°$ of all μ the repartition function of which is an adapted right-continuous process. These sets are convex and compact; it is known ([16]) that the extremal elements of $\underline{\underline{M}}°$ are exactly the Dirac measures on any random variable in $\mathbb{R}_+^2 \cup \{\infty\}$ but, we do not know if the extremal elements of $\underline{\underline{M}}$ are the Dirac measures on any stopping point, as for the one-parameter situation ([14,16]). Anyway for the point at stake, we do restrict ourselves to the subset $\underline{\underline{M}}(L) \subset \underline{\underline{M}}$ of adapted random probability μ the support of which is a.s. contained in $L \cup \{\infty\}$. This subset is convex, closed then compact. It can be proved that $\underline{\underline{T}}(L)$ is exactly the set of all the extremal elements of $\underline{\underline{M}}(L)$ as in ([30]) Theorem 1-5 (where the stopping line considered was of the following type: $L = \{t : |t| = 1\}$) It must be noted that property F4 is crucial to establish this result.

The second step is rather classical. For any $\mu \in \underline{\underline{M}}(L)$ we have $\int 1\!\!1_{[\![L]\!]} Y \, d\mu = \int Y \, d\mu$. Then, if Y is continuous, the linear application $\mu \to E \int 1\!\!1_{[\![L]\!]} Y \, d\mu$ is continuous on $\underline{\underline{M}}(L)$. It attains its maximum at least on one point amongst the extremal elements $\underline{\underline{T}}(L)$. Using the fact that $\underline{\underline{A}}$ is countably generated, this can be proved as in ([14], [30]) by Choquet Theorem.This achieves the proof.

Now, let us come back to the general optimal stopping problem. As for the classical theory ([2],[4],[15]) the notion of Snell envelop is the fundamental tool. It is proved in ([9]) that there exists at least one strong supermartingale J, called the Snell envelop of Y, such that:

$$\forall\, T \in \underline{\underline{T}} : J_T = \operatorname*{esssup}_{S \in \underline{\underline{T}};\, S \geq T} E(Y_S \,/\, \underline{\underline{F}}_T) \;.$$

Notice that it could exist several processes which satisfy the preceding formula, and which are not necessarely undistinguishible ([9]). In the following we construct a process I such that $I_T = J_T$, $\forall\, T \in \underline{\underline{T}}$. This construction is inspired from the classical theory ([15],[33]). A similar has been obtained in ([20]) for the case of two-parameter Markov chains.

<u>Proposition 2-1-2</u>: Let Y be an optional non-negative process, uniformly bounded and such that its trajectories are a.s. lower semi-continuous functions converging to zero at the infinity. Let I be the limiting process of the following increasing sequence of optional processes $(I^n ; n \in \mathbb{N})$ defined by:

$$I^0 = Y \quad \text{and} \quad \forall\, n \in \mathbb{N} : I^{n+1} = \sup_{r \in \mathbb{D}} \{ {}^{o}(I^n_{r+.}) \}$$

where ${}^{o}(I^n_{r+.})$ denotes the optional projection of the process $I^n_{r+.} = (I^n_{r+t} \,;\, t \in \mathbb{R}^2_+)$, and $\mathbb{D}$ is the set of dyadic numbers in $\mathbb{R}^2_+$.

Then I is a strong supermartingale such that

$$\forall\, T \in \underline{\underline{T}} : I_T = J_T \quad \text{a.s.} \quad .$$

Every process having such a property will be called the Snell envelop of Y.

<u>Proof</u>: It is given in full details in ([24]) and we only recall the main steps. We define an operator R on the set of all optional bounded processes as follows

$$\forall\, X \text{ optional bounded} : R(X)_t = \sup_{r \in \mathbb{D}} \{{}^{o}(X_{r+.})_t\} \ , \ \forall\, t \in \mathbb{R}^2_+ \ .$$

Operator R is positive (i.e. $X \geq Y \Rightarrow R(X) \geq R(Y)$), and for any strong supermartingale X, R(X) is a strong supermartingale such that

$\forall\, T \in \underline{\underline{T}} : R(X)_T = X_T$ a.s. .

From the fact that $J \geq Y$, we deduce that

$$\forall\, T \in \underline{\underline{T}} : J_T \geq I^n_T \ , \ \forall\, n \in \mathbb{N} \text{ , and } J_T \geq I_T \text{ a.s. .}$$

On the other hand, it can be proved by a direct computation that:

$$\forall\, T \in \underline{\underline{T}} : E(Y_S \,/\, \underline{\underline{F}}_T) \leq I_T \quad \text{a.s. ,}$$

for any stopping points S and T such that $S \geq T$, and $S - T$ is dyadic a.s.. By using the hypotheses on Y, it is proved in ([24]) that this relation extends to any pair of stopping points. Then

$$\forall\, T \in \underline{\underline{T}} : J_T \leq I_T \quad \text{a.s. .}$$

The proof is completed.

The following general existence result is similar to the one obtained in ([26]) for a discrete parameter set. Nethertheless it involves conditions upon the Snell envelop which limits its domain of application.

<u>Proposition 2-1-3</u>: Let Y be an optional process, bounded and continuous on $\mathbb{R}^2_+ \cup \{\infty\}$, and suppose that its Snell envelop J is also continuous and bounded on $\mathbb{R}^2_+ \cup \{\infty\}$. Then, there exists optimal stopping point. Furthermore, such an optimal solution can be found among the maximal elements of the subset of

stopping points for which the martingale property of the Snell envelop J is preserved, i.e. the subset $\{T \in \underline{\underline{T}} : E(J_0) = E(J_T)\}$.

Proof: If J is continuous and bounded, it is easy to verify by the Zorn lemma that the non empty set $\underline{\underline{T}}_m = \{T \in \underline{\underline{T}} : E(J_0) = E(J_T)\}$ admits maximal elements. Let T be such an element, called here maximal stopping point. By definition, it satisfies the following

$E(J_0) = E(J_T)$, and

$\forall\, S \in \underline{\underline{T}}$ such that $S \geq T$ and $\mathbb{P}(S = T) < 1 : E(J_S) < E(J_0)$.

To prove that T is actually optimal, we must verify that :

$E(J_T) = E(Y_T)$.

For that purpose, let us consider the random set

$H^\lambda = \{(\omega,t) : Y_t(\omega) \geq \lambda J_t(\omega) \quad , t \geq T(\omega)\}$, for $\lambda \in]0,1[$.

This set is optional. Let L^λ be the stopping line début of H^λ . The following formula is proved in ([24]).

$$\forall\, \lambda \in]0,1[\; : E(J_T) = \sup_{S \in \underline{\underline{T}}(L^\lambda)} E(J_S) = \sup_{S \in \underline{\underline{T}}} E(J_S \, 1\!\!1_{\{S \in L^\lambda\}}) \; .$$

where $\underline{\underline{T}}(L^\lambda)$ denotes the set of s.p. S such that $[\![S]\!] \subset [\![L]\!]$.

This formula defines an optimal stopping problem on a stopping line which enters the framework of Proposition 2-1-1. Then, there exists a stopping point S_λ , such that

$E(J_T) = E(J_{S_\lambda})$ and $[\![S_\lambda]\!] \subset [\![L^\lambda]\!]$.

By definition, we have $S_\lambda \geq T$. Because of the maximality of T, this implies $S_\lambda = T$ a.s. , and therefore T belongs to the stopping line L^λ a.s.. By the way the set H^λ and the line L^λ are constructed, this is possible only if T belongs to H^λ a.s.. Then

$\forall\, \lambda \in]0,1[\; : Y_T \geq \lambda J_T$ a.s. .

It follows that

$Y_T = J_T$ a.s. ,

and this achieves the proof.

2-2- Optimal stopping and optional increasing-paths:

In this paragraph, a different approach of the stopping problem is proposed. It reduces to a rather classical distributed control problem of finding an optional increasing path passing by an optimal stopping point. In addition, this method gives a characterization of the Snell envelop, generalizing those of ([23],[24]).

The main idea of this paragraph is resumed in the following result.

Proposition 2-2-1: Let Y be a given two-parameter optional, non-negative process of class (D) . Then,

$$\sup_{T\in \underline{\underline{T}}} E(Y_T) = \sup_{Z\in \underline{\underline{Z}}} \{\sup_{\tau\in \underline{\underline{T}}^Z} E(Y^Z_\tau)\} .$$

Proof: For any o.i.p. Z and any $\underline{\underline{F}}^Z$-stopping time τ, Z_τ is a s.p., and conversely, for any s.p. T, there exists an o.i.p. Z which passes by T, i.e. $T = Z_\tau$ a.s. where τ is a $\underline{\underline{F}}^Z$-stopping time ([35]). Then the set $\underline{\underline{T}}$ can be identified with the set $\{(Z,\tau);\ Z \in \underline{\underline{Z}} \text{ and } \tau \in \underline{\underline{T}}^Z\}$, and that proves the proposition.

The equality in Proposition 2-2-1 shows that the general problem can split up into the following two problems. 1) Find an optimal optional increasing path. 2) Find an optimal stopping time on it. In the particular situation of a Bi-Markov process, the first problem can be assimilated to a distributed control problem. Unfortunately, we do not know how to solve it exactly. Nethertheless it seems that an ε-optimal solution could be constructed following a technique borrowed from the control of alternating processes ([3]).

This approach allows the characterization of the behaviour of the Snell envelop J, on the set on which Y is strictly less than J. The definition of the début of a random set along an o.i.p. will be used.

For every $\lambda \in]0,1[$, set $H^\lambda = \{(\omega,t): Y_t(\omega) \geq \lambda J_t(\omega)\}$, and denote by D_λ^Z the début of H^λ along the o.i.p. Z. Domain H^λ is optional; therefore D_λ^Z is a stopping point. Moreover, the process $J\,1\!\!1_{H^\lambda}$ is non-negative optional and of class (D). We denote by J^λ its Snell envelop. This process is usually called the réduite of J on the set H^λ, ([12]). The following result extends classical properties of réduite.

<u>Proposition 2-2-2</u>: For every stopping point T, one has

$$J_T^\lambda = J_T \quad \text{a.s. .}$$

<u>Proof</u>: It is borrowed from ([15]). J is a strong supermartingale greater than the process $J\,1\!\!1_{H^\lambda}$, then J is necessarely greater than its Snell envelop J^λ. Consequently, for any s.p. S, we get

$$J_S^\lambda \geq 1\!\!1_{\{S \in H^\lambda\}} J_S \geq 1\!\!1_{\{S \in H^\lambda\}} J_S^\lambda \quad , \text{ then}$$

$$J_S^\lambda = J_S \quad \text{a.s. on the set } \{S \in H^\lambda\}.$$

Let I be the strong supermartingale $\lambda J + (1-\lambda)J^\lambda$. Obviously,

$$J_S \geq I_S \quad \forall\, S \in \underline{\underline{T}} \quad .$$

To prove that

$$J_S \leq I_S \quad \forall\, S \in \underline{\underline{T}} \quad ,$$

it is sufficient to verify that

$$Y_S \leq I_S \quad \forall\, S \in \underline{\underline{T}} \quad .$$

On the set $\{S \in H^\lambda\}$, we have $J_S = J_S^\lambda$, then $Y_S \leq I_S$,
on the set $\{S \in H^\lambda\}^c$, we have $Y_S < \lambda J_S$, then $Y_S \leq I_S$.
This achieves the proof.

From this we deduce the formula which characterizes the behaviour of the Snell envelop J, on the domain H^λ . It extends a result given in ([23,24]).

<u>Proposition 2-2-3</u>: For any fixed $\lambda \in]0,1[$, the Snell envelop J of the process Y, satisfies

$$E(J_0) = \sup_{Z \in \underline{\underline{Z}}} E(J_{D_\lambda^Z}) \quad ,$$

where D_λ^Z is the début of the set $\{Y \geq \lambda J\}$ along Z .

Proof: The equality is proved for process J^λ, then it holds for J itself by Proposition 2-2-2. For a given s.p. S, let $\underline{\underline{Z}}^S$ denote the set of all o.i.p. passing a.s. by S, i.e. $\forall\ Z \in \underline{\underline{Z}}^S, \exists\ \sigma \in \underline{\underline{T}}^Z$ such that $S = Z_\sigma$. By definition of J^λ, we have

$$E(J_0^\lambda) = \sup_{S\in \underline{\underline{T}}} E(J_S \, 1\!\!1_{\{S\in H\}}) \quad .$$

Let us prove the following:

$$\forall\ S \in \underline{\underline{T}}\ ,\ \forall\ Z \in \underline{\underline{Z}}^S : \exists\, T \in \underline{\underline{T}} \text{ such that } \quad T \geq D_\lambda^Z \quad \text{and}$$

$$E(J_S \, 1\!\!1_{\{S\in H\}}) = E(J_T \, 1\!\!1_{\{T\in H\}}) \quad .$$

For that purpose, set for any $S \in \underline{\underline{T}}$ and $Z \in \underline{\underline{Z}}^S$:

$$T = S \quad \text{on} \quad \{S \geq D_\lambda^Z\} \quad \text{and} \quad T = \infty \quad \text{on the complementary set.}$$

T is a s.p., due to the fact that $\{S \geq D_\lambda^Z\} \in \underline{\underline{F}}_S$. It is easy to verify

$$\{S\in H\} = \{T \in H\} \subset \{S = T\} \quad .$$

Then, we obtain

$$E(J_S \, 1\!\!1_{\{S\in H\}}) = E(J_T \, 1\!\!1_{\{T\in H\}}) \ .$$

From this formula we deduce the following equalities.

$$E(J_0) = \sup_{S\in\underline{\underline{T}}} E(J_S \, 1\!\!1_{\{S\in H\}}) = \sup_{T\in\underline{\underline{T}}} E(J_T \, 1\!\!1_{\{T\in H\}})$$

$$= \sup_{Z\in\underline{\underline{Z}}} E(J_{D_\lambda^Z} \, 1\!\!1_{\{D^Z \in H\}}) = \sup_{Z\in\underline{\underline{Z}}} E(J_{D_\lambda^Z}) \ .$$

That achieves the proof.

2-3- Optimal stopping of bi-Markov processes :

In this paragraph, we study the optimal stopping problem when the pay-off process Y is a function of the bi-Markov process defined in Chapter 1. We show that the Snell envelop can also be written as a function of X, called the Snell reduite. Under mild assumptions on the model this function is proved to be continuous, and the optimal stopping problem solved.

Let us come back to the probability space $(\Omega, \underline{\underline{A}}, \mathbb{P})$, the filtration $\underline{\underline{F}} = (\underline{\underline{F}}_t\ ; t\in \mathbb{R}_+^2)$, and the family of processes $X = (X^x\ ; x \in E)$ defined in Paragraph 1-1. Note that $\underline{\underline{A}}$ is countably generated if we consider the natural filtrations of the Brownian motions B^1 and B^2.

p-biexcessive majorant of f. Suppose there exists a p-biexcessive function q', greater than f. Then the process J'^x defined by $J'^x_t = e^{-p.t} q'(X^x_t)$, $\forall\, t \in \mathbb{R}^2_+$, is a strong supermartingale which majorizes Y^x. This implies that J'^x majorizes the Snell envelop J^x, and

$$q'(x) = E(J'^x_0) \geq E(J^x_0) = q(x) \;.$$

The proof is completed.

Function q is called the Snell réduite of f, or the p-réduite of f.

The evolution of function q on the subset in E where it majorizes strictly f, is described by the following result.

Proposition 2-3-2: For any $\lambda \in \,]0,1[$, the p-réduite q of f satisfies the following

$$\forall\, x \in E \;:\; q(x) = \sup_{Z \in \underline{\underline{Z}}} E(e^{-p.D^{Z,x}_\lambda} q(X^x_{D^{Z,x}_\lambda}))$$

where $D^{Z,x}_\lambda$ is the entrance point of X^x along Z in the set $\{x \in E \;:\; f(x) \geq \lambda q(x)\}$.

The proof is a straightforward application of Proposition 2-2-3.

Under additional hypotheses on processes X^1 and X^2, the Snell réduite q has better regularity property. In fact, let us consider the following condition on $(X^x \,;\, x \in E)$.

$$(L)\left|\begin{array}{l} \forall\, A > 0 \;,\; \exists\, K > 0 \text{ such that} \\ \forall\, S \in \underline{\underline{T}} \text{ with } |S| \leq A \;:\; E(|X^x_S - X^y_S|) \leq e^{KA}\,|x-y| \;. \end{array}\right.$$

This assumption is easily satisfied in our case, i.e. when processes X^1 and X^2 are defined as solutions of stochastic differential equations. For example, let us suppose that the coefficients b^i_k and σ^i_{jk} in these equations are bounded and lipschitzian. Then, it is well known [13] that, for any $\underline{\underline{M}}^i$-stopping time U such that $U \leq A$, one has

$$E_{\mathbb{P}^i}(|X^{ix^i}_U - X^{iy^i}_U|) \leq e^{KA}\,|x^i - y^i| \qquad ,$$

For a positive real number p and a function $f \in b(E)$, we define a family $Y = (Y^x ; x \in E)$ of pay-off processes by

$$\forall x \in E \ , \ \forall t \in \mathbb{R}_+^2 \ : \ Y_t^x = e^{-p.t} f(X_t^x) \ .$$

Funtion f is called the pay-off function, and p the actualization factor.

Let us denote by J^x the Snell envelop of process Y^x, for $x \in E$.

<u>Proposition 2-3-1</u> : If the pay-off function f is continuous and bounded on E, then there exists a lower semi-continuous function q on E, such that

$$\forall x \in E \ , \ \forall T \in \underline{\underline{T}} \ : \ J_T^x = e^{-p.T} q(X_T^x) \quad \text{a.s.} \ .$$

Moreover function q is the least p-biexcessive majorant of function f.

<u>Proof</u>: In order to apply Proposition 2-1-2, we construct an increasing sequence $(q^n ; n \in \mathbb{N})$ of bounded lower semi-continuous functions on E, as follows:

$$q^0 = f \quad , \text{ and for } n \geq 0 \ : \ q^{n+1} = \sup_{r \in \mathbb{D}} e^{-p.r} P_r q^n \quad .$$

Let q be the limit of this sequence. q is bounded, lower semi-continuous and greater than f, by construction. It is immediate that $,\forall x \in E$, the sequence of processes $((e^{-p.t} q^n(X_t^x) ; t \in \mathbb{R}_+^2) ; n \in \mathbb{N})$ is exactly the sequence $(I^n ; n \in \mathbb{N})$ of Proposition 2-1-2. Then

$$\forall T \in \underline{\underline{T}} \ : \ J_T^x = e^{-p.T} q(X_T) \quad \text{a.s.} \ .$$

By the supermartingale property of J^x, we get, $\forall x \in E$:

$$\forall t \in \mathbb{R}_+^2 \ : \ E(J_0^x) \geq E(J_t^x) \quad \text{i.e.} \quad q(x) \geq e^{-p.t} P_t q(x) \quad .$$

Moreover, it can be proved as in ([15]) (see ([24]) also) that the continuity (more generally the right-continuity) of the pay-off process Y^x implies the right-continuity on $\mathbb{R}_+^2$ of the function $t \to E(J_t^x)$. Then, for any sequence $(t(n) ; n \in \mathbb{N})$, decreasing to zero, we have

$$\forall x \in E \ : \ q(x) = E(J_0^x) = \lim E(J_{t(n)}^x) = \lim e^{-p.t(n)} P_{t(n)} q(x) \ .$$

Hence, q is p-biexcessive. It remains to verify that q is the least

for some K which depends on A, b_k^i and σ_{jk}^i only. By taking the norm on $E = E^1 \times E^2$ defined by $|x| = |x^1| + |x^2|$, $\forall\, x = (x^1, x^2)$, it follows

$$E(|X_S^x - X_S^y|) \leq e^{KA} \, |x-y| \qquad \forall\, S \in \underline{\underline{T}} \quad \text{such that } |S| \leq A/2 \; .$$

That is condition (L).

Proposition 2-3-3: If the pay-off function f is bounded and uniformly continuous on E, and if the bi-Markov family X verifies Hypothesis (L), then the p-réduite q of f is also uniformly continuous on E.

Proof: It is analogous to that of ([31]) for the classical case. Let T be a s.p., a positive constant A and let x, y be two distinct points in E. Let us study the random variable U defined by

$$U = |e^{-p.T} f(X_T^x) - e^{-p.T} f(X_T^y)| \quad .$$

It can be noticed that, for a given s.p. T and a constant A, there exists a s.p. T_A such that $T_A \leq T$ with $T_A = T$ on $\{|T| \leq A\}$ and $|T_A| = A$ on $\{|T| > A\}$, see ([25]). Then, we get the following inequalities

$$\begin{aligned} U &\leq |e^{-p.T} f(X_T^x) - e^{-p.T_A} f(X_{T_A}^x)| \\ &\quad + |e^{-p.T_A} f(X_{T_A}^x) - e^{-p.T_A} f(X_{T_A}^y)| \\ &\quad + |e^{-p.T_A} f(X_{T_A}^y) - e^{-p.T} f(X_T^y)| \\ &\leq 4\, e^{-pA} \|f\| + |e^{-p.T_A} (f(X_{T_A}^x) - f(X_{T_A}^y))| \quad . \end{aligned}$$

Function f being uniformly continuous on E, we have

$$\forall \varepsilon > 0 \; , \; \exists\, \delta \; \text{ such that } |z - z'| < \delta \Rightarrow |f(z) - f(z')| < \varepsilon \quad ,$$

thus

$$E(|f(X_{T_A}^x) - f(X_{T_A}^y)|) \leq \varepsilon + \frac{2}{\delta} \|f\| \, E(|X_{T_A}^x - X_{T_A}^y|) \; .$$

According to Hypothesis (L), we then deduce a majorant of E(U), independently of the s.p. T :

$$E(U) \leq 4\, e^{-pA} \|f\| + \varepsilon + \frac{2}{\delta} \|f\| \, |x-y| \, e^{KA} \quad .$$

Therefore,

$$
\begin{aligned}
|q(x) - q(y)| &= |\sup_{T\in \underline{\underline{T}}} E(e^{-p.T} f(X_T^x)) - \sup_{T\in \underline{\underline{T}}} E(e^{-p.T} f(X_T^y))| \\
&\leq \sup_{T\in \underline{\underline{T}}} E(e^{-p.T} |f(X_T^x) - f(X_T^y)|) \\
&\leq 4\ e^{-p.A} \|f\| + \varepsilon + \frac{2}{\delta} \|f\|\ |x-y|\ e^{KA} .
\end{aligned}
$$

By choosing A, then ε and finally x and y, we can make this majorant as small as we want. This implies the uniform continuity for q.

As a conclusion to this paragraph, we apply these results to the optimal stopping problem.

Proposition 2-3-4: If the pay-off function f is uniformly continuous and bounded on E, and if the bi-Markov family X satisfies Hypothesis (L), then the optimal stopping problems, associated to function f and positive number p, admit solutions.

The proof is an easy consequence of propositions 2-1-3 and 2-3-3.

2-4- Weak harmonicity and variational inequations:

In this paragraph, we characterize, as in (2), the p-réduite q as solution of a system of partial differential inequations with free boundary, also called a system of variational inequalities.

The first result connects the behaviour of the p-réduite with the notion of weak harmonicity.

Proposition 2-4-1: The p-réduite q of a bounded non-negative function f is p-weakly harmonic on the set $\{f < \lambda q\}$, for any $\lambda \in]0,1[$.

The proof is a direct consequence of Proposition 2-2-3 and Definition 1-4-1.

If a function f is sufficiently smooth, then weak harmonicity on a given subset can be expressed locally. The following proposition is proved by using various results of Paragraph 2-1.

Proposition 2-4-2: Let f be a function in $\underline{\underline{D}}(\underline{\underline{L}}^1,\underline{\underline{L}}^2)$, p a positive number, and let A be an open set, then f is both p-biexcessive and p-weakly harmonic on A if and only if

$$\underline{\underline{L}}_p^1 f \leq 0 \quad \text{and} \quad \underline{\underline{L}}_p^2 f \leq 0 \quad \text{on E, and}$$

$$\max(\underline{\underline{L}}_p^1 f, \underline{\underline{L}}_p^2 f) = 0 \quad \text{on A .}$$

Proof: Let us prove the necessary condition. Then, suppose f is p-biexcessive and p-weakly harmonic on an open set A. For $x \in A$ fixed, there exists an open rectangle $B = B^1 \times B^2$ containing x and contained in A. Then f is p-weakly harmonic on B. It is easy to see that the family of exit points $(S_B^{Z,x}, Z \in \underline{\underline{Z}})$ forms a stopping line L. L could be defined also by:

$$L = \{t=(t_1,t_2) : \begin{cases} \text{and } X_{t_2}^{2x^2} \in B^2 \\ t_1 = \inf\{u : X_u^{1x^1} \notin B^1\} \end{cases} \text{or} \begin{cases} \text{and } X_{t_1}^{1x^1} \in B^1 \\ t_2 = \inf\{u : X_u^{2x^2} \notin B^2\}\}\end{cases}$$

According to Proposition 2-1-1, the optimal stopping problem on L associated to f and X^x, admits a solution T. Let Z be an o.i.p. passing by T : $Z_\tau = T = S_B^{Z,x}$. Using Proposition 1-3-2 and Definition 1-4-1, we get

$$f(x) = E(e^{-p.S_B^{Z,x}} f(X^x_{S_B^{Z,x}}))$$

$$= f(x) + E\left(\int_0^\tau (\underline{\underline{L}}_p^1 f(X_u^{xZ})\, \lambda_u^{1Z} + \underline{\underline{L}}_p^2 f(X_u^{xZ})\, \lambda_u^{2Z})\, e^{-pu}\, du\right)$$

By construction the origin 0 never belongs to L. Then it follows that $\tau > 0$ a.s. , and that implies, by continuity, $\underline{\underline{L}}_p^1 f(x) = 0$ or $\underline{\underline{L}}_p^2 f(x) = 0$. The conclusion to be obtained.

Conversely, let us suppose f satisfies the system of the proposition . Obviously f is p-biexcessive, and it remains to prove that f is p-weakly harmonic on A. For $x \in A$ and $\varepsilon > 0$ given, we can construct an o.i.p. Z such that:

$$f(x) - E(e^{-p.S_A^{Z,x}} f(X^x_{S_A^{Z,x}})) \leq \varepsilon \quad .$$

For that purpose, consider the following open sets:

$$B = \{y : \underline{\underline{L}}_p^1 f(y) < \varepsilon\} \quad \text{and} \quad C = \{y : \underline{\underline{L}}_p^2 f(y) < \varepsilon\} \quad .$$

Obviously $B \cup C \supset A$; suppose that $x = (x^1, x^2) \in B$. Let us construct Z as follows. Let $T^1 = (T_1^1, T_2^1)$ be defined by

$$T_1^1 = \inf\{u : (X_u^{1x^1}, x^2) \in B^c\} \quad \text{and} \quad T_2^1 = 0 \quad ,$$

let $T^2 = (T_1^2, T_2^2)$ be defined by

$$T_1^2 = T_1^1 \quad \text{and} \quad T_2^2 = \inf\{u : (X^{1x^1}_{T_1^1}, X_u^{2x^2}) \in C^c\} \quad ,$$

It can be easily verified that T^1 and T^2 are s.p.'s. By iterating that procedure, we construct an increasing sequence $(T^n; n \in \mathbb{N})$ of s.p.'s , which induces an o.i.p. Z, as in Paragraph 1-3.

Everything has been made to insure the following inequality.

$$|E_x(\int_0^\infty (\underline{\underline{L}}_p^1 f(X_u^Z)\ \lambda_u^{1Z} + \underline{\underline{L}}_p^2 f(X_u^Z)\ \lambda_u^{2Z})\ e^{-pu}\ du)| \leq \varepsilon/p \ .$$

Then, we deduce that

$$\sup_{Z \in \underline{\underline{Z}}} E_x(e^{-p.S_A^Z} f(X_{S_A^Z})) \leq f(x) \leq \sup_{Z \in \underline{\underline{Z}}} E_x(e^{-p.S_A^Z} f(X_{S_A^Z})) + \varepsilon/p \ .$$

This leads to the desired conclusion.

As an illustration of these results let us come back to the optimal stopping problem associated to a pay-off function f and an actualization factor $p > 0$. If we can assume that the p-réduite q belongs to domain $\underline{\underline{D}}(\underline{\underline{L}}^1, \underline{\underline{L}}^2)$, then q verifies the following system of partial differential inequations with free boundary.

(S1) $q \geq f$

(S2) $\underline{\underline{L}}_p^1 q \leq 0$ and $\underline{\underline{L}}_p^2 q \leq 0$ on E

(S3) $\max(\underline{\underline{L}}_p^1 q, \underline{\underline{L}}_p^2 q) = 0$ on $\{q > f\}$.

This system (S1,S2,S3) is analogous to the one studied in ([2]) for the classical stopping time problem.

Conversely, we have the following.

Proposition 2-4-3: Let f be a given bounded continuous function on E and let $p > 0$. If System (S1,S2,S3) admits a solution q in $\underline{\underline{D}}(\underline{\underline{L}}^1,\underline{\underline{L}}^2)$, then q is the p-réduite of f.

Proof: If $q \in \underline{\underline{D}}(\underline{\underline{L}}^1,\underline{\underline{L}}^2)$, then q is continuous, and the set $A = \{q > f\}$ is open. Relations (S2) imply that q is p-biexcessive and relation (S3) that q is p-weakly harmonic on A. According to Proposition 2-3-1, the p-réduite of f exists, say q'. Relation (S1) implies that $q \geq q'$. Let us prove that $q' \geq q$. Notice that the set $\{q' = f\}$ contains the set $\{q = f\}$. By Proposition 1-4-1, we have

$$q(x) = H^p_{\{q = f\}}q(x) = \sup_{T \in \underline{\underline{T}}} E_x(1\!\!1_{\{q(X_T) = f(X_T)\}} e^{-p.T} q(X_T))$$

$$= \sup_{T \in \underline{\underline{T}}} E_x(1\!\!1_{\{q(X_T) = f(X_T)\}} e^{-p.T} f(X_T)) \leq \sup_{T \in \underline{\underline{T}}} E_x(e^{-p.T} f(X_T))$$

$$\leq q'(x) .$$

This achieves the proof.

Remark 2-4-1: Under the hypotheses of Proposition 2-4-3 we know, from Proposition 2-1-3, that there exists an optimal stopping point. Moreover, following the method developed in the proof of Proposition 2-4-2, we can construct step by step a tactic passing by a stopping point which is ε-optimal. But we have no idea of an explicit way of obtaining an optimat stopping point.

Remark 2-4-2: System (S1,S2,S3) is partly similar to the one considered in the classical theory of optimal stopping ([2]). The main difference seems to be the existence of a non-linear operator in relation (S3). Such an operator appears commonly in classical distributed control problem, more precisely in the Hamilton-Jacobi-Bellman equation. System (S1,S2,S3) is a free boundary problem which is, to the author's knowledge, open. Nethertheless, the following associated Dirichlet problem i.e., with a smooth fixed boundary, is well known and solved in ([5,21]).

(S'1) $q = f$ on the boundary of a smooth domain A

(S'2) $\mathrm{Max}(\underline{\underline{L}}_p^1 q, \underline{\underline{L}}_p^2 q) = 0$ in A .

Références :

(1) BAKRY, D. : "Théorèmes de section et de projection pour processus à deux indices". Z. Wahr. V. Geb. 55 ; 51-71 ; (1981).

(2) BENSOUSSAN A. - LIONS J.L. : "Applications des inéquations variationnelles au contrôle stochastique"; Dunod, Paris (1978).

(3) BISMUT J.M. : "Contrôle de processusalternants et applications" Z. Warhrs. V. Geb. 47, 241-288 (1979).

(4) BISMUT J.M. - SKALLI B. : "Temps d'arrêt optimal, théorie générale des processus et processus de Markov". Z. f. Wahr. V. Geb. 39, 301-313 (1979).

(5) BREZIS H. - EVANS L.C. : "A variational inequality approach to the Bellman-Dirichlet equation for two elliptic operators". Arch. Rat. Mech. and Anal. 71, 1-14 (1979).

(6) BROSSARD J. - CHEVALIER L. : "Calcul stochastique et inégalités de normes pour les martingales bi-Browniennes. Applications aux fonctions bi-harmoniques". Ann. Inst. Fourier, Grenoble 30, 4. 97-120 (1981).

(7) CAIROLI R. : "Produits de semi-groupes de transition et produits de processus". Publ. Inst. Stat. Paris 15. 311-384 (1966).

(8) CAIROLI R. : "Une représentation intégrale pour fonctions séparément excessives". Ann. Inst. Fourier, Grenoble 18,1. 317-338 (1968).

(9) CAIROLI R. : Enveloppe de Snell d'un processus à paramètre bidimensionnel". Ann. Inst. H. Poincaré 18,1. 47-54 (1982)

(10) CAIROLI R. - J.B. WALSH : "Stochastic Integrals in the plane". Acta Math. 134, 111-183, (1975).

(11) DELLACHERIE C. - MEYER P.A. : "Probabilités et Potentiel". Tomes 1, 2, Hermann, Paris (1975) and (1980).

(12) DELLACHERIE C. - MEYER P.A. : "Probabilités et Potentiel". Tome 3 (to appear).

([13]) DYNKIN E.B. : "Markov processes". Springer Verlag, Berlin (1965).

([14]) EDGAR G.A. - MILLET A. - SUCHESTON L. : "On compactness and optimality of stopping times". Lect. N. in control and Inf. Sc. 38, , Springer Verlag, Berlin (1982).

([15]) EL KAROUI N. : "Les aspects probabilistes du contrôle stochastique". Ecole d'été de St Flour 1979, Lect. N. in Maths 876, 74-239, Springer Verlag, Berlin (1981).

([16]) GHOUSSOUB N. : "An integral representation of randomized probabilities and its applications". Sem. Proba. XVI - Lect. N. in Maths 920, 519-543, Springer Verlag Berlin (1982).

([17]) KOREZLIOGLU H. - LEFORT P. - MAZZIOTTO G. : "Une propriété markovienne et diffusions associées". Lect. N. in Maths 863, 245-274, Springer Verlag, Berlin (1981).

([18]) KRENGEL U. - SUCHESTON L. : "Stopping rules and tactics for processes indexed by directed set". J. of Mult. Anal. Vol 11 199-229 (1981).

([19]) KURTZ T.G. : "The Optional Sampling Theorem for Martingales Indexed by a Directed Set". Annals of Prob. 8 ; 675-681, (1980).

([20]) LAWLER G.F. - VANDERBEI R.J. : "Markov strategies for optimal control problems indexed by a partially ordered set". Preprint.

([21]) LIONS P.L. - MENALDI J.L. : "Optimal control of stochastic integrals and Hamilton - Jacobi-Bellman equations I". SIAM J. Control Opt. 20, 58-95 (1982).

([22]) MANDELBAUM A. - VANDERBEI R.J. : "Optimal stopping and supermartingales over partially ordered sets". Z. f. Wahr. V. Geb. 57, 253-264 (1981).

([23]) MAZZIOTTO G. : "Arrêt optimal d'un bi-Markov et fonctions harmoniques" C.R. Acad Sc. Paris 295. Série I, 173-176 (20/9/82).

([24]) MAZZIOTTO G. : "Sur l'arrêt optimal de processus à deux indices réels". Stoch. Diff. Syst. Led. N. in Control and Inf. Sc. 43, 320-328, Springer Verlag Berlin (1982).

(25) MAZZIOTTO G. : "Arrêt optimal de processus markoviens à deux indices". Prépublication.

(26) MAZZIOTTO G. - SZPIRGLAS J. : "Arrêt optimal sur le plan". Z. f. Wahr. V. Geb. 62, 215-233 (1983).

(27) MERZBACH E. : "Stopping for two-dimensional stochastic processes". Stoch. Pr. and th. Appl. 10, 49-63 (1980).

(28) MEYER P.A. : "Processus de Markov". Lect. N. in Maths 26, Springer Verlag Berlin, (1967).

(29) MEYER P.A. :"Théorie élémentaire des processus à deux indices". Lect. N. in Maths 863, 1-39, Springer Verlag Berlin (1981).

(30) MILLET A. : "On randomized tactics and optimal stopping in the plane". Z. f. Wahr. V. Geb.

(31) NISIO M. : "Some remarks on stochastic optimal controls". 3rd Japan-USSR Symp. Lect. N. in Maths 550, 446-460, Springer Verlag Berlin (1976).

(32) NUALART D. - SANZ M. : "A Markov property for two-parameter gaussian processes". Stochastica 3-1, 1-16, Barcelone (1979).

(33) SHIRYAYEV A.N. : "Optimal stopping rules". Springer Verlag Berlin (1979).

(34) VANDERBEI R.J. : "Towards a stochastic calculus for several Markov processes". Preprint.

(35) WALSH J.B. : "Optional increasing paths". Lect. N. in Maths 863, 172-201, Springer Verlag, Berlin (1981).

(36) WONG E. - ZAKAI M. : "Martingales and stochastic integrals for processes with a multidimensional parameter". Z. f. Wahr. V. Geb. 29, 109-122 (1974).

Gérald MAZZIOTTO
PAA/TIM/MTI
Centre National d'Etude des Télécommunications
38-40, rue du Gal Leclerc
92131 ISSY LES MOULINEAUX
FRANCE

EQUATIONS DU LISSAGE NON LINEAIRE

E. Pardoux
U.E.R. de Mathématiques
Université de Provence
3, Pl. V. Hugo
13331 Marseille Cedex 3

Résumé :

Le but de cette note est d'établir trois couples d'équations, chacun des trois permettant de caractériser la loi conditionnelle d'un problème de lissage non linéaire.

Abstract :

The aim of this note is to state three pairs of equations, each of them caracterizing the conditional law in a non linear smoothing problem.

1. Introduction :

On considère le système différentiel stochastique suivant :

$$dX_t = b(X_t)dt + c(X_t)dW_t + e(X_t)d\tilde{W}_t$$

$$dY_t = h(X_t)dt + dW_t$$

où W_t et $\tilde{W}_t$ sont deux processus de Wiener vectoriels standard indépendants, définis sur un espace de probabilité filtré $(\Omega, \underline{\underline{F}}, \underline{\underline{F}}_t, P)$; on suppose en particulier que W_t et $\tilde{W}_t$ sont des $\underline{\underline{F}}_t$-martingales. On pose $\underline{\underline{G}}_t^s = \sigma\{Y_r - Y_s;\ s \leqslant r \leqslant t\}$, $\underline{\underline{G}}_t = \underline{\underline{G}}_t^o$. Nous allons nous intéresser au problème de lissage non linéaire suivant (problème de lissage à intervalle d'observation fixe): caractériser la loi conditionnelle de X_t, sachant $\underline{\underline{G}}_1$, pour $t \in [0,1]$ (les instants 0 et 1 peuvent être remplacés par n'importe quels instants $t_o < t_1$). Nous avons déjà donné une solution à ce problème dans [8], sous la forme d'un couple de deux Equations aux Dérivées Partielles Stochastiques (EDPS), l'une progressive à résoudre de $s = o$ à $s = t$, l'autre rétrograde de $s = 1$ à $s = t$. Après avoir rappelé ce résultat, nous établirons deux autres couples d'équations, dans chacun desquels une des équations régit l'évolution en t de la "densité conditionnelle non normalisée". Ce travail a été largement motivé par la lecture de l'article d'ANDERSON-RHODES [2].

Précisons maintenant les hypothèses et quelques notations.

On suppose que les processus $\{X_t\}$ et $\{\tilde{W}_t\}$ sont à valeurs dans

$\mathbb{R}^p$, et $\{Y_t\}$ et $\{W_t\}$ dans $\mathbb{R}^d$.

Nous supposerons que tous les coefficients b, c, e et h sont de classe C_b^∞, les fonctions et toutes les dérivées étant bornées , de $\mathbb{R}^p$ à valeurs dans $\mathbb{R}^p$, $\mathbb{R}^{p\times d}$, $\mathbb{R}^{p\times p}$ et $\mathbb{R}^d$ respectivement.

On suppose en outre que $\exists\, \alpha > o$ t.q :

(1.1) $ee^*(x) \geq \alpha I \ , \ \forall x \in \mathbb{R}^p$.

Enfin, si μ_o désigne la loi du v.a. X_o (qui est supposé être $\underline{\underline{F}}_o$-mesurable), on suppose que μ_o admet une densité $p_o(x)$ t.q :

(1.2) $p_o \quad C_b^2(\mathbb{R}^p) \cap H^2(\mathbb{R}^p)^{(2)}, \ p_o(x) > o, \ \forall x \in \mathbb{R}^p$.

On suppose pour fixer les idées que $Y_o = 0$.

On définit les opérateurs aux dérivées partielles :

$$L = \frac{1}{2} \sum_{i,j=1}^{p} a_{ij}(x) \frac{\partial^2}{\partial x_i \partial x_j} + \sum_{i=1}^{p} b_i(x) \frac{\partial}{\partial x_i}$$

$$B_i = \sum_{j=1}^{p} c_{ji}(x) \frac{\partial}{\partial x_j} + h_i \ ; \ i=1\ldots d$$

Grâce aux hypothèses faites sur les a_{ij}, L peut être considéré comme un élément de $\mathcal{L}(H^1(\mathbb{R}^p);H^{-1}(\mathbb{R}^p))^{(1)}$, et $B_i \in \mathcal{L}(H^1(\mathbb{R}^p);L^2(\mathbb{R}^p))$, i=1...d.

Alors les adjoints de ces opérateurs vérifient :

$$L^* \in \mathcal{L}(H^1(\mathbb{R}^p);H^{-1}(\mathbb{R}^p)), \ B_i^* \in \mathcal{L}(H^1(\mathbb{R}^p),L^2(\mathbb{R}^p)), \ i=1,\ldots,d,$$

grâce aux hypothèses faites sur les c_{ij} .

On définit le processus :

$$Z_t = \exp[\int_o^t h(X_s).dY_s - \frac{1}{2}\int_o^t |h(X_s)|^2 ds]$$

et une nouvelle probabilité $\overset{\circ}{P}$ sur $(\Omega,\underline{\underline{G}}_1)$ par :

$$\left.\frac{d\overset{\circ}{P}}{dP}\right|_{\underline{\underline{G}}_1} = (Z_1)^{-1}$$

$$\{Y_t, t \in [0,1]\} \text{ et } \{\tilde{W}_t, t \in [0,1]\}$$

sont deux $\overset{\circ}{P}$ processus de Wiener vectoriels standard indépendants,

---

(1) $H^1(\mathbb{R}^p)$ désigne l'espace de Sobolev des fonctions de $L^2(\mathbb{R}^p)$, dont les dérivées partielles d'ordre 1 au sens des distributions sont aussi des fonctions de $L^2(\mathbb{R}^p)$; $H^{-1}(\mathbb{R}^p)$ est le sous espace de distributions qui s'identifie au dual de $H^1(\mathbb{R}^p)$, lorsqu'on identifie $L^2(\mathbb{R}^p)$ à son dual.

(2) $H^2(\mathbb{R}^p)$ est l'espace des fonctions de $H^1(\mathbb{R}^p)$, dont les dérivées premières appartiennent à $H^1(\mathbb{R}^p)$.

(1.3) $dX_t=(b(X_t)-c\,h(X_t))dt+c(X_t)dY_t+e(X_t)d\tilde{W}_t$

et on a la formule :

(1.4) $$E[f(X_t)/\underline{\underline{G}}_1] = \frac{\mathring{E}[f(X_t)Z_1/\underline{\underline{G}}_1]}{\mathring{E}[Z_1/\underline{\underline{G}}_1]}$$

Pour calculer le membre de gauche de l'égalité (1.4) $\forall f \in C_b(\mathbb{R}^p)$, il suffit de calculer le numérateur du membre de droite $\forall f \in C_b(\mathbb{R}^p)$.

Précisons enfin les notations concernant quatre types d'intégrale stochastique que nous serons amenés à considérer, et que nous définissons ici pour un intégrand $\{\varphi_t\}$ à trajectoires continues à valeurs dans $\mathbb{R}$, en nous plaçant pour simplifier dans le cas d=1:

1. Intégrales progressives

Supposons que $\{\varphi_t, t \in [0,1]\}$ est $\underline{\underline{G}}_t$ adapté.
On définit alors:

a. Intégrale de Ito progressive

$$\int_0^t \varphi_s\, dY_s = \lim_{n\to\infty} \sum_{i=0}^{n-1} \varphi_{t_i^n}(Y_{t_{i+1}^n} - Y_{t_i^n})$$

b. Intégrale de Stratonovitch progressive

$$\int_0^t \varphi_s \circ dY_s = \lim_{n\to\infty} \sum_{i=0}^{n-1} \frac{\varphi_{t_i^n} + \varphi_{t_{i+1}^n}}{2}(Y_{t_{i+1}^n} - Y_{t_i^n})$$

où $t_i^n = \frac{i}{n}\,t$

2. Intégrales rétrogrades

Supposons que $\{\varphi_t, t \in [0,1]\}$ est $\underline{\underline{G}}_1^t$-adapté. On définit alors :

a. Intégrale de Ito rétrograde

$$\int_t^1 \varphi_s \oplus dY_s = \lim_{n\to\infty} \sum_{i=0}^{n-1} \varphi_{\tau_i^n}(Y_{\tau_{i+1}^n} - Y_{\tau_i^n})$$

b. Intégrale de Stratonovitch rétrograde

$$\int_t^1 \varphi_s \circ dY_s = \lim_{n\to\infty} \sum_{i=0}^{n-1} \frac{\varphi_{\tau_i^n} + \varphi_{\tau_{i+1}^n}}{2}(Y_{\tau_{i+1}^n} - Y_{\tau_i^n})$$

où $\tau_i^n = t + \frac{i}{n}(1-t)$.

Nous utilisons la même notation pour les deux types d'intégrales de Stratonovitch, compte tenu de leurs formules de définition identiques, et du fait qu'elles conduisent à la même règle de calcul différentiel, à savoir la règle du calcul différentiel usuel. Par contre, les intégrales stochastiques de Ito progressive et rétrograde conduisent à des règles de calcul différentiel différentes -cf. [8].

Etant donné un espace de Banach X, on notera $M_p^n(0,1;X)$ l'espace de Banach des classes de processus u_t $\underline{\underline{G}}_t$-adaptés à valeurs dans X, qui satisfont :

$$E\int_o^1 \|u_t\|_X^n \, dt < \infty$$

On définit de même $M_r^n(0,1;X)$, avec la seule différence qu'il s'agit de processus $\underline{\underline{G}}_1^t$-adaptés.

Dorénavant, on notera $V \triangleq H^1(\mathbb{R}^p)$.

On définit en outre $\tilde{V} \triangleq H^1(\mathbb{R}^p;\eta(x)dx)$, où $\eta(x) \triangleq (1+|x|^2)^{-p}$, qui est l'espace de Sobolev "avec le poids $\eta(x)$" (cf. [8] page 203).

2. Une première paire d'EDPS

Nous décrivons les résultats de [8]. On considère les EDPS rétrograde et progressive, pour $t \in [0,1]$:

$$(2.1)\quad \begin{cases} dv_t + L\,v_t\,dt + \sum_{i=1}^{d} B_i v_t \oplus dY_t^i = 0 \\ v_1(x) \equiv 1 \end{cases}$$

$$(2.2)\quad \begin{cases} dp_t = L^* p_t\,dt + \sum_{i=1}^{d} B_i^* p_t \, dY_t^i \\ p_o = \text{densité de la loi de } X_o \end{cases}$$

Les équations (2.1) et (2.2) admettent chacune une solution unique, respectivement dans $M_r^2(0,1;\tilde{V})$ et dans $M_p^2(0,1;V)$. En outre, il résulte de (1.1) et (1.2), et de la régularité des coefficients de L et B, que p.s, $\forall t \in [0,1]$, p_t et $v_t \in C^2(\mathbb{R}^p)$, et $p_t(x) > o$, $v_t(x) > o$, $\forall x \in \mathbb{R}^p$.

De plus, $\forall\, t \in [0,1]$,

$$v_t(x) = \overset{o}{E}_{tx}[Z_1^t / \underline{\underline{G}}_1^t]\ (3)$$

et

$$\int p_t(x)\,g(x)\,dx = \overset{o}{E}[g(X_t)\,Z_t / \underline{\underline{G}}_t]$$

pour tout g mesurable et borné, de $\mathbb{R}^p$ à valeurs dans $\mathbb{R}$.

De ces deux formules, on tire, en utilisant la propriété de Markov de $\{X_t\}$ (cf. [8]):

$$\begin{aligned} \overset{o}{E}[f(X_t)Z_1/\underline{\underline{G}}_1] &= \overset{o}{E}{}^{\underline{\underline{G}}_1}[f(X_t)\,Z_t\ \overset{o}{E}{}^{\underline{\underline{G}}_1^t}_{t,X_t}(Z_1^t)] \\ &= \overset{o}{E}[v(t,X_t)\,f(X_t)\,Z_t/\underline{\underline{G}}_1] \\ &= \int p(t,x)\,v(t;x)\,f(x)\,dx \end{aligned}$$

---

(3) $\overset{o}{P}_{tx}$ désigne la solution du problème de martingales associé à (1.3), avec la condition initiale $X_r = x$, $\forall\, r \in [o,t]$.

ceci pour tout f mesurable et borné de $\mathbb{R}^p$ à valeurs dans $\mathbb{R}$. Comparant avec la formule (1.4), on en déduit que

$$p_t(x)v_t(x)\left(\int_{\mathbb{R}^p} p_t(x)v_t(x)dx\right)^{-1}$$

est la densité de la loi conditionnelle de X_t, sachant $\underline{\underline{G}}_1$, i.e. $p(t,x)v(t,x)$ est la "densité conditionnelle non normalisée". On résoud donc le problème de lissage non linéaire en résolvant l'équation en p de 0 à t, et l'équation en v de façon rétrograde, de 1 à t.

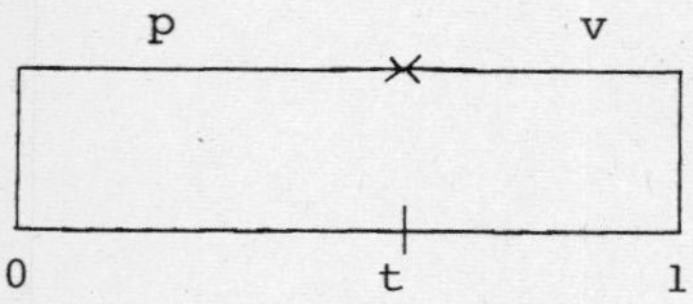

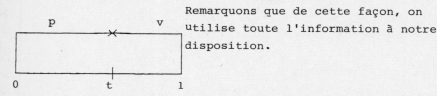

Remarquons que de cette façon, on utilise toute l'information à notre disposition.

La suite de ce travail va être consacrée à la recherche d'équations pour la densité conditionnelle non normalisée $q_t(x)=p_t(x)v_t(x)$, d'abord dans le cas d'indépendance de X et du bruit d'observation (i.e. c=0), puis dans le cas général.

3. Evolution de la densité non normalisée: cas où le bruit d'observation est indépendant du signal

La difficulté pour écrire une équation satisfaite par q provient de ce que q_t est fonction à la fois du passé de Y avant t, et de ses accroissements futurs entre t et 1. Il n'existe donc pas de calcul différentiel stochastique qui nous permette d'écrire directement la différentielle de q_t. Cependant, dans le cas c=0, le résultat s'obtient très simplement, comme on va le voir, en utilisant un argument emprunté à KRYLOV-ROSOVSKII [4]- cf. aussi [8].

Soit $\varphi \in L^\infty(0,1;\mathbb{R}^d)$. On définit :

$$\rho_t(\varphi)= \exp\left[\int_0^t(\varphi(s),dY_s)- \frac{1}{2}\int_0^t|\varphi(s)|^2ds\right]$$

$$\rho^t(\varphi)= \exp\left[\int_t^1(\varphi(s),dY_s)- \frac{1}{2}\int_t^1|\varphi(s)|^2ds\right]$$

$$\rho(\varphi) = \rho_t(\varphi)\,\rho^t(\varphi)$$

On pose :

$$\overline{q}_t(x)= \overset{\circ}{E}(\rho\, q_t(x))$$

$$\overline{p}_t(x)= \overset{\circ}{E}(\rho_t p_t(x))$$

$$\overline{v}_t(x)= \overset{\circ}{E}(\rho^t v_t(x))$$

Il résulte de l'indépendance sous $\overset{\circ}{P}$ de $\underline{\underline{G}}_t$ et $\underline{\underline{G}}_1^t$:

$$\overline{q}_t(x)=\overline{p}_t(x)\overline{v}_t(x)$$

Or on vérifie aisément, en utilisant successivement le calcul de Ito usuel et rétrograde (cf. [8] pages 199-201):

$$\frac{d\bar{p}_t}{dt} = L^*\bar{p}_t + (\varphi(t),h)\bar{p}_t, \quad \bar{p}_o = p_o$$

$$\frac{d\bar{v}_t}{dt} + L\,\bar{v}_t + (\varphi(t),h)\bar{v}_t = 0, \quad \bar{v}_t = 1$$

D'où l'on tire aussitôt :

$$\frac{d\bar{q}_t}{dt} + \bar{p}_t\, L\, \bar{v}_t = \bar{v}_t\, L^*\, \bar{p}_t$$

Cette égalité est vraie $\forall\, \varphi \in L^\infty(0,1;\mathbb{R}^d)$. Mais lorsque φ parcourt cet espace, les v.a.r. ρ, ρ_t et ρ^t parcourent des sous ensembles dont les combinaisons linéaires sont denses respectivement dans $L^2(\Omega,\underline{\underline{G}}_1,\overset{\circ}{P})$, $L^2(\Omega,\underline{\underline{G}}_t,\overset{\circ}{P})$ et $L^2(\Omega,\underline{\underline{G}}_1^t,\overset{\circ}{P})$ (cf. par exemple [9]), d'où :

$$\frac{dq_t}{dt} + p_t\, L\, v_t = v_t\, L^*\, p_t, \quad t \in [0,1]$$

On obtient alors deux équations différentes pour q, suivant que l'on élimine v ou p, en utilisant la relation $q = p\,v$:

$$\frac{dq_t}{dt} + p_t\, L\left(\frac{q_t}{p_t}\right) - \left(\frac{L^* p_t}{p_t}\right) q_t = 0 \tag{3.1}$$

$$\frac{dq_t}{dt} = v_t\, L^*\left(\frac{q_t}{v_t}\right) - \left(\frac{L\, v_t}{v_t}\right) q_t \tag{3.2}$$

Remarquons que l'équation (3.1) est une E.D.P. parabolique rétrograde (i.e. bien posée dans le sens rétrograde du temps), tandis que (3.2) est EDP, parabolique progressive (i.e. bien posée dans le sens usuel du temps). On a la :

<u>Proposition 3.1</u> : $\{p_t, t \in [0,1]\}$ désignant l'unique solution de (2.2), l'équation(3.1), avec la condition finale :

$$q_1 = p_1$$

admet une solution unique parmi les processus $\{q_t, t \in [0,1]\}$ à trajectoires dans $C_f([0,1],L^2(\mathbb{R}^p))$ (4), tels que :

$$q_{\cdot}/p_{\cdot} \in M_r^2(0,1;\tilde{V})$$

---

(4) $C_f([0,1];H)$, où H est un espace de Hilbert désigne l'espace des fonctions définies sur [0,1],à valeurs dans H, qui sont continues, de[0,1]à valeurs dans l'espace H muni de sa topologie faible.

Preuve : Remarquons que la définition des espérances dans (ii) ne necessite pas d'hypothèse d'intégrabilité, puisque les intégrands sont positifs. Il est clair que la quantité $q_t(x)$ définie à la fin du §2 satisfait les conditions de la Proposition. De plus si q, satisfaisant ces mêmes conditions est solution de (3.1) (en un sens faible) alors il résulte de (i) et (ii) que la quantité

$$\frac{\overline{q}_t}{\overline{p}_t} = \frac{\mathring{E}[\rho(\varphi)q_t]}{\mathring{E}[\rho_t(\varphi)p_t]} = \mathring{E}[\rho^t(\varphi)\frac{q_t}{p_t}]$$

est l'unique solution de l'équation en $\overline{v}_t$, donc coïncide avec cette quantité. D'après la latitude de choix de φ, et (i), $\frac{q_t}{p_t}$ est unique, donc aussi q_t .

□

On a un résultat analogue pour l'équation (3.2):

Proposition 3.2 : $\{v_t, t \in [0,1]\}$ désignant l'unique solution de (2.1), l'équarion(3.2),avec la condition initiale :

$$q_o(x) = v_o(x)p_o(x), \quad x \in \mathbb{R}^p$$

admet une solution unique parmi les processus$\{q_t, t\in[0,1]\}$ à trajectoires dans $C_f([0,1];L^2(\mathbb{R}^p))$, tels que :

$$q_{\cdot}/v_{\cdot} \in M_p^2(0,1;V)$$

□

De ces deux Propositions, il résulte deux nouvelles façons de caractériser la densité conditionnelle non normalisée du problème de Lissage.

Considérons le couple d'EDP, défini pour $t \in [0,1]$:

$$(3.3) \quad \begin{cases} dp_t = L^* p_t dt + \sum_{i=1}^{d} h_i p_t dY_t^i, p_o = \text{densité de } \mu_o \\ \frac{dq_t}{dt} + p_t L\,(\frac{q_t}{p_t}) - (\frac{L^* p_t}{p_t}) q_t = 0, \; p_1 = p_1 \end{cases}$$

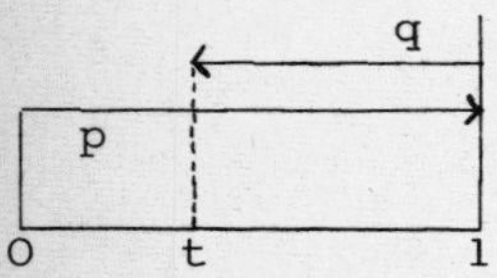

Le système (3.3) permet de résoudre le problème de lissage, en résolvant d'abord l'équation en p, puis, connaissant p, on résoud l'équation en q.

Considérons enfin le système suivant, pour $t \in [0,1]$:

$$(3.4)\quad \begin{cases} dv_t + L\, v_t\, dt + \sum_{i=1}^{d} h_i v_t \oplus d\, Y_t^i,\ v_1 = 1 \\ \dfrac{dq_t}{dt} = v_t\, L\, (\dfrac{q_t}{v_t}) - (\dfrac{Lv_t}{v_t}) q_t\ ,\ q_o(x) = p_o(x) v_o(x). \end{cases}$$

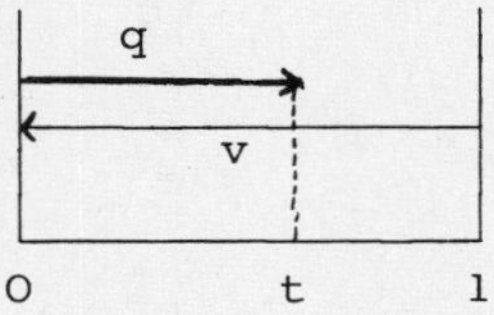

Le système (3.4) permet de résoudre le problème de lissage, en résolvant d'abord l'équation en v, puis, connaissant v, on résoud l'équation en q.

Remarque 3.3: Dans (3.3)[resp.(3.4)], l'équation pour q_t ne contient pas l'observation Y_t. q est fonction de $\{Y_t, t \in [0,1]\}$ par l'intermédiaire de p(resp. de v).
Ceci ne sera plus le cas au §4.

Remarque 3.4: La triple solution que nous donnons au problème de lissage non linéaire peut sembler à première vue parfaitement symétrique par rapport au retournement du temps. Cette impression est en fait largement trompeuse. En effet, les équations de p et de v sont de nature assez différente. p est la densité d'une mesure finie (après normalisation, d'une mesure de probabilité), c'est la solution d'un problème de filtrage. v n'est pas la solution d'un problème de filtrage rétrograde; c'est, à une constante de normalisation près, la dérivée de Radon-Nikodym de la loi du lissage, par rapport à celle du filtrage. Pour faire apparaître un problème de filtrage rétrograde, il faudrait retourner le temps dans l'écriture des différentielles des processus $\{X_t\}$ et $\{Y_t\}$, cf. PARDOUX [10] ANDERSON-RHODES[2].

Remarque 3.5: L'équation (3.1) peut se trouver dans LEONDES-PELLER-STEAR [6], ANDERSON [1] et ANDERSON-RHODES [2]. Les dérivations de ces auteurs nous ont parues soit obscures, soit illicites. Par exemple, [2] utilise le calcul d'Ito, avec des intégrands qui dépendent à la fois du passé et du futur du Wiener par rapport auquel il intègre.

4. Evolution de la densité non normalisée: le cas général.

Nous nous contentons d'esquisser les résultats de cette partie; un exposé plus détaillé fera l'objet d'une autre publication.

4. a Orientation: Si l'on applique la méthode du paragraphe précédent,

on trouve l'équation suivante pour $\overline{q}_t$:

$$\frac{d\overline{q}_t}{dt} + \overline{p}_t L \overline{v}_t = \overline{v}_t L^* p_t + \sum_{i=1}^{d} \varphi_i(t) [\overline{v}_t B_i^* \overline{p}_t - \overline{p}_t B_i \overline{v}_t]$$

Or un calcul simple montre que :

$$(4.1) \quad \overline{v}_t B_i^* \overline{p}_t - \overline{p}_t B_i \overline{v}_t = \widetilde{B}_i^* \overline{q}_t$$

où $\widetilde{B}_i$ est défini par :

$$B_i = h_i + \widetilde{B}_i$$

Donc :

$$\frac{d\overline{q}_t}{dt} + \overline{p}_t L \overline{v}_t = v_t L^* p_t + \sum_{i=1}^{d} \varphi_i(t) \widetilde{B}_i^* \overline{q}_t$$

Intuitivement, l'évolution de q_t est donnée par une égalité de la forme :

$$dq_t = C(p_t, v_t) dt + \sum_{i=1}^{d} \widetilde{B}_i^* q_t \circ dY_t^i$$

Plus précisément, si l'on cherche une équation progressive pour q_t, celle-ci semble devoir prendre la forme :

$$(4.2) \begin{cases} dq_t = A(v_t) q_t + \sum_i \widetilde{B}_i^* q_t \circ dY_t^i \\ q_o(x) = p_o(x) \; v_o(x) \end{cases}$$

Bien sûr, les deux égalités ci-dessus n'ont pas de sens, puisque q_t est adapté à $\underline{\underline{G}}_1$. Cependant, il existe un moyen de contourner cette difficulté. Nous allons expliquer l'idée sur une EDS en dimension finie. A et les B_i étant ici des matrices $n \times n$, on considère l'EDS :

$$dX_t = A X_t dt + \sum_{i=1}^{d} B_i X_t \circ dW_t^i \; , \; X_o = x_o$$

qui est équivalente à :

$$(4.3) \begin{cases} dP_t = \sum_{i=1}^{d} B_i P_t \circ dW_t^i \; , \; P_o = I \\ \dfrac{dY_t}{dt} = P_t^{-1} \circ A \circ P_t Y_t \; , \; Y_o = x_o \\ X_t = P_t Y_t \end{cases}$$

où P_t est un processus à valeurs matrices $n \times n$.
L'intérêt de la formulation (4.3) est qu'elle a encore un sens si l'on remplace x_o et A par des quantités aléatoires, qui peuvent dépendre de toute la trajectoire des $\{W_t^i\}$.

<u>4. b</u> Nous allons maintenant établir les deux équations satisfaites par q_t.

Nous faisons tout d'abord l'hypothèse supplémentaire : $p_o \in L^4(\mathbb{R}^p)$. Alors on a :

$$p \in M_p^4(0,1;L^4(\mathbb{R}^p))$$

En outre : $L^* p \in M_p^2(0,1;L^2(\mathbb{R}^p))$

$$v \in M_r^4(0,1;L^4(\mathbb{R}^p;\eta(x)dx))$$

$$L v \in M_r^2(0,1;L^2(\mathbb{R}^p;\eta(x)dx))$$

Ces estimations se démontrent comme dans [7], I°partie.

Soit Φ_t la solution fondamentale de l'EDPS :

$$du_t = \frac{1}{2}\sum_{i=1}^{p}(\tilde{B}_i^*)^2 u_t\, dt + \sum_{i=1}^{p}\tilde{B}_i^* u_t\, dY_t^i$$

i.e. $\forall u_o \in V$, $u_t = \Phi_t u_o$. Soit

$\xi_o \in L^4(\mathbb{R}^p;\frac{dx}{\eta(x)}) \cap H^2(\mathbb{R}^p;\frac{dx}{\eta(x)})$. Alors le processus $\xi_t = (\Phi_t^{-1})^* \xi_o$ satisfait l'EDPS :

$$d\xi_t = \frac{1}{2}\sum_{i=1}^{p}\tilde{B}_i^2 \xi_t\, dt - \sum_{i=1}^{p}\tilde{B}_i \xi_t\, dY_t^i$$

et on peut vérifier que :

$$\xi \in M^4(0,1;L^4(\mathbb{R}^p;\frac{dx}{\eta(x)})) \cap M^2(0,1;H^2(\mathbb{R}^p;\frac{dx}{\eta(x)}))$$

On a alors, par la formule de Ito pour les processus à valeurs hilbertiennes :

$$d(p_t\xi_t) = [\xi_t L^* p_t + \frac{1}{2}p_t\sum_{i=1}^{d}\tilde{B}_i^2\xi_t - \sum_{i=1}^{d}B_i^* p_t \tilde{B}_i \xi_t]dt +$$

$$+\sum_{i=1}^{d}[\xi_t B_i^* p_t - p_t\tilde{B}_i\xi_t]dY_t^i$$

Avec les notations du §3, on définit :

$(\overline{p\xi})_t = \mathring{E}[p_t\xi_t\rho_t(\varphi)]$. Alors :

$$\frac{d}{dt}(\overline{p\xi})_t = (\overline{\xi L^* p})_t + \frac{1}{2}\sum_{i=1}^{d}(\overline{p\tilde{B}_i^2\xi})_t - \sum_{i=1}^{d}(\overline{B_i^* p\tilde{B}_i\xi})_t +$$

$$+\sum_{i=1}^{d}\varphi_i(t)[(\overline{\xi B_i^*\rho})_t - (\overline{p\tilde{B}_i\xi})_t]$$

D'autre part, toujours avec les notations du §3,

$$\frac{d\overline{v}_t}{dt} + L v_t + \sum_{i=1}^{p}\varphi_i(t)B_i\overline{v}_t = 0$$

En utilisant la relation (4.1) et le même raisonnement qu'au §3, on tire des deux dernières égalités :

$$\frac{d}{dt}(q_t,\xi_t)=(v_t L^* p_t - p_t L v_t,\xi_t)+\frac{1}{2}\sum_{i=1}^{d}((\widetilde{B}_i^*)^2 q_t,\xi_t) -$$

$$- \sum_{i=1}^{d}(\widetilde{B}_i^*(v_t B_i^* p_t),\xi_t)$$

Posons $\widetilde{q}_t = \Phi_t^{-1} q_t$. On a alors :

$$(4.4)\quad \begin{cases} \dfrac{d\widetilde{q}_t}{dt} = \Phi_t^{-1}[v_t L^*[\dfrac{\Phi_t\widetilde{q}_t}{v_t}]-(\dfrac{Lv_t}{v_t})\Phi_t\widetilde{q}_t] + \\ \quad +\dfrac{1}{2}\displaystyle\sum_{i=1}^{d}\Phi_t^{-1}(\widetilde{B}_i^*)^2\Phi_t\widetilde{q}_t - \sum_{i=1}^{d}\Phi_t^{-1}\widetilde{B}_i^*[v_t\widetilde{B}_i^*(\dfrac{\Phi_t\widetilde{q}_t}{v_t})] \\ \widetilde{q}_o(x) = p_o(x)v_o(x) \\ q_t = \Phi_t\,\widetilde{q}_t \end{cases}$$

Remarquons que les opérateurs Φ_t et Φ_t^{-1} peuvent être explicités, en suivant une démarche du type de celle de BISMUT-MICHEL [3] et KUNITA [5].

Nous allons enfin obtenir une "équation rétrograde" pour q_t. Désignons par Ψ_t la solution fondamentale de l'EDPS :

$$dr_t + \frac{1}{2}\sum_{i=1}^{d}(\widetilde{B}_i^*)^2\, r_t\, dt - \sum_{i=1}^{d}\widetilde{B}_i^*\, r_t \oplus dY_t^i = 0$$

i.e. $\forall r_1 \in V,\ r_t = \Psi_t r_1$.

Soit $\theta_1 \in L^4(\mathbb{R}^p; \frac{dx}{\eta(x)}) \cap H^2(\mathbb{R}^p; \frac{dx}{\eta(x)})$.

Alors le processus $\theta_t = (\Psi_t^{-1})^*\,\theta_1$ satisfait l'EDPS :

$$d\theta_t + \frac{1}{2}\sum_{i=1}^{d}\widetilde{B}_i^2\,\theta_t\, dt + \sum_{i=1}^{d}\widetilde{B}_i\,\theta_t \oplus dY_t^i = 0$$

et on peut vérifier que :

$$\theta \in M^4(0,1;L^4(\mathbb{R}^p ; \frac{dx}{\eta(x)}) \cap M^2(0,1;H^2(\mathbb{R}^p; \frac{dx}{\eta(x)}))$$

On a alors :

$$d(v_t\theta_t)+\theta_t L v_t\, dt + \frac{1}{2}\sum_{i=1}^{d}[v_t \widetilde{B}_i^2\,\theta_t + B_i v_t \widetilde{B}_i \theta_t]dt +$$

$$+ \sum_{i=1}^{d}[v_t\widetilde{B}_i\theta_t + \theta_t B_i v_t] \oplus dY_t^i = 0$$

On pose :

$$(\overline{v\theta})_t = \mathring{E}[v_t\ \theta_t\ \rho^t(\varphi)]$$

$$\frac{d}{dt}(\overline{v\theta})_t + (\overline{\eta L v})_t + \frac{1}{2}\sum_{i=1}^{d}(\overline{v\widetilde{B}_i^2\,\theta + B_i\, v\,\widetilde{B}_i\theta})_t +$$

$$+ \sum_{i=1}^{d}\varphi_i(t)\,(\overline{v\,\widetilde{B}_i\ \theta\ \ \theta\ B_i\ v})_t = 0$$

$$\frac{d\bar{p}_t}{dt} = L^* \bar{p}_t + \sum_{i=1}^{d} \varphi_i(t) B_i^* \bar{p}_t$$

D'où l'on tire :

$$\frac{d}{dt}(q_t, \theta_t) + (p_t L v_t - v_t L^* p_t, \theta_t) + \frac{1}{2} \sum_{i=1}^{d} (q_t, \tilde{B}_i^2 \theta_t) + \sum_{i=1}^{d} (\tilde{B}_i^* (p_t B_i v_t), \theta_t) = 0$$

Posons $\hat{q}_t = \Psi_t^{-1} q_t$. On a alors :

$$(4.5) \quad \begin{cases} \dfrac{d\hat{q}_t}{dt} + \Psi_t^{-1} [p_t L (\dfrac{\Psi_t \hat{q}_t}{p_t}) - (\dfrac{L^* p_t}{p_t}) \Psi_t \hat{q}_t] + \\ \quad + \dfrac{1}{2} \displaystyle\sum_{i=1}^{d} \Psi_t^{-1} (\tilde{B}_i^*)^2 \Psi_t \hat{q}_t + \displaystyle\sum_{i=1}^{d} \Psi_t^{-1} \tilde{B}_i^* [p_t B_i (\dfrac{\Psi_t \hat{q}_t}{p_t})] = 0 \\ \hat{q}_1 = p_1 \\ q_t = \Psi_t \hat{q}_t \end{cases}$$

On a des résultats d'unicité pour les systèmes (4.4) et (4.5) analogues à ceux du §3.

BIBLIOGRAPHIE

[1] B.D.O. ANDERSON : Fixed Interval Smoothing for Nonlinear Continuous Time Systems. Information and Control 20, 294-300 (1972).

[2] B.D.O. ANDERSON. I.B. RHODES : Smoothing Algorithms for Nonlinear Finite - Dimensional Systems. Stochastics 9, 139-165 (1983).

[3] J.M. BISMUT-D. MICHEL : Diffusions Conditionnelles II J. Funct. Anal. 45, 274-292 (1982).

[4] N.V. KRYLOV-B.L. ROZOVSKII : On the first integrals and Liouville equations for diffusion processes. in Stochastic Differential Systems, M. Arato, D. Vermes, A. Balakrishnan Eds., Lecture Notes in Control and Information Sciences 36, 117-125, Springer-Verlag (1981).

[5] H. KUNITA : First order stochastic partial differnetial equations, à paraître.

[6] C.T. LEONDES - J.B. PELLER - E.B. STEAR : Nonlinear Smoothing Theory, IEEE Trans. Syst. Scie. and Cyber. SSC-6, 63-71 (1970).

[7] E. PARDOUX : Stochastic PDEs and filtering of diffusion processes. Stochastics 3, 127-167 (1979).

[8] E. PARDOUX : Equations du filtrage non linéaire, de la prédiction et du lissage. Stochastics 6, 193-231 (1982).

[9] E. PARDOUX : Equations of Nonlinear Filtering, and applications to Stochastic Control with Partial Observation. in Non linear Filtering and Stochastic Control, S. Mitter, A. Moro Eds., Lecture Notes in Mathematics 972, 208-248, Springer-Verlag (1982).

[10] E. PARDOUX : Smoothing of a diffusion process conditionned at final time. in Stochastic Differential Systems, M.Kohlmann, N. Christopeit Eds., Lecture Notes in Control and Information Sciences 43, 187-196, Springer-Verlag (1982).

Summary

We consider the following stochastic differential system :

$$dX_t = b(X_t)dt + c(X_t)dW_t + e(X_t)d\tilde{W}_t$$

$$dY_t = h(X_t)dt + dW_t$$

where W_t and $\tilde{W}_t$ are two independent standard Wiener processes, with values in $\mathbb{R}^d$ and $\mathbb{R}^p$ respectively. X_t takes values in $\mathbb{R}^p$, and Y_t in $\mathbb{R}^d$. Definie $\underline{\underline{G}}_1 = \sigma\{Y_t, t \in [0,1]\}$. We consider the fixed interval non linear smoothing problem : caracterize the conditionnal law of X_t, given $\underline{\underline{G}}_1$, for $t \in [0,1]$.

In [8], we gave a solution to this problem, in terms of a pair of Stochastic Partial Differential Equations (SPDEs), one forward to be solved from s=o to s=t, the other backward from s=1 to s=t. After having recalled this result, we establish two other pairs of SPDEs, in which one of the equations governs the evolution of the so-called "unnormalized conditional density" of the smoothing problem.

This work was motivated by the recent paper of ANDERSON-RHODES [2].

APPROXIMATION OF NONLINEAR FILTERING PROBLEMS AND ORDER OF CONVERGENCE.

Jean PICARD

I.N.R.I.A.

Route des Lucioles
Sophia Antipolis
06560 VALBONNE-FRANCE

SUMMARY : In this paper, we consider a filtering problem where the observation is a function of a diffusion corrupted by an independent white noise. We estimate the error caused by a discretization of the time interval ; we obtain some approximations of the optimal filter which can be computed with Monte-Carlo methods and we study the order of convergence.

1 - INTRODUCTION.

Let σ, b and h be "sufficiently regular" real-valued functions defined on $\mathbb{R}$. Define A and B the differential operators :

$$A = \frac{1}{2} \sigma^2(x) \frac{d^2}{dx^2} + b(x) \frac{d}{dx}$$

$$B = \sigma(x) \frac{d}{dx}$$

Let $(\Omega, \underline{F}, \underline{F}_t, P ; 0 \le t \le T)$ be a probability space with a standard Brownian motion (W_t, B_t) and let X_o be a $\underline{F}_o$ measurable variable such that $E|X_o|^r < \infty$ for every r. We consider the filtering problem where the signal process X_t and the observation process Y_t satisfy the equation :

$$\begin{cases} X_t = X_o + \int_o^t b(X_s)ds + \int_o^t \sigma(X_s)dW_s \\ Y_t = \int_o^t h(X_s)ds + B_t \end{cases}$$

Let $\underline{W}_t$ and $\underline{Y}_t$ be the filtrations generated respectively by W and Y. If g is a "regular" function from $\mathbb{R}$ into $\mathbb{R}$, we want to compute :

$$\Pi_t(g) = E\,[g(X_t)|\underline{Y}_t]$$

Let us first recall the Kallianpur-Striebel formula (Kallianpur [1]). Define :

$$L_t = \exp\left(\int_o^t h(X_s)dY_s - \frac{1}{2}\int_o^t h^2(X_s)ds\right)$$

With rather mild hypothesis, $E\,[L_T^{-1}] = 1$; if we define the reference probability $\mathring{P} = L_T^{-1}.\,P$, then Y is a $\mathring{P}$ Brownian motion independent from X, X has the same law under P and $\mathring{P}$ and, with the notation $\mathring{E}^Y[.] = \mathring{E}[.|\underline{Y}_T]$:

$$\Pi_t(g) = \tilde{\Pi}_t(g)/\tilde{\Pi}_t(1)$$

with $\tilde{\Pi}_t(g) = \mathring{E}\,[g(X_t)L_t|\underline{Y}_t] = \mathring{E}^Y\,[g(X_t)L_t]$

The unnormalized filter $\tilde{\Pi}_t$ can also be defined as the measure-valued process solution of the equation :

(1.1) $$\tilde{\Pi}_t(g) = E\,[g(X_o)] + \int_o^t \tilde{\Pi}_s(Ag)ds + \int_o^t \tilde{\Pi}_s(hg)dY_s$$

Now let N be an integer, $\delta = T/N$ and consider the subdivision $t_k = k\delta$, $0 \le k \le N$. We look for an approximation $\tilde{\Pi}_k^\delta$ of $\tilde{\Pi}_{k\delta}$ which involves only the increments $\Delta Y_q = Y_{(q+1)\delta} - Y_{q\delta}$ of the observation process and which is good with respect to the error :

$$\overset{\circ}{E}\,[(\tilde{\Pi}_{k\delta}(g) - \tilde{\Pi}_k^\delta(g))^2]$$

Then, with $\Pi_k^\delta = \tilde{\Pi}_k^\delta / \tilde{\Pi}_k^\delta(1)$, it will be easy to estimate :

$$E\,[|\Pi_{k\delta}(g) - \Pi_k^\delta(g)|]$$

In the case of stochastic differential equations on $\mathbb{R}$, it is possible to construct a discretization scheme which induces an L^2 error dominated by δ (Pardoux-Talay [3]).
So the question is : can we define such a scheme for the nonlinear filtering equation (1.1) ? We will prove that, with some regularity assumptions, the answer is yes.

In section 2, we prove a preliminary result about the representation of random variables by means of stochastic integrals.

In section 3, we study some discretized filters which satisfy :

(1.2) $$\tilde{\Pi}_k^\delta(g) = \overset{\circ}{E}{}^Y\,[g(X_{k\delta})\; L_k^\delta]$$

for some approximation L_k^δ of $L_{k\delta}$. For instance, if we replace the continuous-time signal process X_t by the discrete-time process $X_{q\delta}$, we define :

(1.2a) $$L_k^{\delta.a} = \exp \sum_{q=o}^{k-1} (h(X_{q\delta})\,\Delta Y_q - \tfrac{1}{2}h^2(X_{q\delta})\delta)$$

We can also discretize L_t by :

(1.2b) $$L_k^{\delta.b} = \prod_{q=o}^{k-1} (1 + h(X_{q\delta})\Delta Y_q + \tfrac{1}{2}h^2(X_{q\delta})(\Delta Y_q^2 - \delta))$$

but in this case, L_k^δ is not necessarily positive, so we can prefer :

(1.2c) $$L_k^{\delta.c} = \prod_{q=o}^{k-1} (1 + h_\delta(X_{q\delta})\Delta Y_q + \tfrac{1}{2}h_\delta^2(X_{q\delta})(\Delta Y_q^2 - \delta))$$

where h_δ is the truncated function :

$$h_\delta(x) = (h(x) \wedge \delta^{-1/2}) \vee (-\delta^{-1/2})$$

These three approximations induce L^2 errors dominated by δ (theorem 1).

Then in section 4, we study the normalized filter and in section 5, we describe some other approximations which are obtained from (1.2) by an application of Talay's method [6].

Remark 1 : We suppose for notational convenience that all the processes are real-valued ; nevertheless, one can easily extend the proofs to the multidimensional case. The coefficients σ, b, h may also depend on time.

Other notations :

i) If α is a real-valued function defined on [0,T] :

$$\alpha^* = \sup_t |\alpha_t|$$

ii) If $p \geq 1$, $\|\cdot\|_p$ is the L^p norm in $(\Omega, \underline{\underline{F}}, \overset{\circ}{P})$

iii) $H(M) = \sup \{|h(x)| \ ; \ |x| \leq M\}$

Regularity hypothesis : Subsequently, we will assume :

(H1) All the moments of X_o are finite.

(H2) σ, b, g are K-lipschitz functions for some $K > 0$.

(H3) h is twice continuously differentiable, h' and h" are bounded by K.

(H4) $E\,[\exp((1 + \varepsilon)TH^2(X^*))] < \infty$ for some $\varepsilon > 0$.

Remark on (H4) : Let us first notice :

$$\overset{\circ}{E}[L_T^2] = E\,[\exp(\int_o^T h^2(X_t)dt)] \leq E\,[\exp(TH^2(X^*))]$$

(H4) means that L_T is a little more than square integrable. It is obviously satisfied if h is bounded. In the general case (h has linear growth), it is also satisfied for sufficiently small T if σ is bounded and if $E\,[\exp(aX_o^2)] < \infty$ for some $a > 0$; indeed in this case (Kallianpur [1]) :

$$E\,[\exp(\bar{a}X^{*2})] < \infty \qquad \text{for some } \bar{a} > 0.$$

Remark 2 : A more general case (the signal and the noise were not necessarily independent) was considered in [4]. The basic fact was the following one : the error $|\Pi_T - \Pi_N^\delta|$ is dominated by the error $|L_T - L_N^\delta|$ on the density ; but, for the purpose of this paper, this fact is not precise enough because in general $\|L_T - L_N^\delta\|_2$ is dominated by $\delta^{1/2}$ and not by δ.

2 - A PRELIMINARY RESULT :

Definition : Let C be the space of real-valued continuous functions defined on [0,T], let ρ be an increasing function from $\mathbb{R}_+$ into $\mathbb{R}_+$. A real-valued measurable function Φ defined on $C \times C$ will be said to be ρ-Lipschitz if, for every α, $\bar{\alpha}$ in C :

$$\|\Phi(\alpha,Y)\|_2 < \infty$$

$$\|\Phi(\alpha,Y) - \Phi(\bar{\alpha},Y)\|_2 \leq \rho(\alpha^* \vee \bar{\alpha}^*)\ (\alpha - \bar{\alpha})^*$$

Proposition 1 : Assume (H1), (H2). For each $p > 2$, there exists a constant $\gamma > 0$ (which depends on σ, b, p, T and on the moments of X_0) such that if $\|\rho(X^*)\|_p < \infty$ and if Φ is ρ-Lipschitz, then the variable $\Phi(X,Y)$ can be decomposed in the form :

$$\Phi(X,Y) = \mathring{E}^Y\,[\Phi(X,Y)] + \int_0^T \psi_t\, dW_t \tag{2.1}$$

where ψ_t is a $\underline{\underline{W}}_t \vee \underline{\underline{Y}}_T$ adapted process such that :

$$\|\psi_t\|_2 \leq \gamma \quad \|\rho(X^*)\|_p \tag{2.2}$$

Proof :

First define r by : $1/p + 1/r = 1/2$; since X and Y are $\mathring{P}$ independent :

$$\begin{aligned}\|\Phi(X,Y)\|_2 &\leq \|\Phi(0,Y)\|_2 + \|\Phi(X,Y) - \Phi(0,Y)\|_2\\ &\leq \|\Phi(0,Y)\|_2 + \|\rho(X^*)X^*\|_2\\ &\leq \|\Phi(0,Y)\|_2 + \|\rho(X^*)\|_p\,\|X^*\|_r\\ &< \infty\end{aligned}$$

Therefore, the decomposition (2.1) can be deduced from the well-known Itô representation theorem (Kallianpur [1]) ; moreover :

$$\mathring{E}\,[\psi_t^2] = \frac{d}{dt}\,\mathring{E}[\mathring{E}[\Phi(X,Y)|\underline{\underline{W}}_t \vee \underline{\underline{Y}}_T]^2]$$

Thus, to prove (2.2), it is sufficient to prove that for $0 \leq s \leq t \leq T$:

$$(2.3) \qquad \|\mathring{E}\,[\Phi(X,Y)|\underline{\underline{W}}_t \vee \underline{\underline{Y}}_T] - \mathring{E}\,[\Phi(X,Y)|\underline{\underline{W}}_s \vee \underline{\underline{Y}}_T]\|_2 \leq \gamma(t-s)^{1/2}\|\rho(X^*)\|_p$$

Define $\overline{X}_u$ the solution of :

* if $u \leq t$: $\overline{X}_u = X_{u\wedge s}$

* if $u \geq t$: $\overline{X}_u = X_s + \int_t^u b(\overline{X}_v)dv + \int_t^u \sigma(\overline{X}_v)dW_v$

One easily proves that the left hand side of (2.3) is bounded by :

$$\|\mathring{E}\,[\Phi(X,Y)|\underline{\underline{W}}_t \vee \underline{\underline{Y}}_T] - H\|_2$$

for any $\underline{\underline{W}}_s \vee \underline{\underline{Y}}_T$ measurable variable H ; here we put :

$$H = \mathring{E}\,[\Phi(\overline{X},Y)|\underline{\underline{W}}_s \vee \underline{\underline{Y}}_T] = \mathring{E}\,[\Phi(\overline{X},Y)|\underline{\underline{W}}_t \vee \underline{\underline{Y}}_T]$$

Thus the left hand side of (2.3) is bounded by :

$$\begin{aligned} \|\Phi(X,Y) - \Phi(\overline{X},Y)\|_2 &\leq \|\rho(X^* \vee \overline{X}^*)\|_p\,\|(X - \overline{X})^*\|_r \\ &\leq 2\|\rho(X^*)\|_p\,\|(X - \overline{X})^*\|_r \end{aligned}$$

(because $\|\rho(\overline{X}^*)\|_p \leq \|\rho(X^*)\|_p$)

Now, by a classical argument (Kunita [2]), it follows from (H2) that for some $c_1 > 0$:

$$\mathring{E}\,[\sup_{u\geq t} |X_u - \overline{X}_u|^r\ |\underline{\underline{W}}_t] \leq c_1\,|X_t - X_s|^r$$

Therefore : $\|(X - \overline{X})^*\|_r \leq c_2\,\|\sup_{s\leq u\leq t} |X_u - X_s|\,\|_r$

$$\leq c_3\,(t-s)^{1/2}$$

and (2.3) is proved.

□

Now let us give an example of Lipschitz function : fix N, $\delta \equiv T/N$, suppose that J is a subset of {0, 1,..., N-1} and define :

(2.4) $$\Lambda^J(\alpha,Y) = \prod_{q\in J} \lambda_q(\alpha,Y) \text{ with :}$$

$$\lambda_q(\alpha,Y) = 1 + h(\alpha_{q\delta})\,\Delta Y_q + \frac{1}{2}\,h^2(\alpha_{q\delta})\,(\Delta Y_q^2 - \delta)$$

If J = {0, 1,..., N-1}, we will simply write : $\Lambda \equiv \Lambda^J$.

Proposition 2 : Suppose (H2), (H3). Then the function Λ^J is ρ-Lipschitz for some function ρ of the form :

$$\rho(M) = R(M)\,\exp\left(\frac{1}{2}\,TH^2(M)\right)$$

where R is a polynomial which can be chosen independently of δ and J.

Proof :

$$\mathring{E}\,[\Lambda^J(\alpha,Y)^2] = \prod_{q\in J} \left(1 + h^2(\alpha_{q\delta})\delta + \frac{1}{2}\,h^4(\alpha_{q\delta})\delta^2\right)$$

$$\leq \exp\left(\sum_{q\in J} h^2(\alpha_{q\delta})\delta\right)$$

$$\leq \exp\,(TH^2(\alpha^*)) < \infty$$

Fix $(\alpha,\bar{\alpha}) \in C \times C$, $M = \alpha^* \vee \bar{\alpha}^*$; then

$$\Lambda^J(\alpha,Y) - \Lambda^J(\bar{\alpha},Y) = \sum_{q\in J} \eta_q \qquad \text{with :}$$

$$\eta_q = \left((h(\alpha_{q\delta}) - h(\bar{\alpha}_{q\delta}))\,\Delta Y_q + \frac{1}{2}(h^2(\alpha_{q\delta}) - h^2(\bar{\alpha}_{q\delta}))(\Delta Y_q^2 - \delta)\right) \times$$

$$\prod_{q'\in J;q'<q} \lambda_{q'}(\alpha,Y) \prod_{q'\in J;q'>q} \lambda_{q'}(\bar{\alpha},Y)$$

One can derive from an elementary calculation that :

if $q \neq q'$: $\overset{\circ}{E}\,[\eta_q \eta_{q'}] \le \exp\,(TH^2(M))(H(M)\delta + H^3(M)\delta^2)^2\; K^2(\alpha - \overline{\alpha})^{*2}$

if $q = q'$: $\overset{\circ}{E}\,[\eta_q^2] \le \exp\,(TH^2(M))(\delta + 2H^2(M)\delta^2)\; K^2(\alpha - \overline{\alpha})^{*2}$

Therefore :

$$\overset{\circ}{E}\,[(\Lambda^J(\alpha,Y) - \Lambda^J(\overline{\alpha},Y))^2] \le \exp\,(TH^2(M))\;\gamma(M)\;K^2(\alpha - \overline{\alpha})^{*2}$$

with $\quad \gamma(M) = (H(M) + H^3(M)T)^2 + 1 + 2H^2(M)T$

□

Propositions 1 and 2 immediately imply :

Corollary 1 : With the hypothesis (H1) to (H4), and if f is a real-valued function defined on C such that :

$$|f(\alpha)| \le \overline{\rho}(\alpha^*)$$

$$|f(\alpha) - f(\overline{\alpha})| \le \overline{\rho}(\alpha^* \vee \overline{\alpha}^*)\;(\alpha - \overline{\alpha})^*$$

for some increasing function $\overline{\rho}$ with polynomial growth then :

$$f(X)\Lambda^J(X,Y) = \overset{\circ}{E}{}^Y\,[f(X)\Lambda^J(X,Y)] + \int_0^T \psi_t^J\, dW_t$$

when ψ_t^J is a $\underline{\underline{W}}_t \vee \underline{\underline{Y}}_T$ adapted process such that $\|\psi_t^J\|_2$ is bounded by a number which depends only on T, σ, b, h, ε, $\overline{\rho}$, K and X_o.

3 - THE MAIN RESULT :

In this section, we state the theorem about the speed of convergence of the approximation and we prove it with five lemmas.

Theorem 1 : Assume (H1), (H2), (H3) and (H4). If $\tilde{\Pi}^\delta$ is one of the three approximations (1.2), then, when δ tends to 0 (i.e. when N tends to infinity) :

$$\overset{\circ}{E}\,[(\tilde{\Pi}_T(g) - \tilde{\Pi}_N^\delta(g))^2] = O(\delta^2)$$

Lemma 1 : Suppose that $\tilde{\Pi}^{\delta.a}$, $\tilde{\Pi}^{\delta.b}$ and $\tilde{\Pi}^{\delta.c}$ are respectively the approximations (1.2a), (1.2b) and (1.2c). Then :

i) $\overset{\circ}{E}\,[(\tilde{\Pi}_N^{\delta.a}(g) - \tilde{\Pi}_N^{\delta.b}(g))^2] = O(\delta^2)$

ii) $\overset{\circ}{E}\,[(\tilde{\Pi}_N^{\delta.b}(g) - \tilde{\Pi}_N^{\delta.c}(g))^2] = O(\exp(-c/\delta))$ for some $c > 0$

Proof of i) :

Define :

(3.1) $$\overline{\Lambda}(\alpha,Y) = \prod_q \overline{\lambda}_q(\alpha,Y) \quad \text{with :}$$

$$\overline{\lambda}_q(\alpha,Y) = \exp(h(\alpha_{q\delta})\Delta Y_q - \frac{1}{2}\, h^2(\alpha_{q\delta})\delta)$$

Fix some $\alpha \in C$; then, with the notation (2.4), since $\lambda_q(\alpha,Y)$ and $(\overline{\lambda}_q - \lambda_q)(\alpha,Y)$ are orthogonal :

$$\overset{\circ}{E}\,[(\Lambda - \overline{\Lambda})^2(\alpha,Y)] = \overset{\circ}{E}\,[\overline{\Lambda}(\alpha,Y)^2] - \overset{\circ}{E}\,[\Lambda(\alpha,Y)^2]$$

$$= \prod_q \exp(h^2(\alpha_{q\delta})\delta) - \prod_q (1 + h^2(\alpha_{q\delta})\delta + \frac{1}{2}h^4(\gamma_{q\delta})\delta^2)$$

$$\leq \frac{1}{6}\exp(TH^2(\alpha^*))\ \delta^3 \sum_q h^6(\alpha_{q\delta})$$

(3.2) $$\overset{\circ}{E}\,[(\Lambda - \overline{\Lambda})^2(\alpha,Y)] \leq \frac{\delta^2}{6}\, H^6(\alpha^*)T \exp(TH^2(\alpha^*))$$

Since $\tilde{\Pi}_N^{\delta.a}(g) = \overset{\circ}{E}{}^Y\,[g(X_T)\,\overline{\Lambda}\,(X,Y)]$; $\tilde{\Pi}_N^{\delta.b}(g) = \overset{\circ}{E}{}^Y\,[g(X_T)\,\Lambda\,(X,Y)]$, this implies :

$$\overset{\circ}{E}\,[(\tilde{\Pi}_N^{\delta.a}(g) - \tilde{\Pi}_N^{\delta.b}(g))^2] \leq \frac{\delta^2}{6}\, T\, \overset{\circ}{E}\,[g^2(X_T)H^6(X^*)\exp(TH^2(X^*))]$$

Proof of ii) :

$$\overset{\circ}{E}\,[(\tilde{\Pi}_N^{\delta.b}(g) - \tilde{\Pi}_N^{\delta.c}(g))^2] \leq \overset{\circ}{E}\,[g^2(X_T)\,(L_N^{\delta.b} - L_N^{\delta.c})^2]$$

$$\leq 2\overset{\circ}{E}\,[g^2(X_T)((L_N^{\delta.b})^2 + (L_N^{\delta.c})^2)1_{(H(X^*) > \delta^{-1/2})}]$$

$$\leq 4\overset{\circ}{E}\,[g^2(X_T)\exp(TH^2(X^*))\ 1_{(H(X^*) > \delta^{-1/2})}]$$

It follows from (H4) that :

$$\overset{\circ}{P}\,[H(X^*) > \delta^{-1/2}] \le \exp(-T/\delta)$$

and $\|g^2(X_T)\exp(TH^2(X^*))\|_p < \infty$ for some $p > 1$.

So, if r is defined by $1/p + 1/r = 1$:

$$\overset{\circ}{E}\,[(\tilde{\Pi}_N^{\delta.b}(g) - \tilde{\Pi}_N^{\delta.c}(g))^2] \le 4\,\|g^2(X_T)\exp(TH^2(X^*))\|_p \exp(-\frac{T}{r\delta})$$

□

Lemma 2 : With the notation (2.4), define :

$$I_q = \overset{\circ}{E}{}^Y\,[g(X_T)\,\Lambda\,(X,Y)\int_{q\delta}^{(q+1)\delta}(h(X_s) - h(X_{q\delta}))dY_s]$$

$$I'_q = \overset{\circ}{E}{}^Y\,[g(X_t)\,\Lambda\,(X,Y)\int_{q\delta}^{(q+1)\delta}(h^2(X_s) - h^2(X_{q\delta}))ds]$$

Then : $$\overset{\circ}{E}\,[(\tilde{\Pi}_T(g) - \tilde{\Pi}_N^{\delta.a}(g) - \sum_{q=o}^{N-1}(I_q - I'_q/2))^2] = 0(\delta^2)$$

Proof :

At first define :

$$\beta_q(\alpha,Y) = \int_{q\delta}^{(q+1)\delta}(h(\alpha_s) - h(\alpha_{q\delta}))dY_s$$

(in the stochastic integral, α is considered as a parameter : Stricker-Yor [5])

$$\beta'_q(\alpha) = \int_{q\delta}^{(q+1)\delta}(h^2(\alpha_s) - h^2(\alpha_{q\delta}))ds$$

The inequality

$$|e^u - e^v - (u - v)e^v| \le (e^u + e^v)\,(u - v)^2/2$$

used with $u = \text{Log } L_T$, $v = \text{Log } L_N^{\delta.a}$ provides us with the estimation :

$$|\tilde{\Pi}_T(g) - \tilde{\Pi}_N^{\delta.a}(g) - \overset{\circ}{E}{}^Y [g(X_T)\, L_N^{\delta.a} \operatorname{Log} (L_T/L_N^{\delta.a})]|$$

$$\leq \frac{1}{2} \overset{\circ}{E} [|g(X_T)|(L_T + L_N^{\delta.a}) (\operatorname{Log} \frac{L_T}{L_N^{\delta.a}})^2]$$

The L^2 norm of the right hand side is dominated by δ and :

$$\sum(I_q - I'_q/2) = \overset{\circ}{E}{}^Y [g(X_T)L_N^{\delta.b} \operatorname{Log} (L_T/L_N^{\delta.a})]$$

Therefore with the notations (2.4), (3.1), it is sufficient to prove :

(3.3) $$\|g(X_T)(\Lambda - \bar{\Lambda})(X,Y) \sum\beta'_q(X)\|_2 = 0(\delta)$$

(3.4) $$\|g(X_T)(\Lambda - \bar{\Lambda})(X,Y) \sum\beta_q(X,Y)\|_2 = 0(\delta)$$

(3.3) is easily obtained from (3.2). We only sketch the proof of (3.4) which needs some calculation. We have to study :

$$\overset{\circ}{E} [(\sum\beta_q (\Lambda - \bar{\Lambda}))^2(\alpha,Y)]$$

For $0 \leq q \leq N-1$, we write :

$$\Lambda - \bar{\Lambda} = \bar{\lambda}_q(\prod_{j\neq q} \lambda_j - \prod_{j\neq q} \bar{\lambda}_j) + (\lambda_q - \bar{\lambda}_q) \prod_{j\neq q} \lambda_j$$

From this equality, dropping the "(α,Y)", we prove that :

$$\overset{\circ}{E} [\beta_q^2(\Lambda - \bar{\Lambda})^2] = \overset{\circ}{E} [\beta_q^2 \bar{\lambda}_q^2] \overset{\circ}{E} [(\prod_{j\neq q} \lambda_j - \prod_{j\neq q} \bar{\lambda}_j)^2]$$

$$+ \overset{\circ}{E} [\beta_q^2(\lambda_q - \bar{\lambda}_q)^2] \overset{\circ}{E} [(\prod_{j\neq q} \lambda_j)^2]$$

$$= 0(\delta^3)$$

A similar calculation proves that if $q \neq q'$:

$$\overset{\circ}{E} [\beta_q \beta_{q'}(\Lambda - \bar{\Lambda})^2] = 0(\delta^4)$$

and this yields (3.4).

<u>Lemma 3</u> : $\mathring{E}\,[(\sum I'_q)^2] = 0(\delta^2)$

<u>Proof</u> :

$\|g(X_T)\ \Lambda\ (X,y)\|_2$ is bounded independently of δ, so we can write :

$$g(X_T)\ \Lambda\ (X,Y) = \tilde{\Pi}_N^{\delta}(g) + \int_o^T \psi_t dWt$$

where ψ_t is a $\underline{\underline{W}}_t \vee \underline{\underline{Y}}_T$ adapted process such that $\int_o^T \mathring{E}[\psi_t^2]dt$ is bounded.

$$I'_q = \mathring{E}^Y\,[g(X_T)\ \Lambda\ (X,Y) \int_{q\delta}^{(q+1)\delta} ((q+1)\delta - s)(A(h^2)(X_s)ds + B(h^2)(X_s)dW_s)]$$

$$= \int_{q\delta}^{(q+1)\delta} ((q+1)\delta - s)\ \mathring{E}^Y\,[g(X_T)\ \Lambda\ (X,Y)A(h^2)(X_s) + \psi_s\ B(h^2)(X_s)]ds$$

From Schwarz inequality, for some $c > 0$:

$$|\sum I'_q| \le c\delta \int_o^T (\mathring{E}^Y\,[(g(X_T)\ \Lambda\ (X,Y))^2]^{1/2} + \mathring{E}^Y\,[\psi_s^2]^{1/2})ds$$

$$\|\sum I'_q\|_2 \le c\delta\ (T\,\|g(X_T)\ \Lambda\ (X,Y)\|_2 + (T \int_o^T \|\psi_s\|_2^2\ ds)^{1/2})$$

□

<u>Lemma 4</u> : For every q : $\mathring{E}\,[I_q^2] = 0(\delta^3)$
(that means : $\mathring{E}\,[I_q^2] \le c\delta^3$ where c does not depend on δ, q).

<u>Proof</u> :

We can write I_q with an integration by parts like I'_q but now, the estimate used in the last proof is not precise enough, so we are going to apply the results of section 2.

Fix q, put $J = \{0, 1,\ldots, N-1\} \setminus \{q\}$ and define :

(3.5) $$\xi_q^o = 1\ ;\ \xi_q^1 = \Delta Y_q\ ;\ \xi_q^2 = (\Delta Y_q^2 - \delta)/2$$

$$\overline{I}_q^m = \mathring{E}^Y\,[g(X_T)\ \Lambda^J\ (X,Y)\ \xi_q^m\ h^m(X_{q\delta}) \int_{q\delta}^{(q+1)\delta} (h(X_s) - h(X_{q\delta}))\ dY_s]$$

for $0 \le m \le 2$

Then : $I_q = \overline{I}_q^o + \overline{I}_q^1 + \overline{I}_q^2$

One easily proves $\mathring{E}\,[(\overline{I}_q^1 + \overline{I}_q^2)^2] = O(\delta^3)$ so we have to study $\mathring{E}\,[(\overline{I}_q^o)^2]$.

From corollary 1 :

$$g(X_T)\,\Lambda^J(X,Y) = \mathring{E}^Y\,[g(X_T)\,\Lambda^J(X,Y)] + \int_o^T \psi_t^J\,dW_t$$

where ψ_t^J is a $\underline{\underline{W}}_t \vee \underline{\underline{Y}}_t$ adapted process such that $\|\psi_t^J\|_2$ is bounded. Moreover, ψ_t^J, like Λ_J, is independent of $(Y_s - Y_{q\delta}\ ;\ q\delta \le s \le (q+1)\delta)$.

$$\overline{I}_q^o = \int_{q\delta}^{(q+1)\delta} (Y_{(q+1)\delta} - Y_s)\,\mathring{E}^Y\,[g(X_T)\,\Lambda^J(X,Y)\,Ah(X_s) + \psi_s^J\,Bh(X_s)]\,ds$$

$$|\overline{I}_q^o| \le c\int_{q\delta}^{(q+1)\delta} |Y_{(q+1)\delta} - Y_s|\,(\mathring{E}^Y\,[(g(X_T)\,\Lambda^J(X,Y))^2]^{1/2} + \mathring{E}^Y\,[(\psi_s^J)^2]^{1/2})ds$$

From the independence property :

$$\|\overline{I}_q^o\|_2 \le c\int_{q\delta}^{(q+1)\delta} \|Y_{(q+1)\delta} - Y_s\|_2\,(\|g(X_T)\,\Lambda^J(X,Y)\|_2 + \|\psi_s^J\|_2)ds$$
$$= O(\delta^{3/2})$$

□

<u>Lemma 5</u> : If $q \ne q'$: $\mathring{E}\,[I_q I_{q'}] = O(\delta^4)$

<u>Proof</u> :

Fix $q \ne q'$, put $J = \{0, 1, \ldots, N-1\} \setminus \{q, q'\}$, define ξ_q^j by (3.5) and :

$$\overline{I}_j^{m,n} = \mathring{E}^Y\,[g(X_T)\,\Lambda^J(X,Y)\,h^m(X_{q\delta})\,h^n(X_{q'\delta})\,\xi_q^m\,\xi_{q'}^n \int_{j\delta}^{(j+1)\delta} (h(X_s) - h(X_{j\delta}))dY_s]$$

for $j = q$ or q', $0 \le m,n \le 2$

Then : $I_j = \sum_{m=o}^{2}\sum_{n=o}^{2} I_j^{m,n}$ for $j = q$ or q'.

From corollary 1 :

$$g(X_T)\,\Lambda^J(X,Y) = \mathring{E}^Y\,[g(X_T)\,\Lambda^J(X,Y)] + \int_o^T \psi_t^J\,dW_t$$

where $\|\psi_t^J\|_2$ is bounded so :

$$\overline{I}_j^{m,n} = \overset{\circ}{E}^Y [h^m(X_{q\delta})\ h^n(X_{q'\delta})\ \xi_q^m\ \xi_{q'}^n \int_{j\delta}^{(j+1)\delta} (Y_{(j+1)\delta} - Y_s)\ (g(X_T)\ \Lambda^J(X,Y)Ah(X_s)$$

$$+ \psi_s^J\ Bh(X_s))ds]$$

$$\overset{\circ}{E}\ [\overline{I}_q^{m,n}\ \overline{I}_{q'}^{m',n'}] = \int_{q\delta}^{(q+1)\delta} ds \int_{q'\delta}^{(q'+1)\delta} ds'\ \overset{\circ}{E}\ [(Y_{(q+1)\delta} - Y_s)\ \xi_q^m\ \xi_q^{m'}]$$

$$\overset{\circ}{E}\ [(Y_{(q'+1)\delta} - Y_{s'})\ \xi_{q'}^n\ \xi_{q'}^{n'}]\overset{\circ}{E}[\overset{\circ}{E}^Y[(g(X_T)\ \Lambda^J(X,y)Ah(X_s) + \psi_s^J Bh(X_s))$$

$$h^m(X_{q\delta})h^n(X_{q'\delta})]\ \overset{\circ}{E}^Y\ [(g(X_T)\ \Lambda^J(X,Y)Ah(X_{s'}) + \psi_{s'}^J Bh(X_{s'}))h^{m'}(X_{q\delta})h^{n'}(X_{q'\delta})]]$$

In order to conclude, we remark that the first and the second expectations are dominated by δ and that the third one is bounded so :

$$\overset{\circ}{E}\ [\overline{I}_q^{m,n}\ \overline{I}_{q'}^{m',n'}] = O(\delta^4)$$

□

Then theorem 1 immediately follows from lemmas 1 to 5.

4 - A CONVERGENCE THEOREM FOR THE NORMALIZED FILTER :

The purpose of this section is the proof of a corollary of theorem 1 : we want to replace the unnormalized filter by the normalized one and the reference probability $\overset{\circ}{P}$ by the actual one P ; nevertheless, we study the speed of convergence in L^1 and not in L^2 and we assume a little more restrictive hypothesis : in particular, we want the approximated density L_N^δ to be nonnegative, so we restrict ourselves to the filters (1.2a) and (1.2c).

<u>Corollary 2</u> : Assume (H1), (H2), (H3), (H4) and suppose that $\tilde{\Pi}_k^\delta$ is defined by (1.2a) or (1.2c) ; in the case (1.2c), suppose moreover that g is bounded. Define : $\Pi_k^\delta = \tilde{\Pi}_k^\delta / \tilde{\Pi}_k^\delta(1)$. Then :

$$E\ [|\Pi_T(g) - \Pi_N^\delta(g)|] = O\ (\delta)$$

Proof :

$$\Pi_T(g) - \Pi_N^\delta(g) = \frac{1}{\tilde{\Pi}_T(1)} (\tilde{\Pi}_T(g) - \tilde{\Pi}_N^\delta(g) + \Pi_N^\delta(g)(\tilde{\Pi}_N^\delta(1) - \tilde{\Pi}_T(1)))$$

$$E [|\Pi_T(g) - \Pi_N^\delta(g)|] = \mathring{E} [L_T|\Pi_T(g) - \Pi_N^\delta(g)|]$$

$$= \mathring{E} [\tilde{\Pi}_T(1) \mid \Pi_T(g) - \Pi_N^\delta(g)|]$$

$$\leq \|\tilde{\Pi}_T(g) - \tilde{\Pi}_N^\delta(g)\|_1 + \|\Pi_N^\delta(g)\|_2 \|\tilde{\Pi}_N^\delta(1) - \tilde{\Pi}_T(1)\|_2$$

Therefore, it is sufficient to prove that $\|\Pi_N^\delta(g)\|_2$ is bounded ; define the probability : $P^\delta = L_N^\delta . \mathring{P}$

Then : $\Pi_N^\delta(g) = E_{P\delta} [g(X_T)|\underline{\underline{Y}}_T]$

If g is bounded, it is obviously bounded ; in the case (1.2a) :

$$\|\Pi_N^\delta(g)\|_2^2 = E_{P\delta} [\frac{1}{L_N^{\delta.a}} E_{P\delta} [g(X_T)|\underline{\underline{Y}}_T]^2]$$

$$\leq (E_{P\delta} [\frac{1}{(L_N^{\delta.a})^2}])^{1/2} (E_{P\delta} [E_{P\delta} [g(X_T)|\underline{\underline{Y}}_T]^4])^{1/2}$$

$$\leq (\mathring{E} [\frac{1}{L_N^{\delta.a}}])^{1/2} (E_{P\delta} [g(X_T)^4])^{1/2}$$

Now, one easily proves :

$$\mathring{E} [1/L_N^{\delta.a}] = E [\exp \sum h^2(X_{q\delta})\delta] \leq E [\exp (TH^2(X^*))]$$

$$E_{P\delta} [g(X_T)^4] = \mathring{E} [L_N^{\delta.a} g(X_T)^4] = \mathring{E} [g(X_T)^4] < \infty$$

□

5 - APPROXIMATIONS OF THE DISCRETE-TIME SIGNAL PROCESS :

Up to now, we have studied approximations of the filter which involve an integration with respect to the law of the process $(X_{q\delta}, 0 \leq q \leq N)$. Another step consists of approximating this law for the purpose of simulating it (Talay [6]). Since the discretization of Y induces an error of order δ, it is not worth choosing an approximation of X which induces a smaller error.

So we shall consider the Euler scheme :

$$\begin{cases} \overline{X}_o = X_o \\ \overline{X}_{q+1} = \overline{X}_q + b(\overline{X}_q)\delta + \sigma(\overline{X}_q)\ (W_{(q+1)\delta} - W_{q\delta}) \end{cases}$$

Then define : $\tilde{\Gamma}^\delta_k(g) = \overset{\circ}{E}{}^Y\ [g(\overline{X}_k)\ \overline{L}^\delta_k]$

where $\overline{L}^\delta_k$ is obtained by replacing $W_{q\delta}$ by $\overline{X}_q$ in one of the three definitions (1.2) of L^δ_k.

Theorem 2 : Suppose that σ, b, g and h are four times continuously differentiable and that their derivatives are bounded ; suppose also that h is bounded. Then :

$$\overset{\circ}{E}\ [(\tilde{\Pi}_T(g) - \tilde{\Gamma}^\delta_N(g))^2] = O(\delta^2)$$

Sketch of the proof (details are similar to [6]).

Like lemma 1, we can prove that it is sufficient to study the case (a) :

$$\overline{L}^\delta_k = \exp \sum_{q=o}^{k-1} (h(\overline{X}_q)\Delta Y_q - \frac{1}{2} h^2(\overline{X}_q)\delta)$$

If f(x,Y) is a real-valued function defined on $\mathbb{R} \times C$, define :

$$\mu f(x,Y) = \overset{\circ}{E}{}^Y\ [f(X_{(q+1)\delta},Y)\ |X_{q\delta} = x]$$

$$\overline{\mu} f(x,Y) = \overset{\circ}{E}{}^Y\ [f(\overline{X}_{q+1},Y)\ |\overline{X}_q = x]$$

One can prove that if f(x,Y) is four times continuously differentiable with respect to x, and if for every p, and $1 \leq i \leq 4$, $\|f^{(i)}(x,Y)\|_p$ has polynomial growth when x goes to infinity, then, for every p :

$$\|\mu f(x,Y) - \overline{\mu} f(x,Y)\|_p \leq R(x)\delta^2$$

where R is a polynomial which does not depend on δ.

Then define :

$$v_q(x,Y) = \overset{\circ}{E}{}^{Y} [g(X_T) \frac{L_N^\delta}{L_q^\delta} \quad | X_{q\delta} = x]$$

Some calculations prove that V_q satisfies the above conditions and that the moments of $\overline{X}_q$ are bounded, so :

(5.1) $$\|\mu v_{q+1}(\overline{X}_q,Y) - \bar{\mu} v_{q+1}(\overline{X}_q,Y)\|_p = O(\delta^2)$$

Now define : $Z_q = \overset{\circ}{E}{}^{Y} [\overline{L}_q^\delta \, v_q(\overline{X}_q,Y)]$

We remark :

$$Z_o = \tilde{\Pi}_N^\delta(g) \; ; \; Z_N = \tilde{\Gamma}_N^\delta(g) \quad \text{so :}$$

(5.2) $$\|\tilde{\Pi}_N^\delta(g) - \tilde{\Gamma}_N^\delta(g)\|_2 \le \sum_{q=o}^{N-1} \|Z_{q+1} - Z_q\|_2$$

Now let us study $Z_{q+1} - Z_q$:

$$v_q(x,Y) = \exp\,(h(x)\Delta Y_q - \frac{1}{2} h^2(x)\delta)\; \mu v_{q+1}(x,Y)$$

so : $$Z_q = \overset{\circ}{E}{}^{Y} [\overline{L}_{q+1}^\delta \, \mu v_{q+1}(\overline{X}_q,Y)]$$

(5.3) $$Z_{q+1} - Z_q = \overset{\circ}{E}{}^{Y} [\overline{L}_{q+1}^\delta \, (\bar{\mu} v_{q+1}(\overline{X}_q,Y) - \mu v_{q+1}(\overline{X}_q,Y))]$$

Therefore theorem 2 follows from (5.1), (5.2), (5.3) and theorem 1.

□

Then we can prove an analogue of corollary 2 and obtain, with $\Gamma_k^\delta = \tilde{\Gamma}_k^\delta / \tilde{\Gamma}_k^\delta(1)$:

$$E\,[|\Pi_T(g) - \Gamma_N^\delta(g)|] = O\,(\delta)$$

The filter Γ^δ can be computed by means of a Monte-Carlo method involving simulations of the Markov chain $\overline{X}_q$.

REFERENCES

[1] G. Kallianpur - Stochastic filtering theory, Applications of Mathematics 13, Springer 1980.

[2] H. Kunita - Stochastic differential equations and stochastic flows of diffeomorphisms, Ecole d'Eté de Probabilités de Saint-Flour XII, to appear in Lect. Notes in Math.

[3] E. Pardoux & D. Talay - Approximation and simulation of solutions of stochastic differential equations, to appear in Acta Applicandae Mathematicae.

[4] J. Picard - Problèmes d'approximation en filtrage stochastique, Thèse de 3ème cycle, Université Paris VI, 1982.

[5] C. Stricker & M. Yor - Calcul stochastique dépendant d'un paramètre, Z. Wahrsch. verw. Gebiete, 45, pp. 109-133, 1978.

[6] D. Talay - Efficient numerical schemes for the approximation of expectations of functionals of the solution of S.D.E. and applications, these Proceedings.

ON THE WEAK FINITE STOCHASTIC REALIZATION PROBLEM

G. Picci
CNR-LADSEB and
Istituto di Elettrotechnica
Università di Padova
Italy

J.H. van Schuppen
Mathematical Centre
Amsterdam
The Netherlands

ABSTRACT

The weak finite stochastic realization problem is given a stationary finite valued stochastic process to show existence of and to classify all minimal finite stochastic systems whose output equals the given process in distribution. In this paper the characterization of minimal realizations is investigated and reduced to a factorization problem for positive matrices. The latter problem is discussed and solved in a rather special case.

1. INTRODUCTION

The purpose of this paper is to present a problem formulation of the weak finite stochastic realization problem and to indicate the major current question for this problem: the positive factorization problem.

The weak finite stochastic realization problem is given a finite valued stationary stochastic process to show existence of and to classify all minimal finite stochastic systems whose output equals the given process in distribution. In contrast with this, the strong finite stochastic realization problem is to answer the same question but under the condition that the output process equals the given process almost surely.

The motivation of this problem is the area of control and prediction for systems with point process observations. Examples of practical problems in this area are the control of queues, the prediction of traffic intensities, the estimation of software reliability, and the estimation of certain biomedical signals. These practical problems may be modelled by finite stochastic systems. The prediction and control problems for this class of systems, under the assumption that the parameter values are known, have been considered. Practical application of these results demands the solution of the system identification problem and the stochastic realization problem for finite stochastic systems.

A brief description of the content of the paper follows. A problem formulation and a definition of a finite stochastic system is given in section 2. The characterization of minimal realizations, and the problem of positive factorization is discussed in section 3.

2. PROBLEM FORMULATION

Below a definition is given of a finite stochastic system and the weak finite stochastic realization problem is formulated.

Let (Ω,F,P) be a complete probability space and $T = Z$ be the time index set. The conditional independence relation for a triple of σ-algebra's F_1, F_2, G is defined to satisfy

$$E[x_1 x_2 | G] = E[x_1 | G]E[x_2 | G] \quad \text{a.s.}$$

for all $x_1 \in L^+(F_1)$, $x_2 \in L^+(F_2)$, notation $(F_1 G, F_2) \in CI$.
Here $L^+(F_1)$ is the set of all positive F_1 measurable random variables. The smallest σ-algebra with respect to which a random variable x is measurable is denoted F^x, and that containing the σ-algebra's G,H by $G \vee H$.

2.1. DEFINITION. A finite stochastic system is a collection $\sigma = \{\Omega,F,P,T,Y,B_y,X,B_x,y,x\}$ where $\{\Omega,F,P\}$ is a complete probability space, $T = Z$, Y,X are finite sets called respectively the output space and the state space, B_y, B_x are the σ-algebra's on Y,X generated by all subsets of Y,X, $y : \Omega \times T \to Y$, $x : \Omega \times T \to X$ are stochastic processes called respectively the output process and the state process, such that for all $t \in T$

$$(F_t^{y+} \vee F_t^{x+},\ F^{xt},\ F_t^{x-} \vee F_{t-1}^{y-}) \in CI,$$

where $F_t^{y+} = \sigma(\{y_s, \forall s \geq t\})$, $F_t^{x-} = \sigma(\{x_s, \forall s \leq t\})$.
Notation: $\sigma \in FS\Sigma$.

In a stochastic system one exhibits, besides the externally available output process, the underlying state process. The above defined system is called finite because Y,X are finite sets. The definition stated above has been first given in [7]. In the stochastic automata literature a finite stochastic system is called a stochastic automaton of Mealey-type [6].

2.2. PROBLEM. The weak finite stochastic realization problem (WFSRP) is given a stationary stochastic process on $T = Z$, with values in a finite set Y, and the class of finite stochastic systems, to solve the following subproblems:

a. does there exist a finite stochastic system

$$\sigma = \{\Omega,F,P,T,Y,B_y,X,B_x,y,x\} \in FS\Sigma$$

such that the output process y equals the given process in distribution?; if such a system exists, it is called a weak finite stochastic realization of the given process;

b. classify all minimal weak finite stochastic realizations, where minimal refers to the number of elements in the state space; this involves:

1. the characterization of minimal realizations;
2. the classification as such;
3. the relation between minimal realizations;
4. the construction of an algorithm that constructs all minimal realizations.

The above problem has first been posed by D. Blackwell and L. Koopmans [2], although in a somewhat different form. During the 1960's several contributions to the problem have been given, see the book by A. Paz [6] for references. However, little progress has been made on the problem.

The existence question of the weak finite stochastic realization problem has been solved. The solution is due to A. Heller [4]; see [7] for an alternative proof. However, there is no algorithm that allows one to construct a stochastic realization.

Of the classification subproblem of 2.2 little is known. Let's consider the first question: what are necessary and sufficient conditions for a weak finite stochastic realization to be minimal? This question is unsolved and difficult. To discuss it in more detail attention is restricted to the static problem; this is done in section 3.

3. THE POSITIVE FACTORIZATION PROBLEM

A major question in the weak finite stochastic realization problem is the characterization of minimal realizations. In this section a restricted case of this question is discussed, namely that in which there is only a past and a present and one is asked to construct a state. To be precise, one has the following problem.

3.1. PROBLEM. Assume given two finite sets Y^+, Y^-, and a frequency function $p_0 : Y^+ \times Y^- \to R_+$ (i.e. $\sum_{i \in Y^+, j \in Y^-} p_0(i,j) = 1$).

a. Does there exist a collection $\sigma = \{Y^+, Y^-, X, p\}$ such that
   1. $\sigma = \{Y^+, Y^-, X, p\} \in FP\Sigma$, where X is a finite set, $p : Y^+ \times X \times Y^- \to R_+$ is a frequency function, and with respect to the canonical variables one has $(F^{y^+}, F^X, F^{y^-}) \in CI$;
   2. the restriction of p to $Y^+ \times Y^-$ equals p_0.

   Then σ is called a probabilistic realization of $\{Y^+, Y^-, p_0\}$.
b. Classify all minimal probabilistic realizations of $\{Y^+, Y^-, p_0\}$.
   A probabilistic realization is called minimal if X has the smallest number of elements of all probabilistic realizations.

The existence problem in 3.1 is trivial. The major question is the characterization of minimal realization and their classification.

An equivalent condition for a probabilistic realization is needed and will be given below. First some notation is introduced. The set of the integers is denoted by Z, and for $n \in Z_+$, $Z_n = \{1,2,3,\ldots,n\}$. Furthermore, for $k \in Z_+$, R^k_n denotes the vector space with components in R_+, and $R_+^{k\times m}$ the set of positive matrices. For $k \in Z_+$, with $e_k^T = (1 \ \ 1 \ \ldots \ 1) \in R_+^k$,

$$S^k = \{x \in R_+^k \mid e_k^T x = 1\}$$

will be called the set of stochastic vectors. The set of stochastic matrices, column wise, is defined by

$$S^{k\times m} = \{Q \in R_+^{k\times m} \mid e_k^T Q = e_m^T\}.$$

Given a finite set Y, $\#(Y)$ denotes the number of elements in Y. If Y^+, Y^- are finite sets, $k = \#(Y^-)$, $p : Y^+ \times Y^- \to R_+$ is a frequency function, then define the conditional probability matrix $Q_{y^+|y^-} \in S^{k\times m}$ by

$$Q^{ij}_{y^+|y^-} = p(i,j)/p(j)$$

if this is well defined, and zero otherwise.

3.2. PROPOSITION. Assume given two finite sets Y^+, Y^- with $\#(Y^+) = k$, $\#(Y^-) = m$, and a frequency function $p_0 : Y^+ \times Y^- \to R_+$. The following statements are then equivalent:

a. there exists a probabilistic realization $\{Y^+, Y^-, X, p\} \in FP\Sigma$ with $\#(X) = n$;

b. there exist $n \in Z_+$, $Q_1 \in S^{k\times n}$, $Q_2 \in S^{n\times m}$ such that $Q_{y^+|y^-} = Q_1 . Q_2$;

c. there exist $n \in Z_+$, $\pi \in S^n$, and for all $i \in Z_n$ there exist $p_i \in S^k$, $r_i \in S^m$, such that

$$p_0(i,j) = \sum_{s=1}^n \pi_s \, p_s^i \, r_s^{jT}.$$

Proof. The elementary proof is omitted. □

The characterization 3.2.b., of a factorization of a stochastic matrix into the product of two stochastic matrices, will be used in the sequel. The proof of 3.2 shows that then $Q_1 = Q_{y^+|x}$, $Q_2 = Q_{x|y^-}$.

In the following attention will be restricted to the factorization of positive matrices. It is easily proven that if one has a factorization of a stochastic matrix in a product of two positive matrices, then one can modify the factorization into one with two stochastic matrices.

3.3. DEFINITION. Given $Q \in R_+^{k\times m}$.

a. A positive factorization of Q is a factorization of the form

$$Q = A . B$$

where $A \in R_+^{k\times n}$, $B \in R_+^{n\times m}$ for some $n \in Z_+$;

b. <u>A minimal positive factorization of Q</u> is

1. a positive factorization of Q, say

$Q = A\,.\,B$, with $A \in R_+^{k\times n}$, $B \in R_+^{n\times m}$;

2. if $Q = C\,.\,D$, $C \in R_+^{k\times n_1}$, $D \in R_+^{n_1\times m}$,

is any other positive factorization of Q, then $n_1 \geq n$.

One then calls n the positive rank of Q, notation $n = \text{pos-rank}(Q)$.

3.4. <u>PROBLEM</u>. The <u>minimal positive factorization problem</u> is, given $Q \in R_+^{k\times m}$,

a. to give necessary and sufficient conditions for a positive factorization of Q to be minimal;

b. to classify all minimal positive factorizations of Q;

c. to construct an algorithm that produces all minimal positive factorizations of Q.

3.5. <u>REMARKS</u>.

1. The minimal positive factorization problem is seen to be the restriction of the characterization of minimal weak finite stochastic realizations. This problem is therefore of interest.

2. The problem is also of interest to the area of positive systems, see [5, Chap.6]. The realization problem for positive systems reduces in the static case to the minimal positive factorization problem.

3. In the literature on linear algebra, see [1,3] for some references, the above defined problem is not mentioned. However it is known that the positive rank of a matrix differs in general from the linear rank. In the literature the problem had been posed to give necessary and sufficient conditions for these two rank concepts to be the same. In our opinion this question is uninteresting, the interesting problem being the minimal positive factorization problem.

4. A geometric interpretation of a positive factorization can be given as follows. One has a positive factorization $Q = A\,.\,B$ iff C_1 is contained in C_2, where

$$C_1 = \text{conv.(columns of } Q) \subset R_+^k$$
$$C_2 = \text{conv.(columns of } A),$$

and the right hand side denotes the convex hull generated by the denoted set. Further a positive factorization is minimal iff C_2 is spanned by as few vertices as possible. The minimal positive factorization problem may then be interpreted as the search for a polyhedral cone that contains C_1, lies in R_+^k and has as few vertices as possible.

Let $Q \in R_+^{k\times k}$. For $k = 1,2$, and 3 it is easily shown that the positive rank of Q equals the linear rank of Q. However this is not true for $k = 4$.

3.6. Example. Let

$$Q = \begin{pmatrix} 1 & 0 & 0 & 1 \\ 1 & 1 & 0 & 0 \\ 0 & 1 & 1 & 0 \\ 0 & 0 & 1 & 1 \end{pmatrix} \in R_+^{4\times 4}$$

Then lin - rank(Q) = 3 < 4 = pos - rank(Q).

That in example 3.6 the positive rank of Q is indeed four follows from the following result.

3.7. PROPOSITION. Let $Q \in R_+^{k\times k}$, and denote the columns of Q by q_i, $i \in Z_k$. Assume that

1. $k \geq 4$;
2. $\{q_i, i \in Z_k\}$ lie on different faces of $R_+^{k\times k}$.

Then pos - rank(Q) = k.

The proof is deferred to a future publication. The investigation of the minimal positive factorization problem is being continued.

4. CONCLUDING REMARKS

In this paper the weak finite stochastic realization has been posed. The current open problem is the characterization of minimal realizations. Through reduction it has been shown that this problem is equivalent to the minimal positive factorization problem. The latter problem is currently under investigation.

At the meeting at which this paper has been presented the strong finite stochastic realization problem has also been discussed; see [8] for an exposition.

REFERENCES

1. A. Berman, R.J. Plemmons, Nonnegative matrices in the mathematical sciences, Academic Press, New York, 1979.
2. D. Blackwell, L. Koopmans, On the identifiability problem for functions of finite Markov chains, Ann. Math. Statist., 28(1957)pp.1011-1015.
3. S.L. Campbell, G.D. Poole, Computing nonnegative rank factorizations, Linear Algebra and its Appl.35(1981),pp.175-182.
4. A. Heller, On stochastic processes derived from Markov chains, Ann. Math. Statist., 36(1965),pp.1286-1291.
5. D.G. Luenberger, Introduction to dynamic systems - Theory, models, and applications, Wiley, New York, 1979.
6. A. Paz, Introduction to probabilistic automata, Academic Press, New York, 1971.
7. G. Picci, On the internal structure of finite-state stochastic processes, in "Recent Developments in Variable Structure Systems, Economics, and Biology", Proc. of a U.S.-Italy Seminar, Taormina, Sicily, 1977, Lecture Notes in Econ. and Mathematical Systems, volume 162, Springer-Verlag, Berlin, 1978.
8. J.H. van Schuppen, The strong finite stochastic realization problem - Preliminary results, in "Analysis and Optimization of Systems", A. Bensoussan, J.L. Lions eds., Lecture Notes in Control and Info. Sci., volume 44, Springer-Verlag, Berlin, 1982, pp. 179-190.

CONTROLE LINEAIRE SOUS CONTRAINTE AVEC OBSERVATION PARTIELLE

Monique PONTIER
Université d'Orléans
45046 ORLEANS Cedex

Jacques SZPIRGLAS
CNET/PAA/MTI
38-40 rue du Gl Leclerc
92131 ISSY LES MOULINEAUX

I. INTRODUCTION.

On résout dans ce papier par des méthodes de l'analyse convexe (23) un problème de contrôle linéaire, avec observation partielle, soumis à une contrainte en espérance. On prolonge au cas réparti les idées exploitées dans le cas de l'arrêt optimal ((15),(19),(22)).

On traite le problème suivant : sur un espace de probabilité filtré $(\Omega, \underline{\underline{A}}, \underline{\underline{F}}, \mathbb{P})$ vérifiant les conditions habituelles (9), et sur un intervalle de temps fini $[0,T]$, la dynamique de l'état X d'un système et de son observation Y est décrite par les équations différentielles stochastiques linéaires :

$$(1.1) \qquad dX_t = (A_t X_t + B_t u_t)dt + C_t dV_t \; ; \quad X_o = x_o$$

$$dY_t = H_t X_t dt + G_t dW_t \; ; \quad Y_o = 0.$$

où X, Y et u sont des processus vectoriels de dimensions respectives n, m et k ; V et W sont deux mouvements browniens standard de $\mathbb{R}^\ell$ et $\mathbb{R}^r$; A, B, C, H et G sont des matrices, fonctions continues de t sur $[0,T]$, de dimensions respectives $n \times n$, $n \times k$, $n \times \ell$, $m \times n$ et $m \times r$; x_0 est une variable aléatoire gaussienne de $\mathbb{R}^n$, de moyenne m_o et de variance V_o, indépendante des mouvements browniens **V** et **W**.

Le processus u appartient à l'ensemble des "contrôles admissibles" $\underline{\underline{U}}$ défini, suivant (8), par l'ensemble des processus u tels que :

(1.2) $$u_t = K(t)Z_t,$$

avec Z solution de l'équation stochastique dans $\mathbb{R}^j$:

(1.3) $$dZ_t = \Gamma_t Z_t dt + \Delta_t dY_t \; ; \; Z_o = z_o,$$

où z_o est un vecteur de R^j ; K, Γ et Δ sont des matrices, fonctions continues de t sur $[0,T]$, de dimensions respectives $k \times j$, $j \times j$ et $j \times m$.

Remarque. On aurait pu prendre pour $\underline{\underline{U}}$, comme le fait remarquer Haussmann (14), l'ensemble des processus $u(t,\omega)$ adaptés à la filtration engendrée par le processus d'observation Y, et vérifiant :

$$|u(t,\omega)| \leqq K(1 + \sup_{t \in [0,T]} |Y_t(\omega)|)$$

ou tout autre ensemble $\underline{\underline{U}}$ conduisant au même contrôle optimal.

On se donne enfin les fonctions de coût suivantes :

(1.4) $$J(u) = E(\int_0^T (X_t^* Q_t X_t + u_t^* R_t u_t)dt + X_T^* F X_T),$$

(1.5) $$J'(u) = E(\int_0^T (X_t^* Q_t' X_t + u_t^* R_t' u_t)dt + X_T^* F' X_T),$$

où les matrices Q, Q', F, F', (respectivement R et R') sont des matrices symétriques définies positives de dimension $n \times n$ (resp. $k \times k$), fonctions continues de t sur $[0,T]$.

Le temps n'intervenant pratiquement pas dans les calculs, on l'omettra en général, par la suite.

Un réel positif ε étant donné, on pose :

$$a = \inf_{u \in \underline{\underline{U}}} J'(u) + \varepsilon$$

$$\underline{\underline{U}}^a = \left\{ u \in \underline{\underline{U}} \;/\; J'(u) \leq a \right\}$$

On cherche alors un contrôle u^*, dans $\underline{\underline{U}}^a$, qui minimise le coût J sous la contrainte que J' est majoré par a :

(1.6) $$J(u^*) = \inf_{u \in \underline{\underline{U}}^a} J(u).$$

C'est à dire que l'on cherche à minimiser le coût J dans l'ensemble des contrôles "ε - sous-optimaux" pour le problème sans contrainte associé au coût J'. On étend ainsi le cas avec observation complète (20).

Les problèmes de contrôle linéaire, avec observation partielle ou non, sont classiques. On peut consulter par exemple (1), (2), (4), (5), (8), (11), (13), (14), (17) ou (25). Nous nous référons ici le plus souvent à (8). Par ailleurs, les problèmes de contrôle stochastique avec contrainte ont été peu étudiés, en particulier du point de vue constructif. Ils concernent en général des contraintes trajectorielles sur le contrôle seul ((3),(13),(21)) ou des contraintes portant sur l'état final du système, en moyenne ((13),(16)) ou en probabilité ((6), (18)). Des problèmes d'arrêt optimal avec contrainte ont été aussi traités : (15), (19), (22),(24). Comme l'a fait FRID dans (12) pour le cas discret, nous proposons ici une contrainte portant sur l'ensemble de l'état du système et du contrôle, ainsi que sur l'état final du système.

De manière classique (23), les problèmes d'optimisation sous contrainte se ramènent à la recherche de point-selle d'un lagrangien. Soit donc le lagrangien $\underline{\underline{L}}$ défini sur $\underline{\underline{U}} \times \mathbb{R}^+$ par

$$\underline{\underline{L}}(u,p) = J(u) + p\,(J'(u) - a).$$

On est alors ramené à l'étude du problème sans contrainte associé au coût $J^p = J + pJ'$, qui est ici de même nature que J ou J'. Si u^p est une solution de ce dernier problème, on montre la continuité et la décroissance à l'infini vers $\inf_{u \in \underline{U}} J'(u)$ de l'application $p \to J'(u^p)$. Un tel résultat prouve l'existence en général d'un contrôle optimal pour le problème avec contrainte associée à un réel ε positif ou nul. La forme très particulière du problème entraîne qu'un contrôle optimal pour la contrainte associée à un ε nul est le contrôle optimal classique associé au problème sans contrainte de coût J'. On donne enfin, dans le cas où ε est strictement positif, un théorème constructif d'un contrôle "α - sous - optimal", pour tout réel α strictement positif.

II. METHODES LAGRANGIENNES

Sont rassemblés dans ce paragraphe des résultats généraux, issus des méthodes de l'analyse convexe (23), utiles à la résolution du problème, et qui ne font pas intervenir essentiellement la structure particulière du problème.

On sait qu'une condition suffisante pour un contrôle u* d'être solution d'un problème d'optimisation sous contrainte est l'existence d'un réel p*, positif ou nul, tel que (u*, p*) soit un point-selle du lagrangien $\underline{\underline{L}}$:

$$\underline{\underline{L}}(u,p) = J(u) + p(J'(u) - a) \tag{2.1}$$

c'est-à-dire que, pour tout contrôle u admissible et tout réel positif p, on a :

$$\underline{\underline{L}}(u^*,p) \leq \underline{\underline{L}}(u^*,p^*) \leq \underline{\underline{L}}(u,p^*). \tag{2.2}$$

On déduit immédiatement de ces inégalités la proposition suivante :

Proposition 2.1

Une condition suffisante pour qu'un contrôle u^* soit optimal pour le problème de contrôle avec contrainte posé en (1.6) est que l'une des propriétés suivantes soit vérifiées :

(i) $J(u^*) = \inf_{u \in \underline{U}} J(u)$ *et* $J'(u^*) \leqslant a$;

(ii) *Il existe un réel strictement positif p^* tel que : u^* est optimal pour le problème de contrôle "classique" associé au coût $J^{p^*} = J + p^*J'$ et $J'(u^*) = a$. (On dit alors que u^* "sature" la contrainte).*

Ces deux propriétés expriment simplement que (u^*, p^*) est un point-selle de $\underline{L}$, pour un multiplicateur de Lagrange respectivement nul ou strictement positif.

On est donc conduit à résoudre le problème de contrôle associé à l'équation différentielle (1.1) et au coût J^p, dont on note u^p une solution (qui existe, d'après (8), sous les hypothèses choisies). Si le contrôle u^0 ne vérifie pas la propriété(i)de(2.1), on recherche un réel p^* tel que $J'(u^{p^*}) = a$. Pour ce faire, on montre la proposition suivante :

Proposition 2.2.

Soit, pour tout p, un contrôle u^p optimal pour le problème sans contrainte associé au coût J^p. Alors, l'application définie sur $\mathbb{R}^+$ par :

$$p \to J'(u^p)$$

est décroissante, de limite $\inf_{u \in \underline{U}} J'(u)$ *à l'infini.*

<u>Démonstration</u> : Par définition de u^p, on a pour tout u :

(2.3) $\quad J(u^p) + p\, J'(u^p) \leqslant J(u) + p\, J'(u)$.

En particulier, pour le contrôle u^{p+q}, avec q positif, on a :

$$J(u^p) + p\, J'(u^p) \leqslant J\,(u^{p+q}) + p\, J'\,(u^{p+q}).$$

Inversant dans cette inégalité les rôles de p et p + q, et par transitivité, on obtient :

$$J'(u^{p+q}) \leqslant J'(u^p)$$

Soit ensuite u' un contrôle optimal pour le problème associé au coût J' ; lui appliquant l'inégalité (2.3), il vient :

$$J'(u') \leqslant J'(u^p) \leqslant J'(u') + \frac{J(u') - J(u^p)}{p} \leqslant J'(u') + \frac{J(u')}{p}$$

On peut alors faire tendre p vers l'infini pour conclure la démonstration :

$$\lim_{p \to \infty} J'(u^p) = J'(u') = \inf_{u \in \underline{\underline{U}}} J'(u).$$

III. EXISTENCE D'UN CONTROLE OPTIMAL

Après avoir rappelé quelques résultats sur le contrôle linéaire quadratique, sans contrainte, avec observation partielle, on montre la continuité de l'application :

$$p \to J'(u^p)$$

où u^p est une solution au problème de contrôle linéaire associé au coût J + pJ'. On en déduit l'existence d'un contrôle optimal, sous la contrainte associée à un ε strictement positif. On trouve enfin une solution au cas où le réel ε est nul en faisant tendre p vers l'infini.

La proposition (2.1) conduit naturellement à étudier le problème de contrôle, sans contrainte, associé aux équations différentielles (1.1) et au coût quadratique $J^p = J + pJ'$. D' après (8), ce problème, avec observation partielle, admet pour solution dans $\underline{\underline{U}}$, défini en (1.2) et (1.3), le contrôle "feedback" suivant :

(3.1) $\qquad u^p(t) = -(R + pR')^{-1} B^* \Lambda^p \hat{X}_t^p ,$

où Λ^p est une matrice symétrique, de dimension $n \times n$, solution uniformément bornée sur $[0,T]$ non forcément unique, de l'équation de Riccati :

(3.2) $$- \dot{\Lambda} = Q + pQ' + \Lambda A + A^*\Lambda - \Lambda B(R + pR')^{-1} B^*\Lambda ,$$
$$\Lambda(T) = F + pF'.$$

et le processus $\hat{X}^p$ est défini par :

(3.3) $$\hat{X}_t^p = E(X_t^p / \underline{\underline{G}}_t)$$

où $\underline{\underline{G}}$ est la filtration des observations Y, tandis que les processus X^p et Y sont solution des équations (1.1) où u est remplacé par l'expression (3.1).
Suivant (8), on voit alors que le processus $\hat{X}^p$ est solution de l'équation différentielle :

(3.4) $$d\hat{X} = A^p \hat{X} dt + PH^*(GG^*)^{-1} d\nu_t \quad ; \quad \hat{X}_0 = x_0$$

avec $\qquad A^p = A - B(R + pR')^{-1} B^* \Lambda^p , \quad \nu_t = Y_t - \int_0^t H\hat{X} ds$, mesure d'innovation et P solution de l'équation de Riccati - ne dépendant pas du paramètre p :

(3.5) $$\dot{P} = CC^* - PH^*(GG^*)^{-1} HP + AP + PA^* \quad ; P_0 = V_0$$

Notons aussi que $\tilde{X} = X^p - \hat{X}^p$ est indépendant du contrôle u^p, donc de p.

Avant de montrer la continuité de l'application $p \to J'(u^p)$, il est nécessaire de particulariser la matrice Λ^p et de montrer :

<u>*Lemme 3.1*</u> *On peut choisir pour solution de (3.2) une matrice* Λ_t^p *symétrique, continue en (p,t) sur* $\mathbb{R}^+ \times [0,T]$.

<u>Démonstration</u> : D'après (8), on peut choisir

$$\Lambda_t^p = S_{T-t}^p$$

où S^p est la matrice de covariance de l'erreur d'un filtre de Kalman, et unique solution de l'équation de Riccati :

(3.6) $$\dot{S} = Q + pQ' + A^*S + SA - SB(R + pR')^{-1} B^*S$$
$$S_o = F + pF'$$

De plus, S_t^p est uniformément bornée sur $[0,B] \times [0,T]$ pour tout B réel positif. Alors, les théorèmes généraux sur les équations différentielles (7) s'appliquent et démontrent le lemme.

On peut maintenant montrer la régularité de l'application :

$$p \to J'(u^p)$$

Proposition 3.2

Si u^p est le contrôle défini par les équations (3.1) à (3.5), l'application $p \to J'(u^p)$ est continue sur $\mathbb{R}^+$.

Démonstration. On évalue d'abord $J'(u^p)$, suivant par exemple (8), en remplaçant u^p par l'expression (3.1) dans la définition de J' :

$$J'(u^p) = E(\int_o^T \hat{X}^{p*} M^p \hat{X}^p dt + \hat{X}_T^{p*} F'\hat{X}_T^p) + E(\int_o^T \tilde{X}^* Q'\tilde{X}dt + \tilde{X}_T^* F'\tilde{X}_T).$$

avec

(3.7) $$M^p = Q' + \Lambda^p B(R + pR')^{-1}R'(R + pR')^{-1}B^*\Lambda^p.$$

En effet, puisque X^p se décompose en la somme de deux termes orthogonaux, $\hat{X}^p$ et $\tilde{X}$, on a pour toute matrice M :

$$E(X_t^{p*}MX_t) = E(\hat{X}_t^{p*}M\hat{X}_t^p) + E(\tilde{X}_t^* M\tilde{X}_t)$$

Remarquons ensuite que dans cette expression de $J'(u^p)$, les deux derniers termes ne dépendent pas du paramètre p. Par ailleurs, du fait que $E(\tilde{X}_t^* \tilde{X}_t)$ est P_t matrice de covariance de l'erreur solution de (3.5) , on peut les réécrire :

(3.8) $$\text{Trace } \left(\int_0^T Q'Pdt + F'P_T\right).$$

Quant aux premiers termes, on les évalue en suivant les calculs de (8) pour le contrôle complètement observable et y remplaçant X par $\hat{X}$. Ils s'écrivent alors :

(3.9) $$m_o^* N_o^p m_o + \text{trace } (V_o N_o^p + \int_0^T D_u^* N_u^p D_u du),$$

avec D_u matrice indépendante de p, m_o et V_o moyenne et variance de la variable initiale x_o, et N^p matrice symétrique solution de l'équation :

(3.10) $$\dot{N} + A^{p*}N + NA^p + M^p = o \ ; \ N(T) = F'.$$

Rassemblant (3.8) et (3.9) on obtient enfin :

(3.11) $$J'(u^p) = m_o^* N_o^p m_o + \text{trace } (V_o N_o^p + \int_0^T D_u^* N_u^p D_u \, du + \int_0^T Q'Pdt + F'P_T),$$

où N^p est solution de (3.10) et P solution de (3.5).

La continuité de $p \to J'(u^p)$ est alors conséquence de la régularité en t et p de la matrice Λ_t^p, montrée dans le lemme (3.1). En effet, celle-ci entraîne la continuité en (p,t) des matrices $M^p(t)$ définie en (3.7) et $A^p(t)$ définie en (3.4). Les résultats généraux de (7) vont alors encore s'appliquer à la solution N^p de l'équation différentielle linéaire (3.10). La continuité de N^p permet enfin de conclure, d'après l'expression (3.9), à la continuité de $p \to J'(u^p)$.

<u>Remarque.</u> L'expression de $J^p(u^p)$ peut s'évaluer par les mêmes calculs ; mais dans (3.9), N^p est remplacée par Λ^p, solution de (3.2). On a alors également la continuité des applications

$$p \to J^p(u^p) \quad \text{et} \quad p \to J(u^p).$$

Dans le cas où ε est strictement positif, l'existence d'un contrôle optimal se déduit aisément de cette continuité :

Corollaire 3.3

Pour tout ε strictement positif, il existe un contrôle optimal pour le problème avec contrainte défini en (1.6), associé au réel ε.

Démonstration. Si le contrôle u^o défini par :

$$u^o = -R^{-1}B^* \Lambda^o X^o$$

vérifie la contrainte :

$$J'(u^o) \leqslant \inf_{u \in \underline{\underline{U}}} J'(u) + \varepsilon = a$$

il est solution du problème (1.6), d'après la proposition (2.1). Sinon, grâce à la continuité de $J'(u^p)$ et sa décroissance à l'infini vers $\inf_{u \in \underline{\underline{U}}} J'(u)$ il existe un réel p* tel que :

$$J'(u^{p^*}) = a$$

Le contrôle u^{p^*} est alors optimal (propriété (ii) de (2.1) et le coût associé est donné par :

$$(J + p^*J')(u^{p^*}) - p^*a$$

Le problème pratique est bien sûr que l'on ne connaît que des valeurs approchées de p*, ce qui conduira, dans la dernière partie de ce travail, à construire une solution "α-sous-optimale" pour tout α strictement positif.

Terminons cette partie avec le cas où ε est nul. Notons comme dans la partie 2 par u' le contrôle optimal associé aux équations différentielles (1.1) et au coût J' :

(3.12) $$u' = -R'^{-1} B^* \Lambda' \widehat{X}'$$

avec Λ' solution de l'équation de Riccati :

$$(3.13) \qquad -\dot{\Lambda}' = Q' + \Lambda' A + A^* \Lambda' - \Lambda' B R'^{-1} B^* \Lambda' \; ; \; \Lambda'(T) = F'$$

et le processus $\hat{X}'$ solution de l'équation :

$$(3.14) \qquad d\hat{X}' = (A - R'^{-1} B^* \Lambda')\hat{X}' dt + PH^*(GG^*)^{-1} d\nu_t \; ; \; \hat{X}'_0 = x_0$$

Proposition 3.4

Le contrôle u' défini par les formules (3.12) à (3.14) est optimal pour le problème de contrôle avec contrainte associée au réel ε nul, c'est-à-dire que :

$$J(u') = \inf \left\{ J(u),\ u \in \underline{\underline{U}} \ / \ J'(u) = \inf_{v \in \underline{\underline{U}}} J'(v) \right\}.$$

Remarque. Le contrôle u' réalisant inf J'(v) n'est, à priori, pas unique et donc cette proposition n'est pas triviale.

Démonstration. Remarquons d'abord que, par définition, u' vérifie la contrainte. Montrons ensuite que :

$$J(u') = \inf \left\{ J(u),\ u \in \underline{\underline{U}} \ / \ J'(u) = \inf_{v \in \underline{\underline{U}}} J'(v) \right\}$$

En effet, par définition de u^p et sur l'ensemble des u tels que :

$$J'(u) = \inf_{v \in \underline{\underline{U}}} J'(v),$$

on a :

$$J(u^p) \leqslant J(u) + p(J'(u) - J'(u^p)) \leqslant J(u)$$

Il suffit alors pour conclure de prouver :

$$\lim_{p \to \infty} J(u^p) = J(u')$$

Pour montrer cette convergence, on étudie la continuité de $J(u^p)$ à l'infini en considérant le problème de contrôle associé aux équations (1.1) et au coût J'+qJ (c'est-à-dire que l'on échange les rôles de J et J'). Un contrôle v^q optimal pour ce problème est donné par :

(3.15) $$v^q = -(R' + qR)^{-1} B^* L^q \hat{X}'^q ,$$

où $L^q_t = P^q_{T-t}$, P^q étant l'unique solution d'une équation de Riccati analogue à (3.6) échangeant J et J', et $\hat{X}'^q$ l'unique solution trajectorielle d'une équation analogue à (3.4). Il est alors facile de voir que pour tout réel q strictement positif :

$$v^q = u^{1/q}, \hat{X}'^q = \hat{X}^{1/q}, L^q = q \Lambda^{1/q}$$

lorsque u, $\hat{X}$ et Λ sont définis par les expressions (3.1) à (3.4), homogènes en p.

La proposition 3.2. s'applique à $J(v^q)$ et l'on a, utilisant que v^o n'est autre que u' :

$$\lim_{q \to o} J(v^q) = \lim_{p \to \infty} J(u^p) = J(u')$$

ce qui termine la démonstration.

Remarque. Les mêmes calculs que ceux conduisant à l'expression de $J'(u^p)$ (3.11) permettent d'évaluer le coût optimal J(u') :

(3.16) $$J(u') = m_o^* N'_o m_o + \text{trace } (V_o N'_o + \int_o^T (D^* N' D + QP) dt + F P_T)$$

où N' est l'unique solution de l'équation différentielle linéaire :

(3.17) $$\dot{N}' + A'^* N + NA' + M' = 0 \ ; \quad N'(T) = F$$

avec $A' = A - B R'^{-1} B^* \Lambda'$

et $M' = Q + \Lambda' B R'^{-1} R R'^{-1} B^* \Lambda'$.

IV. SOLUTION APPROCHEE.

Dans le cas où ε est strictement positif, la détermination d'un réel p^* tel que $J'(u^{p^*}) = a$ ne peut se faire exactement, mais on peut l'approcher par dichotomie ou, plus généralement, par discrétisation du paramètre p. La proposition suivante donne un intervalle maximum de discrétisation et la construction d'une solution approchée :

Proposition 4.1

Pour tout ε strictement positif, il existe un réel p_o qui majore p^* :

$$p_o = J(u')/\ \varepsilon,$$

où $J(u')$ est donné par (3.16) et (3.17). De plus, pour tout α strictement positif, on peut construire une suite de réels (p_n) convergeant vers p^* telle que pour un rang assez grand N, et si $u_\alpha = u^{P_N}$:

$$J'(u_\alpha) \leqslant a = \inf_{u \in \underline{\underline{U}}} J(u) + \varepsilon$$

$$J(u_\alpha) \leqslant J(u) + \alpha$$

pour tout contrôle u vérifiant $J'(u) \leqslant a$.

C'est à dire que le contrôle u_α est "α-sous-optimal" pour le problème avec contrainte associée au réel ε.

Démonstration. Soit le point-selle (p^*,u^{p^*}) et le contrôle u', défini par (3.12) à (3.14) qui réalise l'infimum de J' dans $\underline{\underline{U}}$. On a alors :

$$J(u^{p^*}) + p^*\ J'(u^{p^*}) \leq J(u') + p^*\ J'(u').$$

Comme u^{p^*} "sature" la contrainte en ε, il vient :

$$J(u^{p^*}) + p^*\ \varepsilon \leq J(u'),$$

ce qui montre bien la majoration de p^*.

Ensuite, l'application $p \to J'(u^p)$ étant continue et décroissante vers $J'(u')$ on peut, partant de p_o, construire une suite alternante (p_n) convergeant vers p^*. Pour α strictement positif, soit N le premier entier tel que :

$$a - \alpha/p_o < J'(u^{p_N}) \leqslant a.$$

S'il y a égalité, p_N est p^* ; sinon $u_\alpha = u^{p_N}$ est un contrôle admissible vérifiant la contrainte et α-sous-optimal ; en effet, on a pour tout contrôle u :

$$J(u_\alpha) + p_N J'(u_\alpha) \leqslant J(u) + p_N J'(u),$$

soit encore :

$$J(u_\alpha) \leqslant J(u) + p_N(J'(u) - J'(u_\alpha)).$$

Pour les contrôles vérifiant la contrainte, deux cas sont possibles, selon que $J'(u)$ est plus petit ou plus grand que $a - \alpha/p_o$. Pour les premiers, $J'(u)$ étant majoré par $J'(u_\alpha)$, on a nécessairement :

$$J(u_\alpha) \leqslant J(u)$$

Pour les seconds, $J'(u)$ et $J'(u_\alpha)$ appartiennent à l'intervalle :

$$[a - \alpha/p_o,\ a]$$

et leur différence est majorée par α/p_o :

$$J(u_\alpha) \leqslant J(u) + p_N \cdot \alpha/p_o \leqslant J(u) + \alpha,$$

ce qui achève la démonstration.

CONCLUSION.

Ainsi a-t-on montré, pour une contrainte associée à un réel ε positif ou nul, l'existence d'un contrôle optimal. On a de plus une forme explicite dans le cas où ε est nul. Sinon, on construit effectivement une solution approchée. Il faut souligner que la simplicité des résultats repose sur le caractère linéaire quadratique du problème. Néanmoins, on peut noter que tous les résultats de la deuxième partie n'utilisent pas cette structure particulière, et donc s'étendent sans peine à des cas plus généraux de contrôle stochastique réparti en utilisant, par exemple, la modélisation et les théorèmes généraux d'existence de contrôles optimaux de N.El KAROUI (10).

BIBLIOGRAPHIE

(1) *A. BENSOUSSAN, J.L. LIONS* : "Applications des inéquations variationnelles en contrôle stochastique", Dunod 1978.

(2) *J.M. BISMUT* : "Conjugate convex functions in optimal stochastic control", J.Math.Anal. Applic. 44(1973).384-404.

(3) *J.M. BISMUT* :"An example of optimal control with constraints", SIAM J. Control, vol. 12 n°3 (1974), 401-418.

(4) *J.M. BISMUT* : "Linear quadratic optimal stochastic control with random coefficients", SIAM J. Control Optim, vol. 14 n°3 (1976), 419-444.

(5) *J.M. BISMUT* : "On optimal control of linear stochastic equations with a linear quadratic criterion", SIAM J. Control Optim, vol. 15 n°1 (1977), 1-4.

(6) *N. CHRISTOPEIT* : "A stochastic control model with chance constraints", SIAM J. Control Optim , vol. 16 n°5 (1978), 702-714.

(7) *E.A. CODDINGTON, N.LEVINSON* : "Theory of ordinary differential equations", Mc Graw-Hill, New York, 1955.

(8) *M.H.A. DAVIS* : "Linear estimation and stochastic control", Chapman an Hall, Londres,1977.

(9) *C. DELLACHERIE, P.A. MEYER* : "Probabilités et potentiels", tome 1 (1975), tome 2 (1980), Hermann.

(10) *N. EL KAROUI* : "Les aspects probabilistes du contrôle stochastique", Ecole d'été de St.Flour IX-1979, Lect. Notes in Math. n°876, Springer-Verlag, 1981.

(11) *W.H. FLEMING, R.W. RISHEL* : "Deterministic and stochastic optimal control", Springer-Verlag, 1975.

(12) *E.B. FRID* : "On optimal strategies in control problems with constraints", Theory Prob. Appl. vol.XVII n°1 (1972), 188-192.

(13) *U.G. HAUSSMANN* : "Some example of optimal stochastic controls or : the stochastic maximum principle at work", SIAM Review, vol. 23 n°3 (1981), 292-307.

(14) *U.G. HAUSSMANN* : "Optimal control of partially observed diffusions via the separation principle", Lect.Notes in Control and Inf. Sc., n° 43, 302-311.

(15) *D.P. KENNEDY* : "On a constrained optimal stopping problem", J. Appl. Prob. 19 (1982) 631-642.

(16) *H.J. KUSHNER* : "On the stochastic maximum principle with average constraints", J. of Math.Anal. Applic. 12(1965), 13-26.

(17) *R.S. LIPTSER, A.N. SHIRYAYEV* : "Statistic of random processes" tomes 1 et 2, Springer-Verlag 1977.

(18) *M.K. OZGOREN, R.W. LONGMAN, C.A. COOPER* : "Probabilistic inequality constraints in stochastic optimal control theory", J. of Math. Anal. Applic. 66 (1978), 237-259.

(19) *M.PONTIER, J.SZPIRGLAS* : "Arrêt optimal avec contrainte", C.R.A.S. Série A, 17 février 1983.

(20) *M. PONTIER, J.SZPIRGLAS* : "Linear control with constraint", à paraître.

(21) *J.P. QUADRAT* : "Existence de solution et algorithme de résolution numérique de problème de contrôle optimal de diffusion stochastique dégénérée ou non", SIAM control optimization, vol. 18 n°2 (1980), 199-266.

(22) *M.ROBIN* : "On optimal stochastic control problems with constraints", Game theory and related topics, North-Holland, (1979), 187-202.

(23) *R.T. ROCKAFELLAR* : "Convex analysis", Princeton university Press, 1970.

(24) *A.N. SHIRYAYEV* : "Optimal stopping rules", Appl. of Math. n° 8, Springer-Verlag, 1977.

(25) *W.M. WONHAM* : "Linear multivariate control, a geometric approach". L.N. in Economic and Math Systems n°101, Springer-Verlag, 1974.

QUELQUES REMARQUES SUR LES SEMIMARTINGALES GAUSSIENNES ET LE PROBLEME DE L'INNOVATION

C. STRICKER

Cet exposé reprend pour l'essentiel les résultats de notre article [17] mais les démonstrations ont été simplifiées grâce à des discussions fructueuses avec P.A. Meyer et L. Schwartz.

Dans la première partie, nous montrons qu'une semimartingale gaussienne est spéciale et que sa décomposition canonique dans sa filtration naturelle est encore gaussienne. Pour définir la notion de semimartingale, nous adoptons ici le point de vue des mesures vectorielles.

Dans la deuxième partie, nous étudions le problème de l'innovation, notamment dans le cas gaussien.

1 - SEMIMARTINGALES GAUSSIENNES ET MESURES VECTORIELLES.

Afin d'éviter des localisations fastidieuses nous supposerons que l'ensemble d'indice des divers processus et filtrations est $[0,1]$. Soient $(\Omega,\mathcal{F},P)$ un espace probabilisé complet, X un processus gaussien adapté à une filtration $(\mathcal{F}_t)$ possédant la propriété suivante $(*)$: il existe un espace gaussien $\mathcal{H}$ contenant toutes les variables aléatoires $E[X_t|\mathcal{F}_s]$ lorsque s et t parcourent $[0,1]$. La propriété $(*)$ est vérifiée notamment si X est un processus gaussien et si $(\mathcal{F}_t)$ est sa filtration naturelle. On désigne par $\mathcal{E}$ l'ensemble des processus H de la forme : $\sum_{k=1}^{n} H^k 1_{]t_k,t_{k+1}]}$ où H^k est $\mathcal{F}_{t_k}$-mesurable. L'intégrale H.X pour $H \in \mathcal{E}$ est définie de manière classique par : $H.X = \sum_{k=1}^{n} H^k\left(X_{t_{k+1}} - X_{t_k}\right)$. A l'heure actuelle, il existe deux points de vue équivalents de la notion de semimartingale. Dans [15], un processus Y est une semimartingale si $Y = M+A$ où M est une martingale locale et A un processus càdlàg, adapté, à variation finie. Plus récemment, Métivier et Pellaumail [16], puis Dellacherie [5] et Bichteler [2] se sont intéressés à l'aspect mesure vectorielle : un processus càdlàg Y est une semimartingale si l'application $H \to H.Y$ de $\mathcal{E}$ dans L^0 se prolonge en une mesure vectorielle sur la tribu prévisible $\mathcal{P}$ à valeurs dans L^0.

Dans [19] on pourra trouver une démonstration simple de l'équivalence des deux notions de semimartingale. Pour que le processus càdlàg Y soit une semimartingale, il suffit en fait que l'ensemble $K = \{H.Y, |H| \leq 1 \text{ et } H \in \mathcal{E}\}$ soit borné dans L^o. C'est ce point de vue que nous allons adopter ici. Voici le résultat principal de ce paragraphe :

THEOREME 1 : Soit X un processus vérifiant la propriété (*). Si $K = \{H.X, |H| \leq 1 \text{ et } H \in \mathcal{E}\}$ est borné dans L^o, K est aussi borné dans L^2. En outre si la filtration $(\mathcal{F}_t)$ vérifie les conditions habituelles et si le processus X est càdlàg, alors $X = M + A$ où M est une martingale de carré intégrable appartenant à $\mathcal{H}$ et A un processus prévisible à variation de carré intégrable, appartenant aussi à $\mathcal{H}$.

Avant de passer à la démonstration du théorème 1, faisons quelques remarques. Comme K est borné dans L^2, X est en particulier une quasimartingale et on est alors ramené à la situation étudiée par Jain et Monrad [13] car tout processus admet une modification séparable. Toutefois, par souci de complétude, nous démontrerons aussi la deuxième partie du théorème 1 en adaptant la méthode proposée dans notre article [19].

Pour démontrer le théorème 1, nous commençons par centrer notre processus X grâce à la proposition suivante :

PROPOSITION 1 : La fonction $t \mapsto E[X_t]$ est à variation bornée.

DEMONSTRATION : Posons $E[X_t] = m(t)$. Pour toute fonction étagée continue à gauche f(t) sur l'intervalle [0,1], définissons les intégrales élémentaires (se réduisant à des sommes finies) $I(f) = \int_o^1 f(s)\,dm(s)$ et $(f.X)_1 = \int_o^1 f(s)\,dX_s$; la première est l'espérance de la seconde, qui est une v. a. gaussienne. Par hypothèse $K' = \{(f.X)_1, \text{ f étagée et } |f| \leq 1\}$ est un sous-ensemble borné dans L^o, donc dans L^1 puisqu'il est contenu dans l'espace de Hilbert gaussien $\mathcal{H}$. Ainsi $\{I(f), \text{ f étagée et } |f| \leq 1\}$ est borné ; donc m est à variation bornée.

Retranchant la fonction m_t, <u>nous supposons désormais que X est une semimartingale gaussienne centrée</u>. Nous ne restreignons pas la généralité en supposant aussi que $X_o = 0$.

Nous utiliserons à plusieurs reprises un lemme dû à Fernique, dont on trouvera la démonstration dans [7], pp. 8-13. Soit Δ l'ensemble des nombres rationnels dyadiques de [0,1], et soit N une <u>pseudoseminorme</u> borélienne sur $\mathbb{R}^\Delta$, c'est-à-dire une fonction borélienne positive sur $\mathbb{R}^\Delta$, <u>non nécessairement finie</u>, mais satisfaisant à tous les autres axiomes définissant les seminormes. Soit X un processus <u>gaussien centré</u> indexé par Δ ; la fonction N(X) est une variable aléatoire, et l'on a

LEMME 1 : Avec les notations précédentes, ou bien $N(X) = +\infty$ p. s., ou bien il existe $\varepsilon > 0$ tel que $E\left[\exp\left(\varepsilon N^2(X)\right)\right] < +\infty$.

Dans nos applications, en fait X sera indexé par $[0,1]$, et nous conviendrons de noter $N(X)$ la composée de N avec la restriction de la trajectoire à Δ. Nous nous bornerons même à expliciter cette v. a., en laissant au lecteur le soin de décrire la pseudoseminorme utilisée.

On désigne par Λ l'ensemble des subdivisions dyadiques de $[0,1]$ et pour $\lambda = (t_i)$, avec $0 = t_o < t_1 < \ldots < t_k = 1$, on pose : $Q_\lambda(X) = \sum_{i=0}^{k-1} \left(X_{t_{i+1}} - X_{t_i}\right)^2$.

PROPOSITION 2 : L'ensemble $\{Q_\lambda(X),\ \lambda \in \Lambda\}$ est borné dans L^1.

DEMONSTRATION : Montrons d'abord que $X^* = \sup\operatorname{ess}_{t\in[0,1]} |X_t| < +\infty$. Sinon il existerait un ensemble dénombrable D et $\varepsilon > 0$ tels que $P[\sup_{t\in D} |X_t| = +\infty] > \varepsilon$. On pose $T_n = \inf\{t \in \Delta,\ |X_t| \geq n\}$ où Δ est une partie finie de D assez riche pour que $P[|X_{T_n}| \geq n] > \varepsilon$. Or $X_{T_n} - X_o = 1_{]0,T_n]}.X$ appartient à K, ce qui est contradictoire avec l'hypothèse du théorème 1. Ainsi X^* est fini p. s. Comme K est borné dans L^o, ceci entraîne que l'ensemble des v. a. $H_\lambda(X) = \sum_{i=0}^{k-1} X_{t_i}\left(X_{t_{i+1}} - X_{t_i}\right)$ est aussi borné dans L^o. Mais

$$Q_\lambda(X) = \sum_{i=0}^{k-1} \left(X^2_{t_{i+1}} - X^2_{t_i}\right) - 2H_\lambda(X) = X_1^2 - X_o^2 - 2H_\lambda(X),$$

si bien que $\{Q_\lambda(X),\ \lambda \in \Lambda\}$ est borné dans L^o. Si cet ensemble n'était pas borné dans L^1, il existerait une suite (λ_n) de Λ et des nombres réels a_n tels que $N(X) = \sup_n \left(Q_{\lambda_n}(X)/a_n\right)^{1/2}$ soit fini et que $E[N(X)] = +\infty$, ce qui est absurde d'après le lemme de Fernique. Ainsi $\{Q_\lambda(X),\ \lambda \in \Lambda\}$ est borné dans L^1 et même dans tous les L^p.

Remarquons maintenant que si (H^i) est une suite de variables aléatoires $\mathcal{F}_{t_i}$-mesurables bornées par 1, alors :

$$E\Big[\sum_{i=0}^{k-1} H^i\left(X_{t_{i+1}} - X_{t_i} - E[X_{t_{i+1}} - X_{t_i}|\mathcal{F}_{t_i}]\right)\Big]^2$$

$$= E\Big[\sum_{i=0}^{k-1} (H^i)^2\left(X_{t_{i+1}} - X_{t_i} - E[X_{t_{i+1}} - X_{t_i}|\mathcal{F}_{t_i}]\right)^2\Big]$$

$$\leq 2\,E[Q_\lambda(X)].$$

Prenons en particulier pour H^i le signe de $E[X_{t_{i+1}} - X_{t_i}|\mathcal{F}_{t_i}]$. Comme $K = \{H.X,\ H \in \mathcal{E}$ et $|H| \leq 1\}$ est borné dans L^o, nous en déduisons par différence que $E = \{\sum_{i=0}^{k-1} |E[X_{t_{i+1}} - X_{t_i}|\mathcal{F}_{t_i}]|,\ \lambda \in \Lambda\}$ est aussi borné dans L^o. Une nouvelle application du lemme de Fernique entraîne que E est aussi borné dans L^2. En décomposant $H.X$ comme précédemment, nous obtenons que K est borné dans L^2.

Il nous reste maintenant à montrer que la décomposition de X est gaussienne. Les variables aléatoires $M_1^\lambda = \sum_\lambda X_{t_{i+1}} - X_{t_i} - E[X_{t_{i+1}} - X_{t_i} | \mathcal{F}_{t_i}]$ forment un sous-ensemble borné de l'espace de Hilbert gaussien $\mathcal{H}$. Il existe alors une suite (λ_n) de subdivisions dyadiques de plus en plus fines telles que $(M_1^{\lambda_n})$ converge faiblement vers une variable M_1 appartenant à $\mathcal{H}$. Posant $M_t = E[M_1 | \mathcal{F}_t]$ et $A_t = X_t - M_t$, on note que pour t dyadique, $E[M_1^{\lambda_n} | \mathcal{F}_t]$ converge faiblement vers M_t. En outre si (s_k) est une subdivision dyadique de $[0,1]$, si $\lambda_n^k = [s_k, s_{k+1}] \cap \lambda_n$ et si (H_k) est une suite de variables aléatoires $\mathcal{F}$-mesurables, bornées par 1, on a pour toute v. a. Y vérifiant $\|Y\|_{L^2} \leq 1$:

$$E[Y \sum_k H_k (A_{s_{k+1}} - A_{s_k})] = \lim_{n\to\infty} E[Y \sum_k H_k \sum_{\lambda_n^k} E[X_{t_{i+1}} - X_{t_i} | \mathcal{F}_{t_i}]]$$

$$\leq \sup_n \left(E[\sum_{\lambda_n} |E[X_{t+1} - X_{t_i} | \mathcal{F}_{t_i}]|]^2\right)^{1/2}.$$

Ce sup étant fini, A est à variation de carré intégrable. Ainsi nous avons montré que $X = M + A$ où M est une martingale, A un processus prévisible à variation de carré intégrable et (M,A) est gaussien.

REMARQUE : Le lemme de Fernique nous assure qu'il existe $\varepsilon > 0$ tel que $E[\exp \varepsilon([X,X]_1 + \int_0^1 |dA_s|)] < +\infty$. Si on s'intéresse davantage à la décomposition des semimartingales gaussiennes, on peut établir le théorème suivant [17] :

THEOREME 2 : Soit X une semimartingale gaussienne. Alors X appartient à tous les espaces H^p $(p < \infty)$ relatifs à sa filtration naturelle $(\mathcal{F}_t)$. Il existe un ensemble dénombrable D tel que X soit continu dans $D^c \times \Omega$. La partie continue du crochet $[X,X]$ est déterministe. Enfin, si l'on désigne par $X = X_o + M + A$ la décomposition canonique de X dans sa filtration naturelle, les processus X, M, A, M^c, A^c sont tous contenus dans un même espace gaussien.

Faisons maintenant quelques remarques sur la covariance des semimartingales gaussiennes. Nous avons établi la proposition suivante dans [17] mais, par souci de complétude, nous allons la reproduire ici.

PROPOSITION 3 : Si Γ est la covariance d'une semimartingale gaussienne centrée X, il existe une mesure μ sur $[0,1]^2$ telle que pour tout $(s,t) \in [0,1]^2$, $\mu([0,s] \times [0,t]) = \Gamma(s,t)$.

DEMONSTRATION : Comme $X = M + A$ où M est une martingale de carré intégrable et A un processus à variation de carré intégrable, l'application ν de $\mathcal{B}([0,1])$ dans L^2 définie par $\nu(H) = 1_H . X$ est une mesure (vectorielle). Ainsi l'application μ de $(\mathcal{B}([0,1]))^2$ dans $\mathbb{R}$ définie par

$\mu(H \times K) = E[(1_H . X)_1 (1_K . X)_1]$ est une bimesure. On sait [9] que μ est une mesure si et seulement si $C = \sup_{\lambda} \sum_{i,j} |\mu(]t_i, t_{i+1}] \times]t_j, t_{j+1}])|$ est fini lorsque $\lambda = (0 = t_o < t_1 < \ldots < t_n = 1)$ parcourt l'ensemble des subdivisions de $[0,1]$. Si $X = M$, $\mu(H \times K) = E[(1_{H \cap K} . [M,M])_1]$, si bien que $C = E[M,M]_1$. Lorsque $X = A$, on sait [8] qu'il existe une mesure positive bornée γ sur $([0,1], \mathcal{B}([0,1]))$ telle que $|\mu(]s,t] \times]u,v])| \leq \gamma(]s,t])\gamma(]u,v])$; dans ce cas $C \leq \gamma^2([0,1])$. Il reste à examiner le cas "mixte" :

$$C' = \sum_{i,j} |E[(A_{t_{i+1}} - A_{t_i})(M_{t_{j+1}} - M_{t_j})]| \leq \sum_i (E[A_{t_{i+1}} - A_{t_i}]^2)^{1/2} (E[(\varphi^i . M)_1]^2)^{1/2}$$

où $\varphi_i = \sum_j \pm 1_{]t_j, t_{j+1}]}$. Mais $E[(\varphi^i . M)_1]^2 = E[M,M]_1$ et

$\frac{2}{\pi} E[A_{t_{i+1}} - A_{t_i}]^2 = E[|A_{t_{i+1}} - A_{t_i}|]^2$; donc

$C' \leq (\frac{\pi}{2})^{1/2} E[\int_0^1 |dA_s|](E[M,M]_1)^{1/2}$. Par conséquent μ est une mesure et la démonstration de la proposition 3 est achevée.

Emery [6] vient de montrer le joli résultat suivant caractérisant la covariance d'une quasimartingale gaussienne :

THEOREME 3 : Pour que le processus gaussien X soit une quasimartingale, il faut et il suffit qu'il existe une fonction croissante F sur $[0,1]$ telle que pour tout $0 \leq s \leq t \leq 1$ on ait : $\mu(f, 1_{]s,t]}) \leq (\mu(f,f))^{1/2}(F(t) - F(s))$ (*).

Enfin, on peut se poser la question de caractériser les processus gaussiens X qui "intègrent" les fonctions (déterministes) boréliennes bornées. On pourrait être tenté de démontrer que X est une semimartingale. Mais Bakry [1] a fourni un contre-exemple. En revanche si μ est une mesure, alors X "intègre" les fonctions boréliennes bornées sur $[0,1]$ ([10]). Dans ce cas on sait [10] que le crochet $[X,X]$ existe, du moins dans le cas continu. Par ailleurs, suivant Bouleau [3], le lemme de Khintchine montre que si X "intègre" les fonctions déterministes, alors l'ensemble des variables aléatoires $Q_\lambda(X) = \sum_\lambda (X_{t_{i+1}} - X_{t_i})^2$ est borné dans L^1 (et même dans tous les L^p, compte tenu du caractère gaussien de X). Peut-on en conclure que $[X,X]$ existe aussi dans ce cas ?

(*) pour toute fonction étagée f à support dans $[0,s]$.

2 - APPLICATION AU PROBLEME DE L'INNOVATION.

Faisons d'abord quelques remarques sur le problème de l'innovation. Nous nous plaçons dans les conditions habituelles. En particulier nous entendrons par filtration naturelle d'un processus Y la filtration engendrée par ce processus, rendue continue à droite et convenablement complétée. Soit Y une semimartingale par rapport à la filtration $(\mathcal{G}_t)$. Nous dirons que Y appartient à la classe $W(\mathcal{G})$ si Y est continue et si la décomposition canonique de Y est de la forme $B+A$, où B est un $(\mathcal{G}_t)$-mouvement brownien et A est absolument continu par rapport à la mesure de Lebesgue λ sur $[0,1]$. Plus explicitement ([2], VI. 68 et 68 bis), on peut écrire $A_t = \int_0^t Z_s\,ds$ où Z est un processus prévisible par rapport à $(\mathcal{G}_t)$ tel que $\int_0^1 |Z_s|\,ds < +\infty$ p.s..

PROPOSITION 5 : Soit X une semimartingale spéciale par rapport à $(\mathcal{G}_t)$. Si pour tout processus $(\mathcal{G}_t)$-prévisible borné H vérifiant $\int_0^1 |H_s|\,ds = 0$ on a $\int_0^1 H_s\,dX_s = 0$, alors $X = M + A$ où M est une martingale locale et A un processus absolument continu par rapport à la mesure de Lebesgue λ. Si de plus X est continue, alors $[X,X] = [M,M]$ est aussi absolument continu par rapport à λ.

DEMONSTRATION : La décomposition canonique de $H.X$ est : $H.X = H.M + H.A$. Par conséquent, si pour tout processus prévisible borné H vérifiant $\int_0^1 |H_s|\,ds = 0$, on a $\int_0^1 H_s\,dX_s = 0$, il en est de même pour M et A. Il suffit alors d'adapter la démonstration de [2], VI. 68 bis b) pour obtenir que A est absolument continu par rapport à λ. Lorsque X est continue, on obtient de même que $[M,M]$ est aussi absolument continu par rapport à λ.

Grâce à cette proposition, on obtient le

COROLLAIRE : Une semimartingale Y appartient à $W(\mathcal{G})$ si et seulement si $[Y,Y]_t = t$ pour $t \in [0,1]$ et si $H.Y = 0$ pour tout processus $(\mathcal{G}_t)$-prévisible borné H vérifiant $\int_0^1 |H_s|\,ds = 0$. En particulier si Y appartient à $W(\mathcal{G})$, elle appartient aussi à $W(\mathcal{F})$ pour toute sous-filtration $(\mathcal{F}_t)$ à laquelle Y est adaptée.

DEMONSTRATION : Il est évident que si Y appartient à $W(\mathcal{G})$, alors $[Y,Y]_t = t$ et $H.Y = 0$ si H est un processus $(\mathcal{G}_t)$-prévisible borné vérifiant $\int_0^1 |H_s|\,ds = 0$. Réciproquement si $[Y,Y]_t = t$ pour $t \in [0,1]$, Y est continue, la partie martingale locale de Y est nécessairement un $(\mathcal{G}_t)$-mouvement brownien que nous noterons B. Par ailleurs la proposition 5 montre que la partie à variation finie de Y est absolument continue par rapport à la mesure de Lebesgue λ, si bien que Y appartient à $W(\mathcal{G})$. Comme le crochet $[Y,Y]$ ne dépend pas de la sous-filtration $(\mathcal{F}_t)$, on en déduit immédiatement que Y appartient aussi à $W(\mathcal{F})$, compte tenu de la caractérisation ci-dessus.

REMARQUES :

i) On peut noter que la classe $W(\mathcal{G})$ reste invariante par changement de loi dans une même classe d'équivalence car le crochet et les intégrales stochastiques restent invariants.

ii) On peut montrer aisément grâce au processus de Poisson qu'il existe des martingales locales M purement discontinues telles que $H.M = 0$ pour tout processus prévisible borné H vérifiant $\int_0^1 |H_s|\, ds = 0$.

iii) L'hypothèse "$(\mathcal{F}_t)$ sous-filtration de $(\mathcal{G}_t)$" est essentielle dans le corollaire ci-dessus. Voici un contre-exemple qui reprend une idée de Perkins (article à paraître). Soit B un $(\mathcal{F}_t)$-mouvement brownien et $M_t = \sup_{s \le t} B_s$. Le processus de Bessel d'ordre 3 $X = 2M - B$ n'appartient pas à la classe $W(\mathcal{F})$ mais si $(\mathcal{G}_t)$ est la filtration naturelle de X, $X_t = W_t + \int_0^t X_s^{-1}\, ds$ où W est un $(\mathcal{G}_t)$-mouvement brownien. Ainsi X appartient à $W(\mathcal{G})$ sans appartenir à $W(\mathcal{F})$.

Désormais, nous travaillons dans la filtration naturelle $(\mathcal{F}_t)$, et nous désignons par $Y_t = B_t + \int_0^t Z_s\, ds$ la décomposition canonique. On note $(\mathcal{B}_t)$ la filtration naturelle de B. Par hypothèse $\mathcal{B}_t \subset \mathcal{F}_t$. On entend habituellement par <u>problème de l'innovation</u> la question de savoir si on a l'égalité $\mathcal{B}_t = \mathcal{F}_t$ pour tout t. La réponse est négative dans le cas général, comme le montre un contre-exemple célèbre dû à Tsirelson (voir [11] par exemple). Nous allons maintenant considérer le cas particulier où Y est gaussien. Rappelons que $Y_t = B_t + \int_0^t Z_s\, ds$ dans la filtration naturelle $(\mathcal{F}_t)$ de Y. Quitte à retrancher $\int_0^t E[Z_s]\, ds$ à Y_t, nous supposerons désormais que la variable aléatoire Z_s est centrée, ce qui ne changera rien au problème de l'innovation.

LEMME 2 : Pour presque tout s, Z_s appartient à l'espace gaussien $\mathcal{H}$ engendré par Y.

DEMONSTRATION : Le théorème 1 implique que $A_t = \int_0^t Z_s\, ds$ appartient à $\mathcal{H}$. Il est bien connu (c'est une belle conséquence du théorème de convergence des martingales) que si $\sigma_n = (0 = t_0 < t_1 < \ldots < t_k = 1)$ est une suite de subdivisions dont le pas tend vers 0 et si $Z_s^n = \sum_{\sigma_n} (t_{i+1} - t_i)^{-1} (A_{t_{i+1}} - A_{t_i}) 1_{\{t_i \le s \le t_{i+1}\}}$, alors $Z_s^n(\omega)$ tend vers $Z_s(\omega)$ dans $L^1(\lambda)$ pour chaque ω, λ étant la mesure de Lebesgue sur $[0,1]$. Quitte à extraire une sous-suite, on peut supposer que Z^n tend vers Z p.s. pour la loi $P \times \lambda$. Ainsi pour presque tout s

$P[\lim_n Z_s^n = Z_s] = 1$. Comme Z_s^n appartient à $\mathcal{H}$, le lemme en résulte. Quitte à remplacer Z_s par 0 pour les "mauvais" s, on peut supposer en fait que la propriété a lieu pour tout s.

REMARQUE : Voici une démonstration rapide de la réciproque [14]. Si (Z_s) est un processus mesurable, si Z_s appartient à $\mathcal{H}$ pour tout s et si $P[\int_0^1 |Z_s|\, ds < \infty] > 0$, alors $\int_0^1 Z_s\, ds$ existe et appartient à $\mathcal{H}$. En effet supposons d'abord que $\int_0^1 E[Z_s^2]\, ds < +\infty$. Dans ce cas $\int_0^1 Z_s\, ds$ est une v.a. appartenant à L^2. Si U est une v.a. de carré intégrable et orthogonale à $\mathcal{H}$, $E[(\int_0^1 Z_s\, ds)U] = \int_0^1 E[Z_s U]ds = 0$, si bien que U est aussi orthogonale à $\int_0^1 Z_s\, ds$. Donc $\int_0^1 Z_s\, ds$ appartient à $\mathcal{H}$. Le cas général se ramène à cette situation en posant $Z_s^n = \dfrac{nZ_s}{n + \sigma(s)}$ où $\sigma(s)$ est l'écart type de Z. Comme $\int_0^1 E[Z_s^n]^2\, ds < +\infty$, $\int_0^1 Z_s^n ds$ appartient à $\mathcal{H}$ et converge vers $\int_0^1 Z_s\, ds$ sur $\{\int_0^1 |Z_s|\, ds < +\infty\}$ qui a une probabilité positive. Une application simple du lemme de Fernique nous assure que la convergence a lieu p.s. et donc que $\int_0^1 Z_s\, ds$ appartient à $\mathcal{H}$.

La proposition suivante nous donne une représentation des processus absolument continus et adaptés à la filtration naturelle de (B_t).

PROPOSITION 6 : Soit (A_t) un processus absolument continu et adapté à la filtration naturelle de (B_t). Alors le processus (B_t, A_t) est gaussien si et seulement s'il existe une fonction k de $[0,1]^2$ dans $\mathbb{R}$ vérifiant $\int_0^1 (\int_0^1 k^2(s,u)\, du)^{1/2} ds < +\infty$, $k(s,u) = 0$ si $u > s$ et $A_t = \int_0^t (\int_0^s k(s,u)dB_u)ds$ pour tout t.

DEMONSTRATION : Il est évident que si $A_t = \int_0^t (\int_0^s k(s,u)dB_u)ds$ pour tout t, alors (A_t, B_t) est un processus gaussien. Réciproquement, supposons que (A_t, B_t) soit un processus gaussien. Si $\sigma = (0 = t_0 < t_1 < \dots < t_n = 1)$ est une subdivision de $[0,1]$, on pose :

$$H_s^\sigma = \sum_\sigma (t_{i+1} - t_i)^{-1} (A_{t_{i+1}} - A_{t_i}) 1_{\{t_i \le s \le t_{i+1}\}}.$$

Comme A_t appartient à l'espace gaussien engendré par $(B_s,\ s \le t)$, il existe pour tout i une fonction borélienne f^i de $[0,1]$ dans $\mathbb{R}$, de carré intégrable pour la mesure de Lebesgue λ sur $[0,1]$ telle que $A_{t_{i+1}} - A_{t_i} = \int_0^{t_{i+1}} f^i(u)dB_u$. Nous pouvons choisir f^i nulle sur $[t_{i+1}, 1]$. La fonction $k(s,u) = \sum_n (t_{i+1} - t_i)^{-1} f_i(u)\, 1_{\{t_i \le s \le t_{i+1}\}}$ est borélienne, nulle si

$s < t_i < u$ et $H_s^\sigma = \int_0^1 k(s,u)\,dB_u$. Choisissons maintenant une suite de subdivisions σ_n de $[0,1]$ dont le pas tend vers 0. Comme $\sup_n \int_0^1 |H_s^{\sigma_n}|\,ds$ est égal à la variation de A sur $[0,1]$ et que cette variation est intégrable d'après le théorème 1, $E\int_0^1 |H_s^{\sigma_n} - H_s^{\sigma_m}|\,ds$ tend vers 0 lorsque n et m tendent vers $+\infty$.

Mais $E\int_0^1 |H_s^{\sigma_n} - H_s^{\sigma_m}|\,ds = \int_0^1 E\left|\int_0^1 \left(k^n(s,u) - k^m(s,u)\right)dB_u\right|\,ds$

$= \left(\frac{2}{\pi}\right)^{1/2} \int_0^1 \left(\int_0^1 \left(k^n(s,u) - k^m(s,u)\right)^2 du\right)^{1/2} ds$. Ainsi k^n converge vers une fonction borélienne k telle que $A_t = \int_0^t \int_0^1 k(s,u)\,dB_u\,ds$. En vertu des égalités ci-dessus $\int_0^1 \left(\int_0^1 k^2(s,u)\,du\right)^{1/2} ds$ est fini. Par ailleurs, si α_n est le pas de la subdivision σ_n, on a $\int_0^1 \left(\int_0^1 \left(k^n(s,u)\right)^2 1_{\{s<u-\alpha_n\}}\,du\right)^{1/2} ds = 0$, si bien que $k(s,u) = 0$ si $s < u$.

Nous sommes maintenant en mesure d'établir le théorème suivant :

THEOREME 2 : Il existe un processus absolument continu (C'_t) indépendant de $\mathcal{B}_\infty$ et une fonction borélienne $k(s,u)$ de $[0,1]^2$ dans $\mathbb{R}$ vérifiant $\int_0^1 \left(\int_0^1 k^2(s,u)\,du\right)^{1/2} ds < +\infty$, tels que pour tout t :

$$Y_t = B_t + \int_0^t \left(\int_0^s k(s,u)\,dB_u\right) ds + C'_t.$$

DEMONSTRATION : Soit (C''_t) la projection duale prévisible de (C_t) sur $(\mathcal{B}_t)$. Comme (C_t, B_t) est gaussien et que C''_t est la limite faible dans L^2 d'une suite de variables aléatoires de la forme $\sum_{\sigma_n} E[C_{t_{i+1}} - C_{t_i} | \mathcal{B}_{t_i}]$ où (σ_n) est une suite de subdivisions de $[0,t]$ dont le pas tend vers 0, (C''_t, B_t) est aussi un processus gaussien. Par ailleurs $C''_t = \int_0^t Z''_s\,ds$ où (Z''_s) est la projection prévisible de (Z_s). Si on pose $Z'_s = Z_s - Z''_s$ et $C'_t = \int_0^t Z'_s\,ds$, on a $E[Z'_s | \mathcal{B}_s] = 0$; mais B étant une $(\mathcal{F}_t)$ martingale, ceci entraîne que (C'_t) est indépendant de $\mathcal{B}_\infty$. Grâce à la proposition 6 appliquée à C''_t, on obtient la représentation souhaitée de Y. Nous dirons qu'une fonction $k(s,u)$ de $[0,1]^2$ dans $\mathbb{R}$ est un <u>noyau de Volterra</u> si $k(s,u) = 0$ lorsque $s < u$ et si $\int_0^1 \int_0^1 k^2(s,u)\,du\,ds$ est fini.

Rappelons un résultat déjà ancien :

THEOREME 3 : (Hitsuda [8]). Soit Y un processus gaussien. Alors la loi de Y est équivalente à celle d'un mouvement brownien si et seulement s'il existe un mouvement brownien B et un noyau de Volterra k tel que $Y_t = B_t + \int_0^t \left(\int_0^s k(s,u)\,dB_u\right) ds$.

Ce résultat va nous permettre d'établir le théorème suivant :

THEOREME 4 : Soit Y un processus gaussien de la forme $Y_t = B_t + \int_o^t Z_s\, ds$ où (B_t) est un mouvement brownien de la filtration naturelle $(\mathcal{F}_t)$ de Y. Alors la loi de Y est équivalente à celle d'un mouvement brownien si et seulement si toute fonction f de $[0,1]$ dans $\mathbb{R}$, de carré intégrable pour la mesure de Lebesgue λ sur $[0,1]$, est aussi Y-intégrable.

DEMONSTRATION : Rappelons d'abord la définition du mot "Y-intégrable" : Soit Y une semimartingale continue par rapport à une filtration $(\mathcal{G}_t)$ et soit H un processus prévisible. Si $Y = M + A$ est la décomposition canonique de Y, on dit d'après Jacod [12] que H est Y-intégrable si $H.M$ (resp. $H.A$) existe au sens des martingales locales (resp. au sens de l'intégrale de Stieljes). Pour une étude approfondie de cette notion, on pourra consulter [4]. Revenons maintenant aux hypothèses du théorème. On note $L(Y)$ l'ensemble des processus prévisibles qui sont Y-intégrables. Comme $L(Y)$ reste invariant par changement de loi équivalente, il est évident que $L(Y)$ contient $L^2(\lambda)$. Réciproquement, $(f.C)_t = \int_o^t f(s)\, Z_s\, ds$ est un processus gaussien à variation finie, donc à variation intégrable. Si $\sigma(s)$ désigne l'écart-type de Z_s, il en résulte que $\int_o^1 |f(s)|\, \sigma(s)\, ds$ est fini pour toute fonction $f \in L^2(\lambda)$. Donc σ appartient à $L^2(\lambda)$. Avec les notations du théorème 2, on en déduit que $\int_o^1 (Z_s')^2\, ds$ et $\int_o^1 (Z_s'')^2\, ds$ sont des variables aléatoires appartenant à $L^1(P)$. Donc k est un noyau de Volterra. Ainsi d'après le théorème 3, la loi de $Y_t' = B_t + C_t''$ est équivalente à celle d'un mouvement brownien. Or C_t' appartient à l'espace gaussien engendré par $(Y_s,\ s \le t)$. Donc il existe une suite de fonctions boréliennes bornées f^n de $[0,1]$ dans $\mathbb{R}$ telles que $\int_o^t f^n(s)\, dY_s$ converge vers C_t' dans L^2. Mais C_t' est orthogonal à l'espace gaussien engendré par $(Y_s',\ s \le t)$, si bien que $\int_o^t f^n(s)\, dY_s'$ tend vers 0 en probabilité. Or la loi de Y' est équivalente à celle d'un mouvement brownien : donc f^n converge vers 0 dans $L^2(\lambda)$ et $E\,|\int_o^t f^n(s)\, Z_s'\, ds|$ tend aussi vers 0 car

$\left(\int_o^t f^n(s)\, E\,|Z_s'|\, ds\right)^2 \le \int_o^t \left(f^n(s)\right)^2 ds \int_o^t E[Z_s']^2\, ds$. Ainsi $C_t' = 0$ pour tout t. Donc la loi de $Y = Y'$ est équivalente à celle d'un mouvement brownien.

REMARQUE : En examinant de près la démonstration précédente, on peut remplacer dans le théorème 4 la condition $L(Y) \supset L^2(\lambda)$ par : l'ensemble des $f.Y$ est borné dans L^o lorsque f parcourt les fonctions boréliennes élémentaires sur $[0,1]$ vérifiant $\int_o^1 f^2(s)\, ds \le 1$.

Etant donnée une semimartingale $Y_t = B_t + \int_o^t Z_s\, ds$, le théorème précédent montre combien il est important de savoir si $\int_o^1 Z_s^2\, ds < +\infty$. Ceci nous amène à étudier une nouvelle classe de semimartingales : celles de la forme ci-dessus. Plus précisément, si $(\mathcal{G}_t)$ est une filtration, on appelle $W'(\mathcal{G},P)$ l'ensemble des semimartingales appartenant à $W(\mathcal{G})$ dont la partie à variation finie $A_t = \int_0^t Z_s ds$ vérifie la condition $\int_0^1 Z_s^2 ds < +\infty$. Nous nous proposons de caractériser cette classe à partir de propriétés d'intégrabilité. Pour cela nous introduisons d'abord une classe un peu plus large. Soit Y une semimartingale continue, nulle en 0, dont la décomposition canonique $Y = M + A$ possède la propriété :

(1) $d\langle M,M\rangle_t = m_t dt$, $dA_t = a_t dt$ où (m_t) et (a_t) sont prévisibles, m_t étant de plus localement borné et $\int_0^1 a_s^2 ds < +\infty$.

Cette condition entraîne manifestement la condition suivante :

(2) Pour toute suite (H^n) de processus prévisibles élémentaires bornés, telle que $\int_0^1 (H_s^n)^2 ds$ tende vers 0 dans L^o, $(H^n.Y)_1$ tend vers 0 dans L^o.

Voici une conséquence immédiate de (2) :

(3) Si $\mathcal{E}$ désigne l'ensemble des processus prévisibles élémentaires bornés H vérifiant $\int_0^1 H_s^2 ds \leq 1$, alors $\{(H.Y)_1, H \in \mathcal{E}\}$ est borné dans L^o.

Enfin la condition (1) entraîne aussi que :

(4) $L(Y)$ contient l'ensemble des processus prévisibles H vérifiant $\int_0^1 H_s^2 ds < +\infty$.

THEOREME 5 : Les conditions (1), (2), (3) et (4) sont équivalentes.

Comme on a les implications (1) $\Rightarrow$ (2) $\Rightarrow$ (3), (1) $\Rightarrow$ (4) et (4) $\Rightarrow$ (2), il suffit de démontrer que (3) $\Rightarrow$ (1) pour établir le théorème 5. Désormais nous supposerons que la semimartingale continue Y vérifie (3). On désigne par $\mathcal{E}'$ l'ensemble des processus prévisibles bornés H vérifiant $\int_0^1 H_s^2 ds \leq 1$. Il est bien connu que pour tout $H \in \mathcal{E}'$, il existe une suite de processus prévisibles élémentaires bornés (H^n) tels que $\int_0^1 (H_s - H_s^n)^2 ds$ converge vers 0 dans L^o. Ainsi $\{(H.Y)_1, H \in \mathcal{E}'\}$ est aussi borné dans L^o, si bien que lorsque H est un processus prévisible borné vérifiant $\int_0^1 |H_s| ds = 0$, on a $(H.Y)_1 = 0$. D'après la proposition 5, il existe deux processus prévisibles (m_s) et (a_s) tels que $d\langle M,M\rangle_t = m_t d_t$ et $dA_t = a_t dt$, M (resp. A) étant la partie martingale locale (resp. à variation finie) dans la décomposition canonique de Y.

Nous allons d'abord montrer qu'il existe une version de m qui est localement bornée. A cet effet, nous aurons besoin des lemmes suivants.

LEMME 3 : Pour tout processus prévisible H tel que $\int_0^1 |H_s| ds < +\infty$ on a $\int_0^1 |H_s m_s| ds < +\infty$.

DEMONSTRATION : Commençons par montrer que $\{\int_0^1 (H.Y)_s H_s dY_s, H \in \mathcal{E}'\}$ est borné dans L^o. Pour tout H de $\mathcal{E}'$ on pose $T_n = \inf \{t, |(H.Y)_t| \geq n\}$ et

$K^n = (H.Y)^{T_n}$. Fixons $\epsilon > 0$. Comme $\{(H.Y)_1,\ H \in \mathcal{S}'\}$ est borné dans L^o, il existe n tel que pour tout $H \in \mathcal{S}'$, on ait $P[|(H.Y)_1| \geq n] < \frac{\varepsilon}{2}$. D'après les propriétés locales de l'intégrale stochastique, on a :

$P[|\int_o^1 (H.Y)_s H_s \, dY_s| \geq n^2] \leq P[|((K^n H).Y)_1| \geq n^2] + P[T_n < 1] \leq \frac{\varepsilon}{2} + \frac{\varepsilon}{2}$.

Ainsi $\{\int_o^1 (H.Y)_s H_s \, dY_s,\ H \in \mathcal{S}'\}$ est borné dans L^o. Grâce à la formule d'Ito, il en est de même pour $\{(H^2.[M,M])_1,\ H \in \mathcal{S}'\}$. Il est alors clair que pour tout processus prévisible K tel que $\int_o^1 |K_s| \, ds < +\infty$ on a $\int_o^1 |K_s \; m_s| \, ds < +\infty$.

LEMME 4 : Si λ désigne la mesure de Lebesgue sur $[0,1]$, pour presque tout ω la borne supérieure essentielle de $|m_s(\omega)|$ relativement à λ est finie, c'est-à-dire $P[\forall n \in \mathbb{N},\ \lambda\{s, |m_s| \geq n\} > 0] = 0$.

DEMONSTRATION : Raisonnons par l'absurde. Soit H^n l'indicatrice de l'ensemble $\{|m| \geq n\}$. Si $P[\forall n,\ \int_o^1 H_s^n \, ds > 0] > 0$, il existe une suite de réels (γ_n) vérifiant $\gamma_n > 0$ et $P[\forall n,\ \int_o^1 H_s^n \, ds > \gamma_n] > 0$. On pose

$T_n = \inf\{t, \int_o^t H_s^n \, ds > \gamma_n\}$ et $H_t = \Sigma\, n^{-2} \gamma_n^{-1} H^n_{t \wedge T_n}$. Alors $\int_o^1 H_s \, ds < +\infty$

mais $P[\int_o^1 |H_s m_s| \, ds = +\infty] > 0$. Ceci est absurde et le lemme est démontré.

CONSEQUENCE : Soit $T_n = \inf\{t, \lambda\{s \leq t, |m_s| \geq n\} > 0\}$. (T_n) est une suite de temps d'arrêt tendant stationnairement vers 1 d'après le lemme précédent. Ainsi $\tilde{m} = \sum_n (m_\wedge (n+1)) \vee (-n-1)\, 1_{]T_n, T_{n+1}]}$ est une version prévisible de m vérifiant $P[\sup_s |\tilde{m}_s| < +\infty] = 1$. Un lemme bien connu, dû à Lenglart, entraîne alors que $\tilde{m}$ est localement borné. Donc si Y vérifie (3), sa partie martingale locale satisfait à (1). Par différence $\{\int_o^1 H_s \, a_s \, ds,\ H \in \mathcal{S}'\}$ est aussi borné dans L^o et il est clair que si K est un processus prévisible vérifiant $\int_o^1 K_s^2 \, ds < +\infty$, alors $\int_o^1 |K_s \, a_s| \, ds < +\infty$.

Pour achever la démonstration du théorème 5, il reste à établir le lemme suivant que nous énonçons en toute généralité, bien que seul le cas $p = q = 2$ nous serve pour le théorème 5.

LEMME 5 : Soient $p > 1$ et $\frac{1}{p} + \frac{1}{q} = 1$. Si pour tout processus prévisible K tel que $\int_o^1 |K_s|^p \, ds < +\infty$, on a $\int_o^1 |K_s \, a_s| \, ds < +\infty$, alors $\int_o^1 |a_s|^q \, ds < +\infty$.

DEMONSTRATION : Raisonnons à nouveau par l'absurde. Il existe alors une suite d'entiers (n_k) strictement croissante telle que

$$P\left[\forall k, \sum_{n=n_k}^{n_{k+1}} n^q \lambda\{s, n \le |a_s| < n+1\} > k\right] > 0.$$ On pose

$$T_k = \inf\left\{t, \sum_{n=n_k}^{n_{k+1}} n^q \lambda\{s \le t, n \le |a_s| < n+1\} > k\right\}$$

et $$H = \sum_k \frac{1}{k^q} \sum_{n=n_k}^{n_{k+1}} n^{q-1} 1_{\{n \le |a_s| < n+1\} \cap [0, T_k]}.$$ Alors $\int_0^1 H_s^p \, ds < +\infty$

et $P\left[\int_0^1 |H_s a_s| \, ds = +\infty\right] > 0$, ce qui est absurde et le lemme 5 est démontré, ainsi que le théorème 5.

Notons qu'une semimartingale continue Y appartient à $W'(\mathcal{G}, P)$ si et seulement si $d[Y,Y]_t = dt$ et si Y vérifie (1) (ou encore si Y vérifie (1) et $m = 1$). Grâce aux équivalences du théorème 5 et au fait que $[Y,Y]$ est invariant par changement de filtration ou de loi de probabilité équivalente, on a les deux corollaires suivants.

COROLLAIRE 1 : Si Q est une loi absolument continue par rapport à P et si Y vérifie (1) sous P, elle vérifie aussi (1) sous Q. En particulier $W'(\mathcal{G}, P) = W'(\mathcal{G}, Q)$.

COROLLAIRE 2 : Soit Y une semimartingale vérifiant (1). Si $(\mathcal{F}_t)$ est une sous-filtration de $(\mathcal{G}_t)$, vérifiant les conditions habituelles et si Y est adaptée à $(\mathcal{F}_t)$, alors Y possède aussi (1) relativement à la filtration $(\mathcal{F}_t)$. En particulier si Y appartient à $W'(\mathcal{G}, P)$, Y appartient aussi à $W'(\mathcal{F}, P)$.

Ce corollaire est bien connu [14] lorsque $(\mathcal{F}_t)$ est la filtration naturelle de Y et $m = 1$.

Pour terminer, voici une belle application de ce corollaire à l'étude d'un exemple du problème de l'innovation, due à Chitashvili et Toronjadze [21]. Soit $X_t = B_t + \theta t$ où B est un $(\mathcal{G}_t)$-mouvement brownien et θ une v. a. indépendante de B. On note $(\mathcal{F}_t)$ la filtration naturelle de X. On pose :

$$A(s,x) = \begin{cases} 0 & \text{si} \quad s = 0 \\ \int_{-\infty}^{+\infty} a \exp\left(ax - a^2 s/2\right) dF(a) \Big/ \int_{-\infty}^{+\infty} \exp\left(ax - a^2 s/2\right) dF(a). & \end{cases}$$

On vérifie aisément que $A(s,X_s) = E[\theta | \mathfrak{F}_s]$, si bien que la décomposition canonique de X dans la filtration $(\mathfrak{F}_t)$ est (voir [22]) :
$X_t = W_t + \int_0^t A(s,X_s)ds$ où W est un $(\mathfrak{F}_t)$-mouvement brownien. En réalité on a non seulement $\int_0^1 |A(s,X_s)| ds < +\infty$ mais aussi $\int_0^1 A^2(s,X_s)ds < +\infty$ d'après le corollaire ci-dessus. Ainsi X vérifie l'équation différentielle :
$dX_s = dW_s + A(s,X_s)ds$ avec $X_o = 0$. On se restreint aux solutions dont les trajectoires appartiennent à $\{u \in C[0,1], \int_0^1 A^2(s,u(s))ds < +\infty\}$. La proposition suivante est désormais classique.

PROPOSITION : Si X^1 et X^2 sont deux solutions définies sur le même espace probabilisé filtré avec le même mouvement brownien W, alors $X^1 \vee X^2$ et $X^1 \wedge X^2$ sont aussi solutions.

DEMONSTRATION : Comme $X^1 - X^2$ est à variation finie, le temps local en 0 de $X^1 - X^2$ est nul : $L^o(X^1 - X^2) = 0$. La formule de Tanaka entraîne que :

$$X_t^1 \vee X_t^2 = X_t^2 + (X_t^1 - X_t^2)^+ = B_t + \int_0^t A(s,X_s^2)ds$$
$$+ \int_0^t 1_{\{X_s^1 > X_s^2\}} \left(A(s,X_s^1) - A(s,X_s^2)\right)ds$$
$$= B_t + \int_0^t A(s,X_s^1 \vee X_s^2)ds.$$

Ainsi $X^1 \vee X^2$ est aussi solution, de même que $X^1 \wedge X^2$.

On déduit de cette proposition que si X^1 et X^2 ont même loi, $X^1 = X^2$. Pour pouvoir appliquer le théorème de Yamada - Watanabe [23] il reste à montrer que X^1 et X^2 ont même loi. Mais ceci résulte immédiatement du théorème 7.6. de [14], si bien que X est adapté à la filtration naturelle de W et la réponse au problème de l'innovation est positive dans ce cas.

REFERENCES

[1] BAKRY, D. : Une remarque sur les processus gaussiens définissant des mesures L^2. A paraître dans Séminaire de Probabilités XVII. Lecture Notes in M. Springer 1983.

[2] BICHTELER, K. : Stochastic integration and L^p-theory of semimartingales. The Annals of Probability, 1981, vol. 9, n°1, 49-89.

[3] BOULEAU, N. : Sur la variation quadratique de certaines mesures vectorielles. Z. W. 61, 283-290. Springer 1982.

[4] CHOU, C. S., MEYER, P. A., STRICKER, C. : Sur les intégrales stochastiques de processus prévisibles non bornés. Séminaire de Probabilités XIV. Lecture Notes in Math. 721, 128-139. Springer 1980.

[5] DELLACHERIE, C. et MEYER, P. A. : Probabilités et Potentiel B, Théorie des Martingales. Hermann, Paris 1979.

[6] EMERY, M. : Covariance des semimartingales gaussiennes. CRAS Paris, t. 295 (6 Décembre 1982), Série I, 703-705.

[7] FERNIQUE, X. : Régularités des fonctions aléatoires gaussiennes, Ecole d'été de calcul des probabilités. St Flour, IV. Lecture Notes in Math. 480. Springer 1974.

[8] HITSUDA, M. : Representation of Gaussian processes equivalent to Wiener process. Osaka J. Math. 5, 299-312 (1968).

[9] HOROWITZ, J. : Une remarque sur les bimesures. Séminaire de Probabilités XI. Lecture Notes in Math. 581, 59-64. Springer 1977.

[10] HUANG, S. T. and CAMBANIS, S. : Stochastic and multiple Wiener Integrals for Gaussian processes. The Annals of Probability, 1978, vol. 6, n°4, 585-614.

[11] ITO, K., NISIO, M. : On the oscillation function of Gaussian processes. Math. Scand. 22, 209-223 (1968).

[12] JACOD, J. : Calcul stochastique et problèmes de martingales. Lecture Notes in Math. 714. Springer 1974.

[13] JAIN, N. C., MONRAD, D. : Gaussian quasimartingales. Zeitschr. f. Wahrscheinlichkeitstheorie verw. Gebiete 59, 139-159 (1982).

[14] LIPTSER, R. S. and SHIRYAYEV : Statistics of Random Processes I. General Theory. Springer 1977.

[15] MEYER, P. A. : Un cours sur les intégrales stochastiques. Séminaire de Probabilités X. Lecture Notes in M. 511, 245-400. Springer 1976.

[16] METIVIER, M. et PELLAUMAIL, J. : Mesures stochastiques à valeurs dans les espaces L^o. ZW, 40, 101-114. Springer 1977.

[17] STRICKER, C. : Semimartingales gaussiennes, Application au problème de l'innovation. A paraître dans Z. W., Springer 1983.

[18] STRICKER, C. : Quasimartingales, martingales locales, semimartingales et filtrations naturelles. Zeitschr. f. Wahrscheinlichkeitstheorie verw. Gebiete 39, 55-64 (1977).

[19] STRICKER, C. : Caractérisation des semimartingales. A paraître dans le Séminaire XVIII. Lecture Notes in M. Springer 1984.

[20] YOR, M. : Une équation générale du filtrage. Ecole d'été de probabilités de St Flour IX 1979. Lecture Notes in Math. 876. Springer.

[21] CHITASHVILI, R. J. et TORONJADZE, T. A. : On one dimensional stochastic differential equations with unit diffusion coefficients. Structure of solutions. Stochastics 1981, vol. 4, 281-315.

[22] MEYER, P. A. : Sur un problème de filtration. Séminaire de Probabilités VII. Lecture Notes in M. 321, 223-247. Springer 1973.

[23] YAMADA, T. et WATANABE, S. : On the uniqueness of solutions of stochastic differential equations, J. Math. Kyoto Univ., n°11, 1971, 155-167.

[24] YAN, J. A. : Caractérisation d'une classe d'ensembles convexes de L^1 ou H^1. Séminaire de Probabilités XIV, Lecture Notes in M. 784, 220-222. Springer 1980.

C. STRICKER
Laboratoire de Mathématiques
Route de Gray
25030 Besançon Cedex

SUR LES PROPRIETES MARKOVIENNES DU PROCESSUS DE FILTRAGE

Jacques SZPIRGLAS*

On montre dans ce travail, les propriétés markoviennes du processus de filtrage associé à un modèle de filtrage avec observation discontinue (4), (12), généralisant ainsi l'étude de (6). La présence d'une composante discontinue dans le processus d'observation entraine un traitement différent de (6) par l'utilisation du processus de filtrage non normalisé et donc la méthode de la probabilité de référence (14), (12). Celle-ci consiste à construire le modèle de filtrage "signal-observation" sur un espace de probabilité où les relations entre signal et observation sont simples, ici l'indépendance, bien que n'ayant pas de signification physique, puis de changer de probabilité pour obtenir l'espace décrivant le modèle physique. Cette méthode a l'avantage de conduire à une forme explicite du processus de filtrage, la formule de Kallianpur-Striebel, et permet de construire le couple "signal-observation" de façon plus générale que comme solution forte d'équations différentielles stochastiques.

Dans une première partie on construit le modèle de filtrage par la méthode de la probabilité de référence. Dans la seconde partie, on utilise la formule de Kallianpur-Striebel pour montrer que les processus de filtrage normalisé ou non, sont des processus de Markov. Dans la dernière partie, on définit des équations générales du filtrage du type de celles de Zakai. On dégage les différences de traitement par rapport aux équations normalisées, et on en déduit le caractère Fellerien du processus de filtrage normalisé.

On utilisera tout au long de ce papier les notations suivantes. $(E,\underline{\underline{E}})$ désigne un espace de Hausdorff E séparable compact, muni de sa tribu borélienne $\underline{\underline{E}}$. On note b(E) (resp. C(E)) l'ensemble des fonctions boréliennes bornées (resp. continues) sur E, et $(M(E),\underline{\underline{M}}(E))$ (resp. $(\tilde{M}(E),\tilde{\underline{\underline{M}}}(E))$) l'ensemble des lois de probabilité, M(E), (resp. mesures positives bornées, $\tilde{M}(E)$) sur E, munie de sa tribu borélienne $\underline{\underline{M}}(E)$ (resp. $\tilde{\underline{\underline{M}}}(E)$). Notons que C(E) est un espace de Banach séparable pour la métrique de la convergence uniforme, et M(E) est un espace de Hausdorff compact séparable, comme E, pour la topologie de la convergence étroite.

Centre National des Télécommunications, PAA/ATR/MTI
38-40 rue du Général Leclerc, 92131 ISSY LES MOULINEAUX

I/ Construction du modèle de filtrage

Elle se fait en deux étapes; on construit d'abord l'espace de référence, puis par changement de probabilité équivalente, le modèle de filtrage. On rappelle enfin les définitions et certaines propriétés classiques des processus de filtrage.

I-a/ L'espace de référence

Le signal est représenté par un processus de Feller X, continu à droite, limité à gauche (cadlag) à valeurs dans un espace de Hausdorff E, compact séparable, muni de sa tribu borélienne $\underline{\underline{E}}$, de semi-groupe $(P_t;t\geq 0)$ et de générateur infinitésimal $(\underline{\underline{L}},\underline{\underline{D}}(\underline{\underline{L}}))$ défini sur son espace canonique:

$$X = (\Omega^X,\underline{\underline{A}}^X,\underline{\underline{F}}_t^X,\theta_t^X,X_t,(\mathbb{P}_x^X;x\in E))$$

où Ω^X est l'ensemble des fonctions cadlag de $\mathbb{R}_+$ à valeurs dans E. On note ω^X l'élément générique de Ω^X, $\underline{\underline{F}}^X = (\underline{\underline{F}}_t^X;t\geq 0)$ est la filtration engendrée par les applications coordonnées X_t et $\underline{\underline{A}}^X = \underline{\underline{F}}_\infty^X$. $(\theta_t^X;t\geq 0)$ est le semi-groupe d'opérateurs sur Ω^X tel que:

$$X_{t+s} = X_t\circ\theta_s^X \qquad \forall s,t \geq 0 \quad .$$

Pour toute loi initiale de X, μ de M(E) l'ensemble des lois de probabilité sur $(E,\underline{\underline{E}})$, on associe la loi de probabilité $\mathbb{P}_\mu^X$ sur Ω^X définie par:

$$\mathbb{P}_\mu^X = \int_E \mu(dx)\,\mathbb{P}_x^X$$

Le processus d'observation est construit, pour simplifier l'exposé, à partir d'un processus réel Y, à accroissement indépendants (P.A.I) défini sur son espace canonique:

$$Y = (\Omega,\underline{\underline{A}},\underline{\underline{G}}_t,\theta_t,Y_t,\mathbb{P})$$

où Ω est l'ensemble des fonctions cadlag réelles du temps. On note ω l'élément générique de Ω. $\mathbb{P}$ est l'unique loi de probabilité sur Ω qui fait du processus des applications coordonnées Y_t, un P.A.I de décomposition canonique:

$$Y_t = W_t + \int_0^t \int_{\mathbb{R}} x\, \mathbb{1}\{|x|>1\} N(ds,dx) + \int_0^t \int_{\mathbb{R}} x\, \mathbb{1}\{|x|\leq 1\}(N(ds,dx)-ds\nu(dx))$$

où W est un mouvement brownien et N(ds,dx) est la mesure aléatoire des sauts de Y associée à un processus de Poisson homogène de mesure de Levy sur $\mathbb{R}_+ \times \mathbb{R}$, $ds\nu(dx)$, telle que $\nu(\mathbb{R}) < \infty$.

$\underline{\underline{G}} = (\underline{\underline{G}}_t; t\geq 0)$ est la filtration engendrée par le processus Y, complétée pour la probabilité $\mathbb{P}$ et rendue continue à droite(cad), $\underline{\underline{A}} = \underline{\underline{G}}_\infty$.

$(\theta_t; t\geq 0)$ est le semi-groupe d'opérateurs sur Ω tel que:

$$Y_{t+s} - Y_s = Y_t \circ \theta_s \qquad \forall s,t\geq 0 \quad .$$

L'espace de probabilité de référence est défini comme le produit tensoriel des espaces précédents:

$\Omega' = \Omega \times \Omega^X$, $\underline{\underline{A}}' = \underline{\underline{A}} \otimes \underline{\underline{A}}^X$, $\underline{\underline{F}}_t = \underline{\underline{G}}_t \otimes \underline{\underline{F}}_t^X$, $\mathbb{P}_\mu = \mathbb{P} \otimes \mathbb{P}_\mu^X$; $\forall t\geq 0, \forall \mu \in M(E)$.

On redéfinit sur $(\Omega', \underline{\underline{A}}')$, les processus X et Y:

$$X(\omega,\omega^X) = X(\omega^X) \quad \text{et} \quad Y(\omega,\omega^X) = Y(\omega) \quad .$$

On remarque que X et Y sont $\mathbb{P}_\mu$-indépendants pour toute loi μ de M(E).

On note encore $\underline{\underline{G}}$ la filtration sur $(\Omega', \underline{\underline{A}}')$,

$$\underline{\underline{G}} = \underline{\underline{G}} \otimes \{\Omega^X, \emptyset\}$$

On désigne par $\underline{\underline{F}}^\mu$ (resp. $\underline{\underline{G}}^\mu$), la filtration rendue cad, engendrée par $\underline{\underline{F}}$ (resp. $\underline{\underline{G}}$) et les ensembles $\mathbb{P}_\mu$-négligeables de $(\Omega', \underline{\underline{A}}')$. On note $\underline{\underline{A}}^\mu$ la tribu $\mathbb{P}_\mu$-complétée de $\underline{\underline{A}}'$.

I-b/ Le modèle de filtrage

Le modèle décrivant l'évolution réelle du système (X,Y) est obtenu par changement de probabilité équivalente sur cet espace de référence $(\Omega', \underline{\underline{A}}^{\mu}, \mathbb{P}_{\mu})$. On se donne une fonction h réelle continue bornée sur E et une fonction H borélienne bornée sur $E \times \mathbb{R}$, continue sur E telle que:

$$\exists\, \delta > 0,\ 1 + H \geq \delta > 0$$

On suppose de plus que:

$$\int_0^{\infty} h^2(X_s)ds + \int_0^{\infty} \int_{\mathbb{R}} H^2(X_s,x)ds\nu(dx) < \infty \quad \mathbb{P}_{\mu}^X\text{-p.s.},\ \forall\mu \in M(E)$$

Cette hypothèse est vérifiée en horizon fini T (i.e. T remplace ∞) ou si X a une durée de vie finie et h, H sont prolongées par 0 au point cimetière de E.

On définit alors sur $(\Omega', \underline{\underline{A}}^{\mu}, \mathbb{P}_{\mu})$ le processus L^{μ}:

$$L_t^{\mu} = \exp\{M_t^{\mu} - 1/2\int_0^t h^2(X_s)ds\} \prod_{\substack{s \leq t \\ Y_s \neq 0}} (1+\Delta M_s^{\mu}) \exp\{-\Delta M_s^{\mu}\}$$

où M^{μ} est la $(\Omega', \underline{\underline{F}}_t^{\mu}, \mathbb{P}_{\mu})$-martingale:

$$M_t^{\mu} = \int_0^t h(X_s)dW_s + \int_0^t \int_{\mathbb{R}} H(X_{s-},x)(N(ds,dx)-ds\nu(dx))$$

Les hypothèses sur h et H font que, (7),(8), L^{μ} est une martingale relativement à $(\Omega', \underline{\underline{F}}_t^{\mu}, \mathbb{P}_{\mu})$, cadlag, uniformément intégrable telle que

$$\forall\mu \in M(E),\ L_{\infty}^{\mu} >> 0,\ E_{\mathbb{P}_{\mu}}(L_{\infty}^{\mu}) = 1,\ \forall t \geq 0,\ E_{\mathbb{P}_{\mu}}((L_t^{\mu})^2) < \infty \quad .$$

Alors la relation:

$$\mathbb{Q}_{\mu} = L_{\infty}^{\mu}\, \mathbb{P}_{\mu}$$

définit une probabilité $\mathbb{Q}_{\mu}$ sur $(\Omega', \underline{\underline{A}}^{\mu})$ équivalente à $\mathbb{P}_{\mu}$.

Le système (X,Y) défini sur le nouvel espace de probabilité $(\Omega', \underline{\underline{A}}^{\mu}, \mathbb{Q}_{\mu})$, représente le modèle de filtrage désiré. On vérifie que le signal X a même loi sur $(\Omega', \underline{\underline{A}}^{\mu}, \mathbb{Q}_{\mu})$ et $(\Omega', \underline{\underline{A}}^{\mu}, \mathbb{P}_{\mu})$. L'application du théorème de Girsanov (5), montre que le processus Y reste sur $(\Omega',\underline{\underline{A}}^{\mu},\underline{\underline{F}}_t^{\mu},Q_{\mu})$ une semi-martingale de nouvelle décomposition canonique:

$$Y_t = \alpha'_t + W'_t + \int_0^t \int_{\mathbb{R}} x\, 1\!\!1\{|x|>1\} N(ds,dx) + \int_0^t \int_{\mathbb{R}} x\, 1\!\!1\{|x|\leq 1\}(N(ds,dx)-\nu'(ds,dx))$$

où W' est le mouvement brownien sur $(\Omega',\underline{\underline{F}}_t^{\mu},Q_{\mu})$:

$$W'_t = W_t - \int_0^t h(X_s)ds$$

et ν' la mesure de Levy de N(ds,dx) relativement à $(\Omega',\underline{\underline{F}}_t^{\mu},Q_{\mu})$:

$$\nu'(dt,dx) = (1 + H(X_t,x))\, dt\nu(dx)$$

et:

$$\alpha'_t = \int_0^t h(X_s)ds + \int_0^t \int_{\mathbb{R}} x\, 1\!\!1\{|x|\leq 1\} H(X_s,x) ds\nu(dx)$$

Remarque: Le caractère P.A.I des parties martingales de Y n'est pas conservé au contraire du cas continu, par changement de probabilité, puisque la nouvelle mesure de Levy de Y est aléatoire. Ceci explique la différence de traitement avec le cas continu.

I-c/ Les processus de filtrage

Le processus de filtrage de X sachant Y associé à la loi initiale de X, μ, peut être défini de façon générale (13) comme l'unique processus Π^{μ}, à l'indistingabilité près sur $(\Omega',\underline{\underline{A}}^{\mu},Q_{\mu})$, à valeurs dans l'ensemble des lois de probabilité sur E, M(E), tel que:

a) Pour toute fonction f de b(E), l'ensemble des fonctions boréliennes bornées sur E, $\Pi_t^{\mu}(f)$ est une Q_{μ}-projection optionnelle de $f(X_t)$ par rapport à la filtration $\underline{\underline{G}}^{\mu}$.

b) X étant cadlag, Π^{μ} est un processus cadlag pour la topologie de la convergence étroite sur M(E).

En fait, la méthode de la probabilité de référence va

permettre d'expliciter par la formule de Kallianpur-Striebel, le processus de filtrage. En effet grâce à l'indépendance de X et Y sur l'espace de référence, on peut écrire:

$$E_{\mathbb{Q}_\mu}(f(X_t)/\underline{\underline{G}}_t^\mu) = \frac{E_{\mathbb{P}_\mu}(L_t^\mu f(X_t)/\underline{\underline{G}}_t^\mu)}{E_{\mathbb{P}_\mu}(L_t^\mu/\underline{\underline{G}}_t^\mu)} = \frac{E_{\mathbb{P}_\mu}(L_t^\mu f(X_t)/\underline{\underline{G}}_\infty^\mu)}{E_{\mathbb{P}_\mu}(L_t^\mu/\underline{\underline{G}}_\infty^\mu)}$$

Plus précisément, on note K_μ, le noyau markovien de $(\Omega',\underline{\underline{A}}^\mu)$ dans $(\Omega',\underline{\underline{G}}_\infty^\mu)$:

$$K_\mu(\omega_1,\omega_1^X;d\omega_2,d\omega_2^X) = \varepsilon_{\omega_1}(d\omega_2) \otimes \mathbb{P}_\mu^X(d\omega_2^X)$$

où ε_ω désigne la mesure de Dirac en ω . Il est alors facile de vérifier (12), que l'on a la formule de Kallianpur-Striebel:

$$\forall f \in b(E), \forall t \geq 0,\ \Pi_t^\mu(f) = K_\mu(L_t^\mu f(X_t))/K_\mu(L_t^\mu)$$

Ce rapport suggère la définition d'un autre processus, appelé le processus de filtrage non normalisé, $\tilde{\Pi}^\mu$, de X sachant Y, défini sur$(\Omega',\underline{\underline{A}}^\mu, \mathbb{P}_\mu)$ à valeurs dans l'ensemble $\tilde{M}(E)$ des mesures positives bornées sur E, tel que:

$$\forall f \in b(E),\ \forall t \geq 0,\ \tilde{\Pi}_t^\mu(f) = K_\mu(L_t^\mu f(X_t))$$

De plus on peut montrer que $\tilde{\Pi}^\mu$ est $\underline{\underline{G}}^\mu$-optionnel, cadlag pour la topologie de la convergence étroite sur $\tilde{M}(E)$, de carré intégrable pour tout t grâce aux propriétés de L^μ.

Il est classique à présent (12), que le processus de filtrage non normalisé satisfait aux deux équations différentielles stochastiques suivantes dites équations de Zakai:

$\forall \mu \in M(E)$, $\forall f \in \underline{\underline{D}}(\underline{\underline{L}})$, $\forall t \geq 0$, $\mathbb{P}$-p.s.

$$\tilde{\Pi}_t^\mu(f) = \mu(f) + \int_0^t \tilde{\Pi}_s^\mu(\underline{\underline{L}}f)\,ds + \int_0^t \tilde{\Pi}_s^\mu(fh)\,dW_s + \int_0^t\int_{\mathbb{R}} \tilde{\Pi}_{s-}^\mu(fH(.,x))(N(ds,dx)-ds\nu(dx))$$

$\forall\mu \in M(E)$, $\forall f\in C(E)$, $\forall t\geq 0$, $\mathbb{P}$-p.s.

$$\tilde{\Pi}^{\mu}_t(f)=\mu(P_tf)+\int_0^t\tilde{\Pi}^{\mu}_s(hP_{t-s}f)dW_s+\int_0^t\int_{\mathbb{R}}\tilde{\Pi}^{\mu}_{s-}(H(.,x)P_{t-s}f)(N(ds,dx)-ds\nu(dx))$$

II/ Caractère markovien des processus de filtrage

Dans (6) pour le cas continu, Kunita montrait en résolvant les équations du filtrage, le caractère markovien du processus de filtrage. On utilise ici la forme explicite des processus de filtrage, déduite par la méthode de la probabilité de référence,pour étendre ces résultats.

Il faut noter, que l'on définit pour chaque loi initiale μ des processus de filtrage $\tilde{\Pi}^{\mu}$ et Π^{μ}. Il s'agit donc de montrer le caractère markovien des familles $(\tilde{\Pi}^{\mu};\ \mu\in \tilde{M}(E))$ et $(\Pi^{\mu};\mu\in M(E))$. Afin de travailler sur de vrais processus de Markov, on va par des méthodes standard, agrandir l'espace de définition des filtres.

On note $(\tilde{\Omega},\tilde{\underline{\underline{A}}})$ et $(\Omega^*,\underline{\underline{A}}^*)$ les espaces suivants:

$$\tilde{\Omega} = \tilde{M}(E)\times\Omega,\ \tilde{\underline{\underline{A}}} = \tilde{\underline{\underline{M}}}(E)\otimes \underline{\underline{A}} \text{ et } \Omega^* = M(E)\times\Omega,\ \underline{\underline{A}}^* = \underline{\underline{M}}(E)\otimes \underline{\underline{A}}$$

Si $A\in \tilde{\underline{\underline{A}}}$ et $\mu\in \tilde{M}(E)$, on note A^{μ} la coupe en μ de A, i.e.:

$$A^{\mu} = \{\omega/\ (\mu,\omega)\in A\}$$

On munit $(\tilde{\Omega},\tilde{\underline{\underline{A}}})$ d'une filtration $\tilde{\underline{\underline{G}}}$:

$$\tilde{\underline{\underline{G}}}_t = \sigma(A\in \tilde{\underline{\underline{A}}}/A^{\mu}\in \underline{\underline{G}}_t,\ \forall\mu\in \tilde{M}(E))$$

où $\sigma(\ )$ désigne la tribu engendrée par ( ). On définit de façon analogue une filtration $\underline{\underline{G}}^*$, sur $(\Omega^*,\underline{\underline{A}}^*)$. On construit sur $(\tilde{\Omega},\tilde{\underline{\underline{A}}})$ (resp. $(\Omega^*,\underline{\underline{A}}^*)$)la famille de probabilités $(\tilde{\mathbb{P}}_{\mu};\mu\in \tilde{M}(E))$ (resp. $Q^*_{\mu};\mu\in M(E))$ en posant:

$\forall A\in \tilde{\underline{\underline{A}}}$, $\forall\mu\in \tilde{M}(E)$, $\tilde{\mathbb{P}}_{\mu}(A) = \mathbb{P}(A^{\mu})$ (resp. $\forall A\in \underline{\underline{A}}^*$, $\forall\mu\in M(E)$, $Q^*_{\mu}(A)=Q_{\mu}(A^{\mu})$)

On a de même les opérateurs de translation sur $\tilde{\Omega}$ et Ω^*:

$\forall(\mu,\omega)\in \tilde{\Omega}$, $\tilde{\theta}_t(\mu,\omega) = (\mu,\theta_t\omega)$ et $\forall(\mu,\omega)\in \Omega^*$, $\theta^*_t(\mu,\omega) = (\mu,\theta_t\omega)$

On peut alors définir à partir des processus de filtrage $\tilde{\Pi}^{\mu}$ et Π^{μ} de nouveaux processus sur $\tilde{\Omega}$ et Ω^*, en posant:

$$\forall t \geq 0,\ \forall \mu \in \tilde{M}(E),\ \forall \omega \in \tilde{\Omega},\quad \tilde{\Pi}_t(\mu,\omega) = \tilde{\Pi}_t^{\mu}(\omega)$$

$$\forall t \geq 0,\ \forall \mu \in M(E),\ \forall \omega \in \Omega,\quad \Pi_t(\mu,\omega) = \Pi_t^{\mu}(\omega)$$

On a dans la première relation étendu de façon évidente l'opérateur K_{μ} aux μ de $\tilde{M}(E)$.

On a le lemme suivant:

Lemme 1: Les processus $\tilde{\Pi}$ et Π sont respectivement des processus $\tilde{\underline{\underline{G}}}$- et $\underline{\underline{G}}^*$-optionnels.

Démonstration: Il suffit de montrer la mesurabilité faible de ces processus, le reste du lemme découlant de la séparabilité des ensembles $M(E)$ et $\tilde{M}(E)$ et de la continuité à droite des processus de filtrage. Pour cela, on montre qu'il existe une $(\Omega', \underline{\underline{F}}_t^{\mu}, \mathbb{P}_{\mu})$-martingale L indépendante de μ, $\mathbb{P}_{\mu}$-indistinguable de la martingale L^{μ} pour tout μ, et on conclut par application du théorème de Fubini sur la formule de Kallianpur-Striebel.

Soit u une fonction cadlag déterministe à valeurs dans E, telle que:

$$\int_0^{\infty} h^2(u_s)ds + \int_0^{\infty}\int_{\mathbb{R}} H^2(u_s,x)ds\nu(dx) < \infty$$

On définit alors sur $(\Omega, \underline{\underline{G}}, \mathbb{P})$ le processus $L(u)$ en remplaçant formellement dans les formules de définition de L^{μ}, X par u. D'après (9), $L(u)$ est une fonction borélienne de u. Soit alors le processus L, défini sur $(\Omega', \underline{\underline{A}}^{\mu})$:

$$\forall t \geq 0,\ L_t(\omega,\omega^X) = L_t(X(\omega^X))(\omega)$$

On vérifie que L satisfait aux propriétés désirées.

Remarque: Sous deux hypothèses supplémentaires,

$h(X_t)$ est une $(\Omega^X, \underline{\underline{A}}^X, \underline{\underline{F}}_t^X, \mathbb{P}_x^X)$-semi-martingale pour tout x de E,

et $\int_{\mathbb{R}}(x\wedge 1)\nu(dx) < \infty$,

on montre dans (11) que l'on peut choisir une version de L, et donc des processus de filtrage, qui sont des fonctions boréliennes des trajectoires de Y, c'est-à-dire adaptées à la filtration naturelle non complétée de l'observation Y. On utilise une transformation classique (voir (3) par exemple) et des résultats conjoints de (2) et (9) sur le calcul stochastique dépendant d'un paramètre.

Le lemme et la proposition suivants, conséquences du caractère P.A.I homogène de Y sur l'espace de référence, conduira aux propriétés markoviennes des processus de filtrage.

Lemme 2: Le processus L vérifie la propriété suivante:

$$\forall s,t \geq 0,\ \mathbb{P}_\mu\text{-p.s.},\ L_{t+s}(\omega,\omega^X) = L_s(\omega,\omega^X)L_t(\theta_s\omega,\theta_s^X\omega^X)$$

Démonstration: On a: $\forall t,s \geq 0$

$$L_{t+s} = L_s \exp\{\int_s^{s+t} h(X_u)dW_u - 1/2\int_s^{s+t} h^2(X_u)du - \int_s^{s+t}\int_{\mathbb{R}} H(X_u,x)du\nu(dx)\} \prod_{\substack{s<u\leq t\\ \Delta Y_u\neq 0}} (1+H(X_{u-},\Delta Y_u))$$

Il suffit alors de montrer:

$$\left(\int_s^{s+t} h(X_u)dW_u\right)(\omega,\omega^X) = \left(\int_0^t h(X_u)dW_u\right)(\theta_s\omega,\theta_s^X\omega^X)$$

ce qui est vrai pour des fonctions élémentaires

$$h(X_u) = \sum_{i=1}^{i=n} h(X_{t_i})\, 1\!\!1_{]t_i,t_{i+1}]}(u)$$

et par limite pour une fonction $h(X_u)$ arbitraire.

Proposition 3: On a les égalités suivantes:

$\forall t,s\geq 0,\ \forall \mu\in M(E),\ \tilde{\mathbb{P}}_{\mu}\text{-p.s.},\ \tilde{\Pi}_{t+s}(\mu,\omega) = \tilde{\Pi}_t(\tilde{\Pi}_s(\mu,\omega),\theta_s\omega)$

$\forall t,s\geq 0,\ \forall \mu\in M(E),\ Q^*_{\mu}\text{-p.s.},\ \Pi_{t+s}(\mu,\omega) = \Pi_t(\Pi_s(\mu,\omega),\theta_s\omega)$

Démonstration: On s'occupe d'abord du filtre non normalisé.

$\forall t,s\geq 0,\ \forall \mu\in \tilde{M}(E),\ \forall f\in b(E),$

$$\tilde{\Pi}_{t+s}(\mu,\omega)(f) = \int_{\Omega^X}\mathbb{P}^X_{\mu}(d\omega^X)\ L_{t+s}(\omega,\omega^X)\ f(X_{t+s}(\omega^X))$$

D'après le lemme 2:

$$\tilde{\Pi}_{t+s}(\mu,\omega)(f) = \int_{\Omega^X}\mathbb{P}^X_{\mu}(d\omega^X)\ L_s(\omega,\omega^X)\{L_t(\theta_s\omega,.)f(X_t(.))\}(\theta^X_s\omega^X)$$

On applique ensuite la propriété de Markov de X:

$$\tilde{\Pi}_{t+s}(\mu,\omega)(f) = \int_{\Omega^X}\mathbb{P}^X_{\mu}(d\omega^X)\ L_s(\omega,\omega^X)\ \int_{\Omega^X}\mathbb{P}^X_{X_s(\omega X)}(d\omega'^X)L_t(\theta_s\omega,\omega'^X)f(X_t(\omega'^X))$$

Ce qui est exactement, appliqué à f:

$$\tilde{\Pi}_{t+s}(\mu,\omega) = \tilde{\Pi}_t(\tilde{\Pi}_s(\mu,\omega),\theta_s\omega)$$

La formule de Kallianpur-Striebel permet de passer à l'égalité correspondante pour Π. En effet:

$$\Pi_{t+s}(\mu,\omega) = \tilde{\Pi}_t(\tilde{\Pi}_s(\mu,\omega),\theta_s\omega)\ /\ \tilde{\Pi}_t(\tilde{\Pi}_s(\mu,\omega),\theta_s\omega)(1)$$

Or comme L a été choisie indépendamment de μ, on a:

$$\tilde{\Pi}_t(\mu,\omega) = \mu(1)\ \tilde{\Pi}_t(\mu/\mu(1),\omega)$$

et donc divisant numérateur et dénominateur par $\tilde{\Pi}_s(\mu,\omega)(1)$, il vient:

$$\Pi_{t+s}(\mu,\omega) = \tilde{\Pi}_t(\Pi_s(\mu,\omega),\theta_s\omega)\ /\ \tilde{\Pi}_t(\Pi_s(\mu,\omega),\theta_s\omega)(1)$$

C'est-à-dire l'égalité recherchée.

Le caractère markovien des processus de filtrage s'obtient alors facilement:

Proposition 4: Le processus $\tilde{\Pi}$, $\tilde{\Pi} = (\tilde{\Omega},\tilde{\underline{\underline{A}}},\tilde{\underline{\underline{G}}}_t,\tilde{\theta}_t,\tilde{\Pi}_t,(\tilde{\mathbb{P}}_{\mu};\mu\in M(E)))$ est un processus de Markov. Il en est de même du processus Π, $\Pi = (\Omega^*,\underline{\underline{A}}^*,\underline{\underline{G}}^*_t,\theta^*_t,\Pi_t,(Q^*_{\mu};\mu\in M(E)))$.

Démonstration: On commence par $\tilde{\Pi}$. Soit Γ un borélien de $\tilde{M}(E)$, on a:

$$\tilde{\mathbb{P}}_{\mu}(\tilde{\Pi}_{t+s} \in \Gamma \,/\, \tilde{\underline{\underline{G}}}_s) = \tilde{\mathbb{P}}(\tilde{\Pi}_{t+s}(\mu,.) \in \Gamma \,/\, \underline{\underline{G}}_s) = \mathbb{P}(\{\tilde{\Pi}_t(\tilde{\Pi}_s(\mu,\omega),.) \in \Gamma\} \circ \theta_t / \underline{\underline{G}}_s)$$

Comme $\tilde{\Pi}_s(\mu,\omega)$ est $\underline{\underline{G}}_s$-mesurable, et Y un P.A.I homogène;

$$\tilde{\mathbb{P}}_{\mu}(\tilde{\Pi}_{t+s} \in \Gamma \,/\, \tilde{\underline{\underline{G}}}_s) = \tilde{\mathbb{P}}_{\tilde{\Pi}_s(\mu,\omega)}(\tilde{\Pi}_t \in \Gamma)$$

Ce qui est la propriété de Markov pour $\tilde{\Pi}$.

On vérifie ensuite la propriété de Markov pour Π . Soit Γ un borélien de M(E), on a:

$$Q^*_{\mu}(\Pi_{t+s} \in \Gamma \,/\, \underline{\underline{G}}^*_s) = Q_{\mu}(\Pi_{t+s}(\mu,.) \in \Gamma \,/\, \underline{\underline{G}}_s)$$

Par changement de probabilité et par définition de $\tilde{\Pi}$, il vient:

$$Q^*_{\mu}(\Pi_{t+s} \in \Gamma \,/\, \underline{\underline{G}}^*_s) = E_{\mathbb{P}}(\tilde{\Pi}_{t+s}(\mu,.)(1)\, 1\!\!1\{\Pi_{t+s}(\mu,.) \in \Gamma\} / \underline{\underline{G}}_s) / \tilde{\Pi}_s(\mu,\omega)(1)$$

Grâce à la proposition 3, et la propriété de P.A.I homogène de Y, il vient:

$$Q^*_{\mu}(\Pi_{t+s} \in \Gamma \,/\, \underline{\underline{G}}^*_s) = E_{\mathbb{P}_{\Pi_s(\mu,\omega)}}(\tilde{\Pi}_t(1)\, 1\!\!1\{\Pi_t \in \Gamma\}) = Q^*_{\Pi_s(\mu,\omega)}(\Pi_t \in \Gamma)$$

La propriété de Feller du processus de filtrage sera montrée dans la prochaine partie, à l'aide des équations du filtrage.

III/ Equations du filtrage

La forme des équations non normalisées du filtrage, ou équations de Zakai rappelée au I, conduit à définir comme dans le cas continu (10), des équations générales qui auront comme solution unique, dans le cadre d'un problème de filtrage, le processus non normalisé du filtrage.

On considère donnés un espace de Hausdorff compact séparable muni de sa tribu borélienne $(E,\underline{\underline{E}})$, une fonction h réelle continue sur E, une fonction borélienne bornée H sur $E\times\mathbb{R}$, continue sur E et telle que $1+H$ soit minorée par une constante strictement positive, une mesure positive bornée sur $\mathbb{R}$, ν, et un semi-groupe de Feller sur $C(E)$, $(P_t;t\geq 0)$, de générateur infinitésimal $(\underline{\underline{L}},\underline{\underline{D}}(\underline{\underline{L}}))$.

Soit alors un ensemble $(\Omega,\underline{\underline{A}},\underline{\underline{G}},\mathbb{P},Y,u_o,u_t)$, vérifiant les propriétés suivantes:

(i) $(\Omega,\underline{\underline{A}},\underline{\underline{G}},\mathbb{P})$ est un espace de probabilité filtré vérifiant les conditions habituelles.

(ii) Le processus Y est la somme sur $(\Omega,\underline{\underline{A}},\mathbb{P})$ d'un mouvement brownien W et d'un processus de Poisson homogène N de mesure de Levy $dt\nu(dx)$ relativement à la filtration $\underline{\underline{G}}$.

(iii) u_o est une variable aléatoire sur $(\Omega,\underline{\underline{A}},\mathbb{P})$, $\underline{\underline{G}}_o$-mesurable à valeurs dans $\tilde{M}(E)$, de carré intégrable i.e. $E((u_o(1))^2) < \infty$.

(iv) $(u_t;t\geq 0)$ est un processus $\underline{\underline{G}}$-adapté à valeurs dans $\tilde{M}(E)$, cadlag pour la topologie de la convergence étroite de carré intégrable pour tout t, i.e. $\forall t\geq 0$, $E((u_t(1))^2) < \infty$ de valeur initiale u_o.

<u>Définition 5</u>: On dit qu'un ensemble $(\Omega,\underline{\underline{A}},\underline{\underline{G}},\mathbb{P},Y,u_o,u_t)$ vérifiant les conditions (i) à (iv) ci-dessus, est solution de l'équation(Z-1), respectivement (Z-2), si de plus:

(Z-1) $\forall f\in\underline{\underline{D}}(\underline{\underline{L}})$, $\forall t\geq 0$, $\mathbb{P}$-p.s.

$$u_t(f) = u_o(f)+\int_0^t u_s(\underline{\underline{L}}f)ds+\int_0^t u_s(fh)dW_s+$$
$$+\int_0^{t}\int_{\mathbb{R}} u_{s-}(H(.,x)f)\{N(ds,dx)-ds\nu(dx)\}$$

respectivement,

(Z-2) $\forall f \in C(E)$, $\forall t \geq 0$, $\mathbb{P}$-p.s.,

$$u_t(f) = u_\circ(P_t f) + \int_0^t u_s(hP_{t-s}f)\,dW_s + \int_0^t \int_{\mathbb{R}} u_{s-}(H(.,x)P_{t-s}f)\,\{N(ds,dx) - ds\nu(dx)\}$$

Nous ne montrerons que ce qui diffère sensiblement du cas continu normalisé de (6) ou (10). Comme dans (10), ces deux équations sont équivalentes entre elles dans la mesure où toute solution de l'une est solution de l'autre. On a aussi la proposition suivante.

Proposition 6: Il y a unicité trajectorielle des solutions de l'équation (Z-2).

Démonstration: Si on a deux solutions de (Z-2), u et u', définies sur le même espace de probabilité $(\Omega, \underline{\underline{A}}, \underline{\underline{G}}, \mathbb{P})$ avec le même P.A.I Y et la même condition initiale $u_\circ$, on montre que u et u' sont indistinguables. On remarque d'abord que si u est solution de (Z-2), sa norme $u_t(1)$ est une $(\Omega, \underline{\underline{G}}, \mathbb{P})$-martingale. En effet:

$$u_t(1) = u_\circ(1) + \int_0^t u_s(h)\,dW_s + \int_0^t \int_{\mathbb{R}} u_{s-}(H(.,x))\,\{N(ds,dx) - ds\nu(dx)\}$$

$u_t(1)$ est donc déjà une martingale locale, qui est par hypothèse de carré intégrable pour tout t, d'où c'est une vraie martingale. On en déduit alors, ce qui permettra de faire comme dans (6):

$$\forall T < \infty \quad , \sup_{t \leq T} E((u_t(1))^2) \leq E((u_T(1))^2) < \infty$$

La démonstration de l'unicité repose alors sur l'application du lemme de Gronwall. On note:

$$\forall f \in C(E), \ \rho_t(f) = E((u_t(f) - u'_t(f))^2)$$

On a, si $\|f\| = \sup_E |f(x)|$ et T est fixé:

$$\rho_t(f) \leq 2\|f\|^2 \left(E(u_T(1))^2 + E(u'_T(1))^2 \right)$$

D'où pour une constante c,

$$\rho_t(f) \le c\,\|f\|^2$$

u et u' étant solutions de (Z-2), il vient:

$$\forall t \le T,\ \rho_t(f) = \int_0^t \rho_s(hP_{t-s}f)ds + \int_0^t\int_{\mathbb{R}} \rho_s(H(.,x)P_{t-s}f)ds\nu(dx)$$

D'où, reportant la première majoration dans l'égalité précédente, on a:

$$\rho_t(f) \le ckt\|f\|^2$$

avec $k = (\|h\|^2 + \|H\|^2)(1 + \nu(\mathbb{R}))$

Après n itérations, il vient:

$$\rho_t(f) \le c\|f\|^2(kt)^n/n!$$

D'où, $\forall t \le T$, $\forall f \in C(E)$, $\rho_t(f) = 0$ et $u_t(f) = u'_t(f)$ $\mathbb{P}$-p.s..

Comme u et u' sont cadlag, C(E) séparable et T arbitraire, u et u' sont indistinguables.

Remarque: L'hypothèse, $\forall t \ge 0$, $E((u_t(1))^2) < \infty$ est nécessaire à la définition des solutions des équations non normalisées du filtrage.

On montre comme dans (10) la proposition suivante.

Proposition 7: *Un ensemble* $(\Omega, \underline{\underline{A}}, \underline{\underline{G}}, \mathbb{P}, Y, u_0)$ *étant donné, vérifiant les propriétés (i) à (iii) ci-dessus, il existe une solution* $(\Omega, \underline{\underline{A}}, \underline{\underline{G}}, \mathbb{P}, Y, u_0, u_t)$ *à l'équation (Z-2).*

Comme dans (4), on peut montrer le lemme suivant.

Lemme 8: *Si* $(u^\mu; \mu \in \tilde{M}(E))$ *est une famille de solutions de l'équation (Z-2) associées aux lois initiales* μ *de* $\tilde{M}(E)$, *on a:*

$$\forall f \in C(E),\ \lim_{\mu_n \to \mu} E_{\mathbb{P}}((u_t^{\mu_n}(f) - u_t^{\mu}(f))^2) = 0$$

pour toute suite μ_n *de* $\tilde{M}(E)$ *convergeant étroitement vers* μ.

On en déduit la propriété de Feller du processus de filtrage.

Proposition 9: *Le processus de filtrage* Π ,

$\Pi = (\Omega^*, \underline{A}^*. \underline{\underline{G}}^*_t, \theta^*_t, \Pi_t, (Q^*_\mu; \mu \in M(E)))$, *est un processus de Feller, cadlag à valeurs dans l'ensemble des lois de probabilité sur E.*

Démonstration: Soit $F(\mu)$ une fonction continue sur $M(E)$ construite de la façon suivante:
Soient $n \in \mathbb{N}$, $f_1, \ldots, f_n \in C(E)$ et une fonction $F(x_1, \ldots, x_n)$ sur $\mathbb{R}^n$, telle que la fonction $\phi(k, x_1, \ldots, x_n) = kF(x_1/k, \ldots, x_n/k)$ soit lipchitzienne de rapport α en $k, x_1, \ldots, x_n$. Alors on pose:

$$F(\mu) = F(\mu(f_1), \ldots, \mu(f_n)).$$

On montre que $E_{Q^*_\mu}(F(\Pi_t))$, est continue sur $M(E)$. En effet, par changement de probabilité, il vient:

$$\forall \mu, \nu \in M(E), \ |E_{Q^*_\mu}(F(\Pi_t)) - E_{Q^*_\nu}(F(\Pi_t))| \leq \sum_{i=0}^{i=n} E_{\mathbb{P}}(|\Pi^\nu_t(f_i) - \Pi^\mu_t(f_i)|)$$

en convenant que $f_0 = 1$. L'application du lemme 8 entraine le résultat.

Il est alors facile de voir, $M(E)$ étant compact, que le théorème de Stone-Weierstrass s'applique et que l'ensemble de telles fonctions réelles sur $M(E)$, est dense pour la convergence uniforme dans l'ensemble de toutes les fonctions réelles continues sur $M(E)$. On en déduit par limite uniforme, que si $G(\mu)$ est continue sur $M(E)$, $E_{Q^*_\mu}(G(\Pi_t))$ est aussi continue sur $M(E)$.

Comme Π est cadlag pour la topologie de la convergence étroite;

$$\lim_{t\to 0} E_{Q^*_\mu}(G(\Pi_t)) = G(\mu)$$

On a ainsi montré que Π est un processus de Feller

Bibliographie

(1) P. BREMAUD, M. YOR :"Changes of filtration and probability measures" Z.f.Wahr.V.Geb. 45 (1978) 269-295.

(2) E. CINLAR, J. JACOD, P. PROTTER, M.J. SHARPE :"Semi-martingales and Markov Processes" Z.f.Wahr.V.Geb. 54 (1980) 161-219.

(3) M.H.A. DAVIS :"On a Multiplicative Functional Transformation Arising in Non Linear Filtering Theory" Z.f.Wahr.V.Geb. 54 (1980) 125-140.

(4) I. GERTNER :"An alternative approach to non linear filtering" Stoch. Proc. Appl. 7 (1978) 231-246.

(5) J. JACOD, J. MEMIN :"Caractéristiques locales et conditions de continuité absolue pour les semi-martingales" Z.f.Wahr.V.Geb. 35 (1976) 1-37.

(6) H. KUNITA :"Asymptotic Behaviour of the Non Linear Filtering Error of Markov Processes" J. Mult. Anal. 1 N°4 (1971) 365-393.

(7) D. LEPINGLE, J. MEMIN :"Sur l'intégrabilité uniforme des martingales exponentielles" Z.f.Wahr.V.Geb. 42 (1978) 175-203.

(8) R.S. LIPSTER, A.N. SHIRYAYEV :"Statistic of Random Processes" Appl. of Math. N°5 Springer-Verlag 1977.

(9) C. STRICKER, M. YOR :"Calcul stochastique dépendant d'un paramètre" Z.f.Wahr.V.Geb. 45 (1978) 109-133.

(10) J. SZPIRGLAS :"Sur l'équivalence d'équations différentielles stochastiques à valeurs mesures intervenant dans le filtrage markovien non linéaire" Ann. Inst. H. Poincaré, Vol XIV N°1 (1978) 33-59.

(11) J. SZPIRGLAS :"Contributions aux théories du filtrage et du contrôle stochastiques" Thèse Univ. Paris IV, 21 Juin 1982.

(12) J. SZPIRGLAS, G. MAZZIOTTO :"Modèle général de filtrage non linéaire et équations différentielles stochastiques associées" Ann. Inst. H. Poincaré, Vol XV N°2 (1979) 147-173.

(13) M. YOR :"Sur les théories du filtrage et de la prédiction" Sem. Prob. XI, Lect. Notes in Math. N°581, Springer-Verlag 1977.

(14) M. ZAKAI :"On the Optimal Filtering of Diffusion Processes" Z.f.Wahr. V.Geb. 11 (1969) 230-249.

EFFICIENT NUMERICAL SCHEMES FOR THE APPROXIMATION OF EXPECTATIONS OF FUNCTIONALS OF THE SOLUTION OF A S.D.E., AND APPLICATIONS

Denis TALAY
I N R I A
Sophia-Antipolis
06560 VALBONNE
FRANCE

Summary : This paper presents results appearing in Talay [8], and show their possible applications in non linear filtering in particular. We develop a method to approximate expectations of functionals of the solution of a S.D.E. ; this method is efficiently implementable on a computer, and is based on both Monte-Carlo methods and discretizations of S.D.E. Convergence is shown and bounds of the error are given.

INTRODUCTION

Let (X_t) be the solution of the Stochastic Differential Equation :

$$\text{(E)} \qquad X_t = X_o + \int_o^t b(X_s)\, ds + \int_o^t \sigma(X_s)\, dW_s \quad , \; 0 \le t \le T$$

Where :

- $X_t \in R^d$, $W_t \in R^m$
- X_o is a random variable independent of (W_t), whose the moments are finite.

The aim of this article is to propose methods for the numerical approximation of such quantities as :

$$\text{(1)} \qquad E \{ g(X_t) \exp \{ \int_o^t \alpha(X_s)\, ds + \int_o^t \beta(X_s)\, dW_s \} \}$$

where :

- $g : R^d \to R$ is a regular function, of growth at most polynomial
- $\alpha : R^d \to R$, $\beta : R^d \to L(R^m, R)$ are regular bounded functions.

First, let us remark that the problem (1) can be written again in a simpler way :

Let $Z_t = (Z_t^{(1)}, Z_t^{(2)}, Z_t^{(3)})$ be defined by :

$$Z_t = (X_t, \exp \int_o^t \alpha(X_s)\, ds \; , \; \exp \int_o^t \beta(X_s)\, dW_s)$$

(Z_t) satisfies the S.D.E. :

$$dZ_t^{(1)} = b(Z_t^{(1)})\,dt + \sigma(Z_t^{(1)})\,dW_t$$

$$dZ_t^{(2)} = \alpha(Z_t^{(1)}).\,Z_t^{(2)}\,dt$$

$$dZ_t^{(3)} = \frac{1}{2}\{\,Z_t^{(3)}\ \beta(Z_t^{(1)})\beta(Z_t^{(1)})^*\,\}dt + Z_t^{(3)}\ \beta(Z_t^{(1)})\,dW_t$$

Let $\rho : R^d \times R \times R$ be defined by : $\rho(x, y, z) = g(x).y.z$

Then we want to approximate :

$$E\,\rho(\,Z_T^{(1)}\,,\,Z_T^{(2)}\,,\,Z_T^{(3)}\,)$$

and the problem is (2) :

(2) "GIVEN (X_t) THE SOLUTION OF A S.D.E., AND ρ ANY "REGULAR" FUNCTION OF GROWTH AT MOST POLYNOMIAL, DEVELOP A GOOD METHOD WHICH CAN BE PERFORMED EFFICIENTLY ON COMPUTERS TO APPROXIMATE :

$$E\,\rho(X_T) \text{ "}$$

In part 1, we describe methods and we study their order of convergence.
In part 2, we apply these results to the numerical approximation of the solution of a parabolic type P.D.E., possibly of large dimension.
In part 3, we are interested in the numerical treatment of Kallianpur-Striebel Formula in non linear filtering theory.

<u>IMPORTANT REMARK</u> : most proofs of results of part 1 are only sketched ; for more details and comments, please refer to Talay [8].

PART 1 : EFFICIENT METHODS

1 - 1 Principle of the methods

Let τ be a random variable and f a borelian function, such that $E|f(\tau)|$ is finite.

To compute the expectation of $f(\tau)$, $Ef(\tau)$, one uses a "Monte-Carlo method" : one simulates N independent realizations of τ, denoted by $\tau^{(i)}$ $(1 \leq i \leq N)$, and one computes the sum :

$$\frac{1}{N} \sum_{i=1}^{N} f(\tau^{(i)})$$

according to the strong law of large numbers, this sum converges to $Ef(\tau)$. (For improvements of the method, see Rubinstein [7] for example).

So, in view of solving the problem (2) of the introduction, we should like to simulate N realizations of X_t. Unfortunately, in most cases, we do not know the law of X_t. Hence we discretize (E) to get an approximate solution of it, $(\overline{X}_t)$, whose we simulate N independent realizations, $\overline{X}_t^{(i)}$, $1 \leq i \leq N$, and we do compute :

$$\frac{1}{N} \sum_{i=1}^{N} f(\overline{X}_T^{(i)})$$

The error of the method satisfies the following inequality :

$$\left|E\, f(X_T) - \frac{1}{N} \sum_{i=1}^{N} f(\overline{X}_T^{(i)})\right| \leq \left|E\, f(X_T) - E\, f(\overline{X}_T)\right|$$

$$+ \left|E\, f(\overline{X}_T) - \frac{1}{N} \sum_{i=1}^{N} f(\overline{X}_T^{(i)})\right| = \varepsilon_1 + \varepsilon_2$$

To minimize ε_2, one only needs to choose a sufficient large N.

The error ε_1 depends on the step of the discretization of $[0,T]$ which is used, and denoted by h.

So our aim is to produce an algorithm of discretization which has the following properties :

1) the error ε_1 satisfies the inequality :
$$\varepsilon_1 \leq C.h^r$$
where : C is a constant and r is large enough.
2) in view to minimize the computation time (remember that we must get N independent trajectories of $(\bar{X}_t)$), the complexity of the algorithm is reduced.
3) in the same view, the laws of the random variables occuring in the algorithm are simple to simulate ; a good choice would be discrete laws.

1 - 2 <u>What about usual discretization schemes</u> ?

Let $h = \frac{T}{M}$, $M \in \mathbb{N}$, be the step of the discretization of $[0,T]$:

$$\{ 0 = t_o , h = t_1, \ldots, T = t_M \}$$

Suppose the dimensions d and m equal to 1. Let us consider the scheme :

$$\bar{X}_o^{(h)} = X_o$$

$$\begin{aligned} \bar{X}_{p+1}^{(h)} &= \bar{X}_p^{(h)} + \sigma(\bar{X}_p^{(h)}) (W_{t_{p+1}} - W_{t_p}) \\ &+ \{ b(\bar{X}_p^{(h)}) - \frac{1}{2} \sigma(\bar{X}_p^{(h)}) \sigma'(\bar{X}_p^{(h)}) \} h \\ &+ \frac{1}{2} \sigma(\bar{X}_p^{(h)}) \sigma'(\bar{X}_p^{(h)}) (W_{t_{p+1}} - W_{t_p})^2 \end{aligned}$$

This scheme is the best one within a large class of schemes for the mean-square approximation and the trajectorial approximation of (X_t) (see Pardoux & Talay [6]).

Nevertheless, generally we only get :

$$|E\, f(X_T) - E\, f(\overline{X}_T^{(h)})| \leq C.h$$

Thus, for $b \equiv 0$, $\sigma \equiv \mathrm{Id}$, $f : x \to x^2$;

$$E\, f(\overline{X}_T^{(h)}) - E\, f(X_T) = C.h + O(h^2)$$

Such a speed of convergence is too low.

1 - 3 The "Monte-Carlo scheme"

Again, $d = m = 1$.

Milshtein [4] has constructed a new scheme and given some good reasons to expect its speed of convergence would be of order h^2.

Our proof of this statement is given below ; it appears that it works also for the following version of the Milshtein's scheme, which is better from a computational point of view (from now on, every time no confusion is possible, we shall write $\overline{X}_p$ instead of $\overline{X}_p^{(h)}$).

$$\overline{X}_o = X_o$$

$$\begin{aligned}\overline{X}_{p+1} = \overline{X}_p &+ \sigma(\overline{X}_p).R_{p+1}\sqrt{h} \\ &+ \{\, b(\overline{X}_p) - \tfrac{1}{2}\,\sigma(\overline{X}_p)\,\sigma'(\overline{X}_p)\,\}\, h \\ &+ \tfrac{1}{2}\,\sigma(\overline{X}_p)\,\sigma'(\overline{X}_p)\,(R_{p+1})^2\, h \\ &+ \tfrac{1}{2}\,\{b(\overline{X}_p)\,\sigma'(\overline{X}_p) + b'(\overline{X}_p)\,\sigma(\overline{X}_p) + \tfrac{1}{2}\,\sigma^2(\overline{X}_p)\,\sigma''(\overline{X}_p)\}R_{p+1}\, h^{3/2} \\ &+ \tfrac{1}{2}\,\{\, b(\overline{X}_p)\, b'(\overline{X}_p) + \tfrac{1}{2}\,\sigma^2(\overline{X}_p)\, b''(\overline{X}_p)\,\}\, h^2\end{aligned}$$

where :

1) $R_o = 0$

2) R_{p+1} is a random variable independent of the σ-field generated by $\{\, R_j,\ 0 \leq j \leq p\,\}$

3) R_{p+1} satisfies :

$E(R_{p+1}) = E(R_{p+1})^3 = E(R_{p+1})^5 = 0$

$E(R_{p+1})^2 = 1$

$E(R_{p+1})^4 = 3$

$E(R_{p+1})^6 < +\infty$

For example, the law of R_{p+1} is a discrete law such that :

$P\{R_{p+1} = \sqrt{3}\} = \frac{1}{6}$

$P\{R_{p+1} = -\sqrt{3}\} = \frac{1}{6}$

$P\{R_{p+1} = 0\} = \frac{2}{3}$

We denote this law by $\mathcal{R}$.

Remark : it is clear that $R_{p+1}\sqrt{h}$ plays the part of $W_{t_{p+1}} - W_{t_p}$.

1 - 4 Accuracy of the Monte-Carlo scheme (d=m=1)

We wish to prove : "for any given function f in a large subset of the space of continuous functions, it exists a real C not depending on h such that, if $(\overline{X}_t^{(h)})$ is the approximate solution of (E), piecewise constant on each interval $[t_p = ph; t_{p+1} = (p+1)h[$, and given by the Monte-Carlo scheme :

(1-4 ; a) $$|E\,f(X_T) - E\,f(\overline{X}_T^{(h)})| \leq C.h^2 \text{ "}$$

The main tool of the proof is the Kolmogorov's backward equation associated to the diffusion process (X_t) and the function f :

(KBE) $$\begin{cases} \frac{\partial v}{\partial t}(t,.) + Lv(t,.) = 0, \; 0 \leq t < T \\ v(T,.) = f(.) \end{cases}$$

This tool has been suggested to the author by A. Bensoussan ; the idea is to substitute (1-4 ; b) to (1-4 ; a) :

(1-4 ; b) $$|\,Ev(T,X_T) - Ev(T,\overline{X}_T^{(h)})\,| \leq C.h^2$$

Before giving the main lines of the proof, we state 2 lemmas.

Lemma 1-4-1

Let us suppose :

(H1) b, σ are continuous functions and their derivatives up to order 6 are continuous bounded functions.

(H2) f and its derivatives up to order 6 are continuous functions such that :

$$\exists M_1 \in R, \ \exists r \in N / \forall x \in R : |f^{(i)}(x)| \leq M_1(|x|^r + 1), \ 0 \leq i \leq 6$$

Then the P.D.E. (KBE) has a solution ; the solution we exhibit has the following properties :

(i) v is 6 times continuously differentiable with respect to x ; moreover :

$$\exists s \in N, \forall T, \exists M_2 \in R / \forall x \in R, \forall t \in [0,T],$$

$$\left| \frac{\partial^i}{\partial x^i} v(t,x) \right| \leq M_2 \left(|x|^s + 1\right), \ 0 \leq i \leq 6$$

(ii) v and its derivatives with respect to x are continuous in (t,x).

Sketch of the proof of lemma 1-4-1

Let us first remark that we do not assume : σ bounded, nor $\sigma(x) \geq \sigma_o > 0$, nor f bounded. Then we must use a probabilistic interpretation of (KBE).

For fixed (t,x), let $(X_\Theta^{t,x})$ be the process defined by :

$$X_\Theta^{t,x} = x + \int_t^\theta b(X_u^{t,x})\, du + \int_t^\theta \sigma(X_u^{t,x})\, dW_u$$

Under the above assumptions, $(X_\Theta^{t,x})$ is 6 times continuously differentiable with respect to x (see Kunita [3]).

Computing the derivatives of $(X_\Theta^{t,x})$ with respect to x, we get the following inequalities :

. $\forall T,\ \exists M_3 \in R \ / \forall n \in N, \forall t \in [0,T], \forall \Theta \in [t,T], \forall x \in R$:

$$E \mid X_\Theta^{t,x} \mid \leq M_3 \ (1+x^{2n})$$

. The processes :

$\frac{\partial^i}{\partial x_i} X_\Theta^{t,x} (1 \leq i \leq 6)$

have all their moments bounded uniformly in (Θ, t, x) over $[0,T] \times [0,T] \times R$.

Let v be defined by :

$(t,x) \in [0,T] \times R \to v\ (t,x) = E\ f(X_T^{t,x})$

It is known that v is solution of the P.D.E. (KBE) (see Gikhman & Skorokhod [1] for example). The properties (i), (ii) follow from the above properties of $(X_\Theta^{t,x})$. $\blacksquare$

Easy computations lead to

<u>Lemma 1.4.2.</u>

Under the assumption (H1) of Lemma 1-4-1 and the following :

(H1') $\exists M_4 \in R \ / \ \forall x \in R : |\sigma^2(x)\ \sigma''(x)| \leq M_4\ (|x| + 1)$

$$|\sigma^2(x)\ b''(x)| \leq M_4\ (|x| + 1)$$

one gets :

$\forall T, \forall n \in N,\ \exists M_5 \in R \ / \forall h = \frac{T}{M},\ M \in N, \forall p = 0,\ 1,\ \ldots,\ \frac{T}{h}$

$$E\ |\overline{X}_p^{(h)}|^n \leq M_5$$

Now we can state the main result of this paragraph :

Theorem 1-4-3

Let us suppose (H1), (H1') and (H2) of the two previous lemmas. Then the "Monte-Carlo scheme" satisfies the estimation :

$$\forall T,\ \exists C \in R \ / \forall h = \frac{T}{M},\ M \in N,\ |E\, f(X_T) - E\, f(\overline{X}_T^{(h)})| \leq C.h^2$$

Sketch of the proof of Theorem 1-4-3

We introduce the following notation : for two integrable random variables X and Y:

$$\text{we write } X \overset{E}{=} Y \text{ instead of : } E(X) = E(Y).$$

Let v(t,x) be defined as previously. The two lemmas permit to justify the following inequalities (after some work which is not mentioned here) :

$$\begin{aligned} v(t_{p+1}, \overline{X}_{p+1}^{(h)}) \overset{E}{=} & v(t_p, \overline{X}_p^{(h)}) + h \left(\frac{\partial v}{\partial t}(t_p, \overline{X}_p^{(h)}) + Lv(t_p, \overline{X}_p^{(h)}) \right) \\ & + \frac{1}{2} h^2 \left(L^2 v (t_p, \overline{X}_p^{(h)}) + \frac{\partial^2 v}{\partial t^2}(t_p, \overline{X}_p^{(h)}) \right. \\ & \left. + 2 \frac{\partial}{\partial t} Lv (t_p, \overline{X}_p^{(h)}) \right) + r_{p+1}^{(h)} \end{aligned}$$

With : $r_{p+1}^{(h)}$ such that : $\exists\, C \ / \forall h : |r_{p+1}^{(h)}| \leq C.\, h^3$, $p=0,\dots, M-1$.

$$\begin{aligned} v(t_{p+1}, \overline{X}_{p+1}^{(h)}) & \overset{E}{=} v(t_p, \overline{X}_p^{(h)}) + r_{p+1}^{(h)} && \text{(since v is solution of (KBE))} \\ & \overset{E}{=} v(0, \overline{X}_o^{(h)}) + 0(h^2) && \text{(by recurrence on p)} \\ & \overset{E}{=} v(0, X_o) + 0(h^2) && \text{(since } \overline{X}_o = X_o\text{)} \\ & \overset{E}{=} v(T, X_T) + 0(h^2) && \text{(by definition of v)} \end{aligned}$$

In particular, for $t_{p+1} = T$, we get :

$$v(T, \overline{X}_T^{(h)}) \overset{E}{=} v(T, X_T) + 0(h^2)$$

That is to say :

$$E\, f(\overline{X}_T^{(h)}) = E\, f(X_T) + 0(h^2)$$

∎

1 - 5 The "MCRK Scheme"

We should like to avoid the assumption (H1') which is restrictive, and the computation of the second derivatives of b and σ (in order to reduce the computation time).

To this end, we introduce a new scheme which is of Runge-Kutta type (see the midpoint method for the numerical analysis of Ordinary Differential Equations, e.g. in Henrici [2]).

First, we introduce notations :

$$t_{p+} = t_p + \frac{h}{2}$$

Then the "MCRK scheme" is defined by :

$$\overline{X}_o = X_o$$

$$\overline{X}_{p+} = \overline{X}_p + \sigma(\overline{X}_p)\, R^1_{p+1} \frac{\sqrt{h}}{\sqrt{2}} + \{b(\overline{X}_p) - \frac{1}{2}\sigma(\overline{X}_p)\,\sigma'(\overline{X}_p)\} \frac{h}{2}$$
$$+ \frac{1}{2}\sigma(\overline{X}_p)\,\sigma'(\overline{X}_p)\ (R^1_{p+1})^2 \cdot \frac{h}{2}$$

$$\overline{X}_{p+1} = \overline{X}_p + \{2\,\sigma(\overline{X}_p)\, R^1_{p+1} \frac{\sqrt{h}}{\sqrt{2}} + 2\,\sigma(\overline{X}_{p+})\, R^2_{p+1} \frac{\sqrt{h}}{\sqrt{2}} - \sigma(\overline{X}_p)\, R_{p+1} \sqrt{h}\}$$
$$+ \{b(\overline{X}_{p+}) - \frac{1}{2}\sigma(\overline{X}_{p+})\,\sigma'(\overline{X}_{p+})\}\, h$$
$$+ \{\sigma(\overline{X}_p)\,\sigma'(\overline{X}_p)\ (R^1_{p+1})^2 \frac{h}{2} + \sigma(\overline{X}_{p+})\,\sigma'(\overline{X}_{p+})\ (R^2_{p+1})^2 \frac{h}{2}$$
$$- \frac{1}{2}\sigma(\overline{X}_p)\,\sigma'(\overline{X}_p)\ (R_{p+1})^2\, h\}$$

where :

. $R^1_o = 0$

. R^1_{p+1} is independent of the σ-field generated by

$$\{R^1_j,\ R^2_j,\ 0 \le j \le p\}$$

. R^2_{p+1} is independent of the σ-field generated by

$$\{R^1_j,\ R^2_j,\ 0 \le j \le p\} \cup \{R^1_{p+1}\}$$

. $R_{p+1} = (R^1_{p+1} + R^2_{p+1})/\sqrt{2}$

. R^1_{p+1} and R^2_{p+1} have the discrete law $\mathcal{R}$, defined in 1-3.

Remark :

$R^1_{p+1}\frac{\sqrt{h}}{\sqrt{2}}$ (resp. $R^2_{p+1}\frac{\sqrt{h}}{\sqrt{2}}$) plays the part of $W_{t_{p+}} - W_{t_p}$ (resp. $W_{t_{p+1}} - W_{t_{p+}}$)

The following theorem can be proved :

Theorem 1-5-1

Under the only assumptions (H1) and (H2) of 1-4, the "MCRK scheme" satisfies the estimation :

$$\forall T,\ \exists C \in R \ / \ \forall h = \frac{T}{M},\ M \in N : \left|E f(X_T) - E f(\overline{X}_T^{(h)})\right| \le C.h^2$$

1 - 6 The multidimensional case

The generalization of the Monte-Carlo scheme is not efficient :

1) at each step, it would be necessary to compute such quantities as :

$$\sum_{r,s,j,k} \frac{\partial^2 \sigma_{ij}}{\partial x_r \partial x_s}(\overline{X}_p)\ \sigma_{rk}(\overline{X}_p)\ \sigma_{sk}(\overline{X}_p)$$

2) We cannot avoid an assumption of type (H1').

So we recommend the "MCRK scheme", defined below (again we introduce a new notation :
$\overset{i}{\Sigma}$ means that the sum is executed with respect to all the indices different from i, p, p+1) :

$$\overline{X}_o^{(i)} = X_o^{(i)}$$

$$\overline{X}_{p+}^{(i)} = \overline{X}_p^{(i)} + \overset{i}{\Sigma}\, \sigma_{ij}\, (\overline{X}_p)\, R_{i+1}^{1(j)} \cdot \frac{\sqrt{h}}{\sqrt{2}} + b_i(\overline{X}_p)\, \frac{h}{2}$$

$$+ \overset{i}{\Sigma}\, \frac{\partial\sigma_{ij}}{x_r}\, (\overline{X}_p)\, \sigma_{rl}\, (\overline{X}_p)\, S_{p+1}^{1(lj)}\, \frac{h}{2}$$

$$\overline{X}_{p+1}^{(i)} = \overline{X}_p^{(i)} + 2\, \overset{i}{\Sigma}\, \sigma_{ij}\, (\overline{X}_p)\, R_{p+1}^{1(j)}\, \frac{\sqrt{h}}{\sqrt{2}}$$

$$+ 2\, \overset{i}{\Sigma}\, \sigma_{ij}\, (\overline{X}_{p+})\, R_{p+1}^{2(j)}\, \frac{\sqrt{h}}{\sqrt{2}}$$

$$- \overset{i}{\Sigma}\, \sigma_{ij}\, (\overline{X}_p)\, R_{p+1}^{(j)}\, \sqrt{h}$$

$$+ b_i\, (\overline{X}_{p+}) \,.\, h$$

$$+ 2\overset{i}{\Sigma}\, \frac{\partial\sigma_{ij}}{\partial x_r}\, (\overline{X}_p)\, \sigma_{rl}(\overline{X}_p)\, S_{p+1}^{1(lj)} h$$

$$+ 2\overset{i}{\Sigma}\, \frac{\partial\sigma_{ij}}{\partial x_r}\, (\overline{X}_{p+})\, \sigma_{rl}(\overline{X}_{p+})\, S_{p+1}^{2(lj)} h$$

$$- \overset{i}{\Sigma}\, \frac{\partial\sigma_{ij}}{\partial x_r}\, (\overline{X}_p)\, \sigma_{rl}(\overline{X}_p)\, S_{p+1}^{(lj)}\, h$$

where, if F_2 is the family defined by :

$$F_2 = \bigcup_{p=o}^{M} \bigcup_{j=1}^{m} \{ \{R_p^{1(j)} \frac{\sqrt{h}}{\sqrt{2}} ,\, R_p^{2(j)} \frac{\sqrt{h}}{\sqrt{2}} ,\, R_p^{(j)} \sqrt{h}\}$$

$$\cup \bigcup_{l=1}^{m} \{ S_p^{1(lj)} h,\, S_p^{2(lj)} h,\, S_p^{(lj)} h\}\}$$

and :

$$F_1 = \bigcup_{p=o}^{M} \bigcup_{j=1}^{m} \{\{ W^{(j)}_{t_{p+}} - W^{(j)}_{t_p} , W^{(j)}_{t_{p+1}} - W^{(j)}_{t_{p+}} , W^{(j)}_{t_{p+1}} - W^{(j)}_{t_p} \}$$

$$\bigcup \bigcup_{l=1}^{m} \{ \int_{t_p}^{t_{p+}} (W^{(l)}_s - W^{(l)}_{t_p}) \, dW^{(j)}_s ,$$

$$\int_{t_{p+}}^{t_{p+1}} (W^{(l)}_s - W^{(l)}_{t_{p+}}) \, dW^{(j)}_s ,$$

$$\int_{t_p}^{t_{p+1}} (W^{(l)}_s - W^{(l)}_{t_p}) \, dW^{(j)}_s \}\},$$

then F_2 is "Monte-Carlo equivalent" to F_1 in the following sense :

<u>Definition 1-6-1</u>

The family $F_2 = \{Y_n, n \leq N\}$ of random variables is "Monte-Carlo equivalent" to the family $F_1 = \{X_n, n \leq N\}$ iff for any sets $\{i_1, i_2, \ldots i_5 \,;\, i_1+i_2+\ldots i_5 = 5\}$ and $\{n_1, \ldots, n_5 \,; 1 \leq n_1, \ldots, n_5 \leq N\}$ of integers we have the property :

"$E[X_{n_1}^{i_1} X_{n_2}^{i_2} \ldots X_{n_5}^{i_5}] = 0$ or $C.h$ or $C.h^2$ for one C independent of h

implies :

$E(Y_{n_1}^{i_1} Y_{n_2}^{i_2} \ldots Y_{n_5}^{i_5}) = E(X_{n_1}^{i_1} X_{n_2}^{i_2} \ldots X_{n_5}^{i_5})$"

Then we choose a convenient family of discrete random variables , F_2, and we state :

Theorem 1-6-2

Let us suppose :

(H1) b_i, σ_{ij} are 6 times differentiable, and all their partial derivatives are continuous and bounded.

(H2) f is 6 times differentiable ; f and each partial derivative of f belongs to
$\{\phi : R^d \to R / \exists\, C \in R,\ \exists\, s \in N ;\ |\phi(x)| \leq C(||x||^s + 1),\ \forall x \in R^d\}$

Then the "MCRK scheme" satisfies :

$$\forall T,\ \exists\, C \in R \,/\, \forall h = \frac{T}{M},\ M \in N :\quad |E\, f(X_T) - E\, f(\bar{X}_T^{(h)})| \leq C.h^2$$

PART 2 : APPLICATION TO THE NUMERICAL ANALYSIS OF PARABOLIC PDE

We are given some real functions a_{ij} on R^d such that the matrix $(a_{ij}(x))_{1 \leq i,j \leq d}$ is symetric and non-negative definite for all $x \in R^d$; then let $(\sigma_{ij}(x))$ be such that :

$$\sigma(x)\ \sigma(x)^* = a(x)$$

We are given also some real functions $b_i(x)$, ϕ on R^d, and we suppose that the functions b_i, σ_{ij} (resp. ϕ) satisfy the assumptions (H_1) (resp.(H_2)) of Theorem 1-6-1.
Let us now consider the operator :

$$L = \sum_{i=1}^{d} b_i \frac{\partial}{\partial x_i} + \frac{1}{2} \sum_{i,j=1}^{d} a_{ij} \frac{\partial^2}{\partial x_i \partial x_j}$$

and the backward P.D.E. :

$$\begin{cases} \dfrac{\partial v}{\partial s}(s,x) + Lv\,(s,x) + \alpha(s,x)\, v(s,x) = 0, \quad s < T \\[2ex] v(T,x) = \phi(x) \end{cases}$$

where :

$\alpha : [0,T] \times R^d \to R$ satisfies the same hypothesis as b_i, and is bounded.

Let $(Z_\Theta^{s,x})$ the solution of the S.D.E. :

$$Z_\Theta^{s,x} = x + \int_s^\Theta b(Z_u^{s,x})\, du + \int_s^\Theta \sigma(Z_u^{s,x})\, dW_u \ , \ \Theta \geq s$$

where (W_t) is a Brownian motion on any probability space.

Feynman-Kac's formula gets :

$$\forall (s,x) \in [0,T] \times R^d \ , \ v(s,x) = E\left(\phi(Z_T^{s,x}) \exp \int_s^T \alpha(\Theta, Z_\Theta^{s,x})\, d\Theta\right)$$

Now let us define X_t by :

$$X_t = (X_t^{(1)}, X_t^{(2)}) = (Z_t^{s,x}, \exp \int_s^T \alpha(\Theta, Z_\Theta^{s,x})\, d\Theta)$$

(X_t) is solution of S.D.E. whose the coefficients satisfy the assumptions (H1) of Theorem 1-6-2 ; moreover the function f defined by :

$(x, y) \in R^d \times R \to f(x,y) = \phi(x).y$

satisfies the assumption (H2).

Accordingly to this theorem and the paragraph 1-1, a sufficiently large number of simulations of X_T given by the MCRK scheme with a step h permit to get an approximation of v(s,x) of order h^2.

PART 3 - APPLICATION TO A NUMERICAL TREATMENT OF KALLIANPUR-STRIEBEL FORMULA

3 - 1 Account of the situation

Let (X_t) be an N-dimensional inobserved process such that :

(3-1;a) $X_t = X_o + \int_o^t b(X_s)ds + \int_o^t \sigma(X_s)dW_s$

Suppose we observe a D-dimensional process (Y_t) solution of :

(3-1;b) $Y_t = Y_o + \int_o^t H(X_s)\, ds + \int_o^t dV_s$

where :

(W_t, V_t) is NxD-dimensional Wiener process (the underlying σ-field on the canonical space of trajectories is the σ-field generated by the process (X_t, Y_t)).

We suppose that b, σ, H and their derivatives are bounded regular functions.

Under these assumptions, the solution of the martingale problem associated to (3-1.a) (resp. to the system {(3-1;a) , (3-1;b)}) exists uniquely ; we denote it by $\overline{P}$ (resp. P) and $\overline{E}$ (resp. E) is the expectation computed under the law $\overline{P}$ (resp. P).

By the usual change of reference probability law, for any "regular" function Ψ, we have the Kallianpur-Striebel formula

$$E\{(\Psi(X_t)| F_t^Y\} = \frac{\overline{E}(\Psi(X_t)M_t)}{\overline{E}(M_t)}$$

where :

. F_t^Y is the σ-field generated by $\{Y_s, s \leq t\}$

. $M_t = \exp \{ \int_0^t H(X_s).\, dY_s - \frac{1}{2} \int_0^t |H(X_s)|^2\, ds \}$

For simplicity, <u>let us now suppose that (Y_t) is a real valued process.</u>

Then :

$$M_t = \exp \{ H(X_t)\, Y_t - \int_0^t Y_s.\, LH\, (X_s)\, ds - \int_0^t Y_s\, \nabla H(X_s)\, .\, \sigma(X_s)\, dW_s$$
$$- \frac{1}{2} \int_0^t |H(X_s)|^2\, ds \}$$

Define a new probability, $\tilde{P}$, by :

$$\frac{d\tilde{P}}{d\overline{P}} = \exp \{ -\int_0^t Y_s\, .\nabla H(X_s)\, .\, \sigma(X_s)\, dW_t$$
$$- \frac{1}{2} \int_0^t Y_s^2\, (\nabla H(X_s),\ \sigma\sigma^*\, (X_s)\, \nabla H(X_s))\, ds \}$$

(this new change of probability appears "naturally" in the search of the robust form of the filtering equations, see Pardoux [5], § 4).

Then :

$$\overline{E}(\Psi(X_t)\ M_t)) = \tilde{E}(\Psi(X_t)\ \exp\{H(X_t)Y_t - \int_o^\tau Y_s.LH(X_s)ds$$

$$- \frac{1}{2}\int_o^t (H(X_s))^2\, ds$$

$$+ \frac{1}{2}\int_o^t Y_s^2(\nabla H(X_s), \sigma\sigma^*(X_s)\nabla H(X_s))ds\})$$

Define Z_t by :

$$Z_t = \exp\ \{ -\int_o^t Y_s.LH(X_s)\ ds - \frac{1}{2}\int_o^t |H(X_s)|^2\ ds$$

$$+ \frac{1}{2}\int_o^t Y_s^2\ \ (\nabla H(X_s),\ \sigma\sigma^*(X_s)\nabla H(X_s))ds$$

Under the law $\tilde{P}$:

$$d\begin{pmatrix} X_t \\ Z_t \end{pmatrix} = \begin{pmatrix} b(X_t) - Y_t\ \sigma\sigma^*(X_t)\nabla H(X_t) \\ - Y_t.LH(X_t) - \frac{1}{2}|H(X_t)|^2 + \frac{1}{2}Y_t^2\ \ (\nabla H(X_t),\ \sigma\sigma^*(X_t)\nabla H(X_t)) \end{pmatrix} dt$$

$$+ \begin{pmatrix} \sigma(X_t) \\ 0 \end{pmatrix} d\tilde{W}_t$$

So, considering $t \to Y_t(\omega)$ as a deterministic function (remember that we are supposed to observe a trajectory of Y_t), we want to compute :

$$\tilde{E}\ \{\Psi(Xt)\ Z_t\ \exp\ \{H(X_t)\ Y_t\}\}$$

Thus our problem can be formulated as follows :

"Let $t \to U_j(t)$ $(1 \leq j \leq J)$ be some continuous real-valued fonctions (without no more regularity).
Let (b_{ij}) a d x J matrix, σ_{ij} a d x r matrix,
and (W_t) a R^r - brownian motion on a probability space (Ω, F_t, P).
Let (X_t) the solution of the S.D.E. :
$dX_t = b(X_t)U_t\ dt + \sigma(X_t)dW_t$
Let f be a regular function, possibly of polynomial growth.
We want to approximate $Ef(X_T)$".

3.2 Description of the scheme

Unfortunately, we could not succeed to develop a scheme ot "MCRK" type ; the scheme is of "Monte-Carlo" type, and thus the second derivatives of b and σ will appear.

Anyway, for an easier reading, we shall describe the scheme using random variables with Gaussian laws, but, in practice, we replace this family of random variables by any family of discrete random variables which is "Monte-Carlo equivalent" to it (see definition 1-6-1) : (*).

$$\bar{X}_o^{(i)} = X_o$$

$$\bar{X}_{p+1}^{(i)} = \bar{X}_p^{(i)} + \Sigma^i \sigma_{ij}(\bar{X}_p)\Delta W_{p+1}^{(j)} + \Sigma^i b_{ij}(\bar{X}_p) \int_{t_p}^{t_{p+1}} U_s^{(j)} ds$$

$$+ \Sigma^i \frac{\partial\sigma_{ij}}{\partial x_k}(\bar{X}_p)\,\sigma_{kl}(\bar{X}_p) \int_{t_p}^{t_{p+1}} (W_s^{(1)} - W_{t_p}^{(1)})\, dW_s^{(j)}$$

$$+ \Sigma^i \frac{\partial\sigma_{ij}}{\partial x_k}(\bar{X}_p)\, b_{kl}(\bar{X}_p) \int_{t_p}^{t_{p+1}} \int_{t_p}^{s} U_\Theta^{(1)}\, d\Theta\, dW_s^{(j)}$$

$$+ \Sigma^i \frac{\partial b_{ij}}{\partial x_k}(\bar{X}_p)\,\sigma_{kl}(\bar{X}_p) \int_{t_p}^{t_{p+1}} (W_s^{(1)} - W_{t_p}^{(1)})\, U_s^{(j)}\, ds$$

$$+ \frac{1}{4} \Sigma^i \frac{\partial^2\sigma_{ij}}{\partial x_k \partial x_l}(\bar{X}_p)\,\sigma_{km}(\bar{X}_p)\,\sigma_{lm}(\bar{X}_p)\, h\Delta W_{p+1}^{(j)}$$

$$+ \Sigma^i \frac{\partial b_{ij}}{\partial x_k}(\bar{X}_p)\, b_{kl}(\bar{X}_p) \int_{t_p}^{t_{p+1}} \int_{t_p}^{s} U_\Theta^{(1)}\, d\Theta\, U_s^{(j)}\, ds$$

$$+ \frac{1}{4} \Sigma^i \frac{\partial^2 b_{ij}}{\partial x_k \partial x_l}(\bar{X}_p)\,\sigma_{km}(\bar{X}_p)\,\sigma_{lm}(\bar{X}_p)\, h \int_{t_p}^{t_{p+1}} U_s^{(j)}\, ds$$

Now we consider the P.D.E. :

$$\begin{cases} \frac{\partial v}{\partial t}(t,x) + \sum_{i,j} b_{ij}(x) U_t^{(j)} \frac{\partial v}{\partial x_i}(t,x) + \frac{1}{2} \sum_{i,j,k} \sigma_{ik}(x)\sigma_{jk}(x) \frac{\partial^2 v}{\partial x_i \partial x_j}(t,x) = 0 \\ v(T,x) = f(x) \end{cases}$$

(*) $\Delta W_{p+1}^{(j)}$ denote $W_{t_{P+1}}^{(j)} - W_{t_p}^{(j)}$

We do the assumptions :

(H1) b_{ij}, σ_{ij} are continuous bounded functions and their partial derivatives of order 1,2,...6 are continuous bounded functions.

(H2) see assumption (H2) of theorem 1-6-2.

Then, slightly modifying the proof of theorem 1-4-3, we can prove (see [8]) :

$$\forall\, T,\ \exists\, C, \forall h \ / \ |E\, f(X_T) - Ef\,(\overline{X}_T^{(h)})| \leq C.h^2$$

REFERENCES

(1) I. GIKHMAN & A. SHOROKHOD - "Introduction à la théorie des processus aléatoires" - Ed. Mir - (1980)

(2) P. HENRICI - "Discrete Variable Methods in Ordinary Differential Equations" - Wiley (1962).

(3) H. KUNITA - Stochastic Differential Equations and Stochastic flows of Diffeomorphims - Cours de l'Ecole d'Eté de Probalites de Saint-Flour 1982 - To appear in "Lecture Notes in Mathematics", Springer - Verlag.

(4) G.N. MILSHTEIN - A Method of Second Order Accuracy integration of S.D.E. - Theory Proba. Appl. 23, pp 396 - 401 (1976).

(5) E. PARDOUX - Equations du Filtrage non linéaire, de la Prédiction et du Lissage - Stochastics, vol. 6, pp 193-231 (1982).

(6) E. PARDOUX & D. TALAY - Approximation and Simulation of Solutions of S.D.E. - to appear in Acta Applicandae Mathematicae (1984).

(7) R.Y. RUBINSTEIN - "Simulation and the Monte-Carlo method" Wiley (1981)

(8) D. TALAY - "Discretisation d'une E.D.S et calcul approché d'espérances de fonctionnelles de la solution". (à paraître).

DISTRIBUTIONS-VALUED SEMIMARTINGALES and APPLICATIONS TO CONTROL AND FILTERING

by

A.S.Ustunel*

Introduction

In this work we try to give a partial review of the recent results about the stochastic integration on the nuclear spaces and their applications to some problems of finite and infinite dimensional stochastic analysis without any pretention to be complete or exhaustive.In the first part we tried to make realize the reader that the concept of semimartingale is more natural in the context of nuclear spaces than in the context of Hilbert spaces.Afterwards the general definitions, the essential results about the characterization of semimartingales and the integration by parts formula are announced.In the second section we apply these results to show the hypoellipticity of some stochastic partial differential operators,the uniqueness and the hypoellipticity of the heat equation in the space of the tempered distributions,the characterization of the C^∞ or analytic flows of the semimartingales,a generalization of the generalized Itô-Stratonovitch formula and as a subproduct a pathwise Girsanov formula and the reduction of the unnormalized density equation (whose drift being a pseudodifferential operator) into a deterministic evolution equation with random coefficients (this result is first established by M.H.A. Davis for the second order elliptic case).

The references are not at all complete,especially about the results in the filtering and control theory we refer to the references of this volume.

I. Motivations and Preliminaries

In the following we denote by $(\Omega,\mathcal{F},\mathcal{F}_t,P)$ a complete probability space satisfying "the usual conditions"(cf. 2).If $p\geq 1$ and H is a Hilbert space we denote by $S^p(H)$ the Banach space of the special semimartingales with values in H ,euipped with the norm

$$\| x\|_p = \| [m,m]_\infty^{1/2} + \int_0^\infty |da_s| \|_{L^p}$$

for $x=m+a$ .If $H=\mathbb{R}$ we write simply S^p.

Suppose that $\tilde{X}:H\longrightarrow S^1$ is a linear,continuous mapping.We ask whether there exists a semimartingale $\hat{X}$ with values in H'(i.e. the continuous dual of H) such that $\tilde{X}(h)=\langle\hat{X},h\rangle$ for any $h\in H$ .In general the answer is no.However ,if $T:H\longrightarrow H$ is a nuclear(or Hilbert-Schmidt) mapping,then $\tilde{X}\circ T:H\longrightarrow S^1$ is also nuclear and can be represented as

$$\tilde{X}\circ T = \sum_{i=1}^\infty \lambda_i f_i\otimes x_i$$

where $(f_i)\subset H'$, $(x_i)\subset S^1$ are bounded and $(\lambda_i)\in l^1$.If we define $X_t(\omega)$ by

$$\hat{X}_t(\omega) = \sum_i \lambda_i x_t^i(\omega) f_i$$

then $\hat{X}$ is an element of $S^1(H)$ and we have

$$\tilde{X}\circ T(h) = \langle\hat{X},h\rangle$$

for any $h\in H$.

The above example says that if any linear continuous mapping from H into a Banach space were a nuclear mapping then any cylindrical semimartingale on H would be in fact a semimartingale with values in H'. Unfortunately,for a Hilbert space this is impossible unless it is of finite dimension.However,we can imagine a family of Hilbert spaces $(H_n;n\in\mathbb{N})$ such that $H_n\supset H_{n+1}$,for any $n\in\mathbb{N}$ and the injections $H_{n+1}\hookrightarrow H_n$ are nuclear(or Hilbert-Schmidt).We shall denote by F the set

$$F = \bigcap_{n=1}^\infty H_n$$

and say that $f_k\rightarrow 0$ in F if $f_k\rightarrow 0$ in each H_n.With this topology F becomes a Fréchet space with the property that,any continuous,linear mapping S from F into a Banach space X is nuclear since it can be decomposed as $S=\hat{S}\circ i_n$ where i_n is the injection $F\hookrightarrow H_n$ and $\hat{S}:H_n\rightarrow X$ is continuous.Of course the (continuous)dual of F can be identified with

$F' = \bigcup_{n=1}^{\infty} H'_n$ where the injections $H'_n \hookrightarrow H'_{n+1}$ are nuclear. Consequently, if F' is equipped with the inductive limit topology, then, any continuous linear mapping from F' into a Banach space is also a nuclear mapping. Since F is metrizable, the concept of a stochastic process, measurability, etc., are well defined. However F' is not metrizable (metrizability of F' implies that it is finite dimensional). Hence we have to extend the definition of a process to such spaces, called nuclear spaces. In order to include $\mathcal{D}(\mathbb{R}^d)$, $\mathcal{D}'(\mathbb{R}^d)$, etc., we shall give a general setting:

We shall assume that F is a complete, locally convex nuclear space whose dual F' is also nuclear under the strong topology, denoted by F'_β. If U is a neighbourhood (of zero) in F, we denote by F(U) the Banach space which is the completion of $F/p_U^{-1}(0)$ with respect to the gauge function p_U of U (which is a norm on $F/p_U^{-1}(0)$) and k(U) denotes the canonical mapping from F into F(U). If B is an absolutely convex, closed and bounded subset of F we denote by F[B] the space spanned by B with the norm p_B (i.e., the gauge function of B). The nuclearity of F is equivalent to the existence of a neighbourhood base $\mathcal{U}_h(F)$ such that, for any $U \in \mathcal{U}_h(F)$, F(U) is a separable Hilbert space and there exists $V \in \mathcal{U}_h(F)$ $V \subset U$ such that the mapping $k(U,V): F(V) \longrightarrow F(U)$ defined by $k(U,V) \circ k(V) = k(U)$, is a nuclear mapping (cf. [5], [3]). Let us note that F(U) plays the role of H_n in the countable case and its dual can be identified with $F'[U°]$, U° being the polar of U. Evidently $F'[U°]$ can be injected into F' and the injection is nuclear. Let us remark finally that the set

$$\mathcal{K}_h(F') = \{U°: U \in \mathcal{U}_h(F)\}$$

is a fundemental system of compact sets in F' and the similar notations can be defined for F' just by interchanging the notations.

Now we can give the following

<u>Definition I.1</u>

i) Let X be the set of stochastic processes $\{X^U;\ U \in \mathcal{U}_h(F'_\beta)\}$, where X^U is a stochastic process with values in F'(U). We say that X is a projective system (of processes) if for any $V, U \in \mathcal{U}_h(F'_\beta)$, $V \subset U$,

$k(U,V) \circ X^V$ and X^U are undistinguishable.

ii) We say that X is a semimartingale, martingale, etc., if any element X^U of X is a semimartingale, martingale, etc., with values in $F'(U)$.

iii) We say that X possesses the property π if any $X^U, U \in \mathcal{U}_h(F'_\beta)$, possesses the property π in $F'(U)$.

iv) Suppose that there exists a stochastic process X' with values in F' such that, for any $U \in \mathcal{U}_h(F'), t \geq 0$,

$$k(U)(X'_t) = X^U_t \quad \text{a.s.}$$

Then we say that X has a limit X' in F'.

Now we can announce the result for which we have made all these definitions :

Theorem I.1

Suppose that $\tilde{X}: F \to S°$ is a linear, sequentially continuous mapping, where S° is the non-locally convex Fréchet space of the scalar semimartingales. Then there exists a unique projective system of semimartingales $\{X^U; U \in \mathcal{U}_h(F')\}$ such that, for any $f \in F[U°]$,

$$\tilde{X}(f) = \langle X^U, f \rangle$$

The proof of this result is given in [13]. The essential difficulty comes from the fact that S° is not locally convex. If we replace S° by S^1, which is a Banach space then, using the closed graph theorem and the fact that F is bornological, we obtain the following

Theorem I.2

Suppose that $\tilde{X}: F \to S^1$ is a linear, sequentially continuous mapping. Then there exists $K \in \mathcal{K}_h(F')$ and a semimartingale $\hat{X}$ with values in $F'[K]$ such that, for any $f \in F$,

$$\langle i_K(\hat{X}), f \rangle = \tilde{X}(f),$$

where i_K is the injection $F'[K] \hookrightarrow F'$.

If F'_β is metrizable, we can also show by the help of Girsanov's Theorem (cf. [14]) :

Theorem I.3

Suppose that F'_β is metrizable. Then, for any finite time interval, there exists a probability Q equivalent to P such that the hypothesis of Theorem I.1 under P implies the hypothesis of Theorem I.2 under Q on that time interval and consequently the conclusion of Theorem I.2 holds.

Remark

This result extends a recent theorem which says that any probability measure on a nuclear Fréchet space has a Hilbert space support.

Having defined the concept of projective system of semimartingales, we have to prove that the previsible processes can be "integrated" with respect to these objects. The first thing to note is the fact that for scalar or operator valued previsible processes the stochastic integrals can be defined directly by Theorem I.1. It remains to define the stochastic integrals of F-valued previsible processes. Suppose that H is a bounded, weakly previsible process with values in F. Then it is absorbed by some $F[U^\circ]$, $U \in \mathcal{U}_h(F'_\beta)$, as a bounded, (strongly) previsible process. Then we can define the stochastic integral of H with respect to X^U, the element of X corresponding to U (cf. [4]). Afterwards, it is not hard to show that the resulting (scalar) semimartingale is independent of the choice of U and this integral is denoted by

$$\int_0^t \langle H_s, dX_s \rangle \quad \text{(cf. [12])}.$$

One of the most important features of this stochastic integral is the integration by parts formula :

Theorem I.4

Suppose that Z is an element of $S^1(F)$ (i.e., it satisfies the properties announced in Theorem I.2 when F'_β is replaced by F) and X is a projective system of semimartingales on F' with a limit X'. Then $\{\langle Z_t, X'_t \rangle ; t \geq 0\}$ has a modification which is a semimartingale and one has

$$\langle Z_t, X'_t \rangle = \int_0^t \langle Z_{s-}, dX_s \rangle + \int_0^t \langle dZ_s, X_{s-} \rangle + [\![Z, X]\!]_t$$

where $[\![Z, X]\!]$ is a cadlag process of finite variation.

For the proof of this result the reader is referred to [12]. Let us simply remark that, by Theorem I.2, the stochastic integrals are well defined and in case F or F'_β is metrizable, by the help of Theorem I.3 we can drop the hypothesis that Z is in $S^1(F)$.

II. Applications

Let us look at the simplest example : Suppose that z is a finite dimensional semimartingale (say in $\mathbb{R}^d$), then, δ_z, defined by

$$\langle \delta_{z_t}, f\rangle = f(z_t) \quad , f \in \mathcal{D}(\mathbb{R}^d)$$

defines a semimartingale with values in $\mathcal{E}'(\mathbb{R}^d)$ (i.e., the space of distributions with compact support). In fact , we have

Proposition II.1

i) Let $T \in \mathcal{D}'(\mathbb{R}^d)$ and define X_t by

$$\langle X_t, f\rangle = \langle T * \delta_{z_t}, f\rangle \quad , f \in \mathcal{D}(\mathbb{R}^d).$$

Then X is a semimartingale with values in $\mathcal{D}'(\mathbb{R}^d)$ satisfying the following stochastic partial differential equation:

$$X_t = X_o - \int_o^t \partial_i X_{s-}\, dz_s^i + 1/2 \int_o^t \partial_i\, \partial_j X_{s-}\, d\langle z^{c,i}, z^{c,j}\rangle_s +$$

$$+ \sum_{s\leq t}{}' (X_s - X_{s-} + \sum_i \partial_i X_{s-}\, \Delta z_s^i),$$

where ∂_i denotes $\partial/\partial x_i$. Moreover, for any initial condition X_o, this equation has one and only one solution.

ii) Denote by $P_t(\omega, D_x)$ the stochastic partial differential operator of the right hand side of the above equation. Suppose that $Z \in S^\circ(\mathcal{D}'(\mathbb{R}^d))$ satisfies the equation

$$Z - PZ = h$$

with $h \in S^\circ(\mathcal{E}(\mathbb{R}^d))$ (i.e. the space of semimartingales with values in the space of C^∞-functions on $\mathbb{R}^d$). Then Z is also an element of $S^\circ(\mathcal{E}(\mathbb{R}^d))$

Proof: The proof follows from the integration by parts formula: If X is any solution, using Theorem I.4, one can show that

$$\langle X_t * \delta_{-z_t}, f\rangle = \langle T, f\rangle \quad \text{a.s.},$$

for any $f \in \mathcal{D}(\mathbb{R}^d)$, hence δ_{-z} is a first integral of the equation. For the hypoellipticity, the proof is similar. II.Q.E.D.

Remark: The first part of Proposition II.1 has a geometric interpretation:Suppose that there are two reference frames R1 and R2 such that R2 moves with respect to R1 following the trajectories of the semimartingale z.Then ,any distribution T in R2 seems to an observer in R1 as a ditribution valued semimartingale satisfying the above stochastic partial differential equation.The integration by parts formula permits us to show that this coordinate transformation is reversible.

Corollary II.1

The following evolution equation has a unique solution in $\mathcal{S}'(\mathbb{R}^d)$ (i.e.,the space of tempered distributions):

$$du/dt = \frac{1}{2} \Delta u \quad , \quad u_o \in \mathcal{S}'(\mathbb{R}^d) .$$

Proof:

Taking the Fourier transform of u , we have

$$d\hat{u}_t = -1/2 \; |x|^2 \hat{u}_t \; dt .$$

Define $Z_t(x,\varpi)$ by

$$Z_t(x,\varpi) = \exp(ix.W_t(\varpi) + t/2 |x|^2)$$

where W is the standard Wiener process with values in $\mathbb{R}^d$ .Then $\hat{X}=\hat{u}.Z$ satisfies

$$d\hat{X}_t = ix.\hat{X}_t \; dW_t - 1/2 |x|^2 \; \hat{X}_t \; dt$$

but this is the Fourier transform of the equation

$$dX_t = - \partial_i X_t \; dW_t^i + 1/2 \; \Delta X_t dt \quad .$$

II.Q.E.D.

Similarly we have the following

Corollary II.2

Suppose that

$$du/dt = 1/2 \; \Delta u + h_t$$

where h_t is C^∞ for any $t \geq o$.Then u_t belongs to $\mathcal{E}(\mathbb{R}^d)$ for any $t \geq o$.

One of the main problems of finite dimensional stochastic analysis is the construction of the infinitely differentiable flows of the semimartingales.The following theorem gives a necessary and sufficient condition for this :

Theorem II.1

a) Let X be a random field on $[0,1] \times \mathbb{R}^d$ such that, for any $x \in \mathbb{R}^d$, $X(x)$ is a semimartingale in S^1. Suppose that the following conditions are satisfied:

i) For any bounded, scalar, previsible process h, the mapping

$$x \longmapsto E\int_0^1 h_s \; dX(x)_s$$

is C^∞ with values in $\mathbb{R}^d$.

ii) For any $H \in L^\infty(\Omega,\mathcal{F},P)$, $t \in [0,1]$,

$$x \longmapsto E(H\,X(x)_t)$$

is C^∞ with values in $\mathbb{R}^d$.

Then there exists an $\mathcal{E}(\mathbb{R}^d) \tilde{\otimes} \mathbb{R}^d$ - valued (i.e., the completed projective tensor product) semimartingale $\hat{X}$ such that, for any $x \in \mathbb{R}^d$, $\hat{X}(x)$ and $X(x)$ are undistinguishable.

b) If $\hat{X}$ is any $\mathcal{E}(\mathbb{R}^d) \tilde{\otimes} \mathbb{R}^d$-valued semimartingale, then, there exists a probability measure Q, equivalent to P, under which the conditions (i) and (ii) are satisfied.

c) (a) and (b) hold if one replaces C^∞-condition with analyticity in a region of $\mathbb{C}^d$.

Proof of (a) and (c) follows from the Radon-Nikodym property of nuclear spaces. (b) is a consequence of Theorem I.3. (cf. 17).

The following result is a straightforward application of Theorem II.1 (cf. [15]):

Corollary II.3

Let O be an open subset of $\mathbb{R}^k$, $b: \mathbb{R}^d \times O \to \mathbb{R}^d$, $\sigma: \mathbb{R}^d \times O \to \mathbb{R}^d \otimes \mathbb{R}^d$ are C^∞-vector fields with bounded derivatives. Then there exists an $\mathcal{E}(O) \tilde{\otimes} \mathbb{R}^d$-valued semimartingale π such that, for any $z \in O$, $\pi(z)$ is the solution of the following Itô equation:

$$dx_t(z) = b(x_t(z),z)\,dt + \sigma(x_t(z),z).dW_t$$

$$\pi_0 \in \mathcal{E}(\mathbb{R}^k) \tilde{\otimes} \mathbb{R}^d, \quad x_0 = \pi_0(z).$$

Let us suppose that z is a semimartingale with values in O. Then we may ask if one can developpe $\pi_t(\omega, z_t(\omega))$. In fact, using the integration by parts formula we have a generalization of the so-called generalized Itô-Stratonovitch formula (cf. [17]):

Corollary II.4

For any semimartingale z with values in O, we have

$$\pi_t(z_t) = \pi_0(z_0) + \int_0^t b(\pi_s(z_s), z_s)\, ds + \int_0^t \sigma(\pi_s(z_s), z_s).dW_s +$$
$$+ \int_0^t D_z \pi_s(z_{s-})\, dz_s +$$
$$+1/2 \int_0^t D_z^2 \pi_s(z_{s-}).\, d\langle z^c, z^c\rangle_s +$$
$$+ \int_0^t (D_z \sigma(\pi_s(z_s), z_s) + D_x \sigma(\pi_s(z_s), z_s) D_z \pi_s(z_s))\, d\langle W, z^c\rangle_s$$
$$+ \sum_{0<s\leq t} (\pi_s(z_s) - \pi_s(z_{s-}) - D_z \pi_s(z_{s-})\, \Delta z_s)$$

where D is the Jacobian and D^2 is the Hessian operators and $\langle W, z^c\rangle$ is the operator valued process corresponding to Doob-Meyer decomposition.

Remark: In case the parameter is the initial condition of the diffusion, this result has been proved by using the approximation of the integrals in [7] and [10]. For analytic proofs in Sobolev spaces cf. [9] and [19].

As a particular case, let us look at the flow π defined by the following diffusion process:

$$dx_t = \sigma(x_t)\, dW_t$$
$$x_0(x) = x \in \mathbb{R}^d .$$

Let b be a smooth vector field on $\mathbb{R}^d$ with values in $\mathbb{R}^d$. We know that $x \mapsto \pi_t(x)$ is a diffeomorphism (cf. [7], [9]) whose inverse is also a semimartingale in $\mathcal{E}(\mathbb{R}^d)\tilde{\otimes}\mathbb{R}^d$ (cf. [14]). Let us define $\theta_t(x)$ as the solution of the following equation:

$$\theta_t(x) = x + \int_0^t (D\pi_s)^{-1}(\theta_s(x))((\sigma.D\sigma + b)\circ\pi_s)(\theta_s(x))\, ds .$$

Then it is easy to see that $\pi_t(\theta_t(x))$ is the solution of

$$dy_t = b(y_t)\, dt + \sigma(y_t)\, dW_t$$
$$y_0 = x .$$

A possible name for this formula may be "the pathwise Girsanov formula" and it suggests that one can study the optimal control problems on the paths of the diffusion processes and this method seems interesting from the numerical point of view.

In the filtering of diffusion processes, the unnormalized probability density function p_t satisfies the following stochastic partial differential equation:

$$dp_t = Lp_t dt + hp_t . dW_t ,$$

where L is a pseudodifferential operator, h is a bounded mapping with values in $\mathbb{R}^d$. Let us suppose that everything is smooth and define M as

$$M_t(\omega,x) = \exp(-h(x).W_t(\omega) + |h(x)|^2 t/2).$$

Of course M is a semimartingale with values in $\mathcal{E}(\mathbb{R}^d)$ (cf. [15]).

Proposition II.1

Let K_t be the (random) operator defined by

$$K_t f = M_t . L(f.M_t^{-1}) , \quad f \in \mathcal{D}(\mathbb{R}^d).$$

Then, $y_t = M_t . p_t$ satisfies the following equation:

$$\frac{\partial y_t}{\partial t} = K_t y_t - |h|^2 y_t$$

$$y_o = p_o .$$

Proof:

Let f be a test function. Using the integration by parts formula we have

$$\begin{aligned}\langle p_t, fM_t\rangle = p_o(f) + \int_o^t \langle Lp_s, fM_s\rangle ds + \int_o^t \langle h_i p_s, fM_s\rangle dW_s^i - \\ - \int_o^t \langle p_s, h_i fM_s\rangle dW_s^i - \\ - \int_o^t \langle |h|^2 p_s, fM_s\rangle ds \\ = p_o(f) + \int_o^t \langle M_s Lp_s, f\rangle ds - \int_o^t \langle |h|^2 M_s p_s, f\rangle ds .\end{aligned}$$

II.Q.E.D.

References

[1] A.Badrikian:Séminaire sur les Fonctions Aléatoires et les Mesures Cylindriques.Lect.Notes in Math.Springer,Vol.139,1970.

[2] C.Dellacherie and P.A.Meyer:Probabilités et Potentiel,Vol.I,II. Hermann,Paris,1975 and 1980.

[3] A.Grothendieck:"Produits tensoriels topologiques et espaces nucléaires". Mem.Amer.Math.Soc.6(1955).

[4] M.Métivier and J.Pellaumail:Stochastic Integration.Academic Press, New York,1980.

[5] H.H.Schaefer:Topological Vector S paces.GRTM,Springer,1970.

[6] L.Schwartz:Théorie des Distributions.Hermann,Paris,1973.

[7] J-M.Bismut:Mécanique Aléatoire.Lect.Notes in Math.Springer,Vol.866, 1981.

[9] A.P.Carverhill and K.D.Elworthy:"Flows of stochastical dynamical systems:The functional analytic approach".Preprint,1982.

[10] H.Kunita:"Some extensions of Itô's formula".Sém.de Proba.XV,Lect. Notes in Math.Vol.850,Springer,1980.

[11] A.S.Ustunel:"Formule de Feynman-Kac stochastique".C.R.Acad.Sci. Paris 292,Ser.A-B(1981),p.595-597.

[12] A.S.Ustunel:"Stochastic integration on nuclear spaces and its applications".Ann.Inst.H.Poincaré,Sect.A,XVIII(2),p.165-200, (1982).

[13] A.S.Ustunel:"A characterization of semimartingales on nuclear spaces".Z.Wahrsch.verw.Gebiete,60(1982)p.21-39.

[13'] A.S.Ustunel:"An erratum to "A characterization of semimartingales on nuclear spaces"after Laurent Schwartz " .To appear in Zeit.Wahrsch.verw.Gebiete.

[14] A.S.Ustunel:"Some applications of stochastic integration in infinite dimensions".Stochastics 7(1982),p.255-288.

[15] A.S.Ustunel:"Some applications of stochastic calculus on the nuclear

spaces to the nonlinear stochastic problems".Nonlinear Stochastic Problems,Ed.by R.S.Bucy & J.M.F.Moura,Nato ASI Series,D.Reidel Publ.Comp.,1983.

[16] A.S.Ustunel:"A generalization of Itô's formula".J.Funct.Anal.,47 (1982),p.143-152.

[17] A.S.Ustunel:"Analytic semimartingales and their boundary values". J.Funct.Anal.(1983),p.142-158.

[18] A.S.Ustunel:"Additive processes on nuclear spaces".Preprint,1983.

[19] A-S.Sznitman:"Martingales dépendant d'un paramètre:Une formule d'Itô".Z.Wahrsch.verw.Gebiete,60,p.41-70(1982).

---

*Member of CNET,PAA/ATR/SST.

Corresponding adress:2,Bd.Auguste Blanqui,75013 Paris,France.

Lecture Notes in Control and Information Sciences

Edited by A. V. Balakrishnan and M. Thoma

Vol. 22: Optimization Techniques
Proceedings of the 9th IFIP Conference on Optimization Techniques,
Warsaw, September 4–8, 1979
Part 1
Edited by K. Iracki, K. Malanowski, S. Walukiewicz
XVI, 569 pages. 1980

Vol. 23: Optimization Techniques
Proceedings of the 9th IFIP Conference on Optimization Techniques,
Warsaw, September 4-8, 1979
Part 2
Edited by K. Iracki, K. Malanowski, S. Walukiewicz
XV, 621 pages. 1980

Vol. 24: Methods and Applications in Adaptive Control
Proceedings of an International Symposium
Bochum, 1980
Edited by H. Unbehauen
VI, 309 pages. 1980

Vol. 25: Stochastic Differential Systems – Filtering and Control
Proceedings of the IFIP-WG7/1 Working Conference
Vilnius, Lithuania, USSR, Aug. 28 – Sept. 2, 1978
Edited by B. Grigelionis
X, 362 pages. 1980

Vol. 26: D. L. Iglehart, G. S. Shedler
Regenerative Simulation of Response Times in Networks of Queues
XII, 204 pages. 1980

Vol. 27: D. H. Jacobson, D. H. Martin, M. Pachter, T. Geveci
Extensions of Linear-Quadratic Control Theory
XI, 288 pages. 1980

Vol. 28: Analysis and Optimization of Systems
Proceedings of the Fourth International Conference on Analysis and Optimization of Systems
Versailles, December 16–19, 1980
Edited by A. Bensoussan and J. L. Lions
XIV, 999 pages. 1980

Vol. 29: M. Vidyasagar,
Input-Output Analysis of Large-Scale Interconnected Systems – Decomposition, Well-Posedness and Stability
VI, 221 pages. 1981

Vol. 30: Optimization and Optimal Control
Proceedings of a Conference Held at Oberwolfach, March 16–22, 1980
Edited by A. Auslender, W. Oettli, and J. Stoer
VIII, 254 pages. 1981

Vol. 31: Berc Rustem
Projection Methods in Constrained Optimisation and Applications to Optimal Policy Decisions
XV, 315 pages. 1981

Vol. 32: Tsuyoshi Matsuo,
Realization Theory of Continuous-Time Dynamical Systems
VI, 329 pages, 1981

Vol. 33: Peter Dransfield
Hydraulic Control Systems – Design and Analysis of Their Dynamics
VII, 227 pages, 1981

Vol. 34: H.W. Knobloch
Higher Order Necessary Conditions in Optimal Control Theory
V, 173 pages, 1981

Vol. 35: Global Modelling
Proceedings of the IFIP-WG 7/1 Working Conference Dubrovnik, Yugoslavia,
Sept. 1–5, 1980
Edited by S. Krčevinac
VIII, 232 pages, 1981

Vol. 36: Stochastic Differential Systems
Proceedings of the 3rd IFIP-WG 7/1 Working Conference
Visegrád, Hungary, Sept. 15–20, 1980
Edited by M. Arató, D. Vermes, A.V. Balak
VI, 238 pages, 1981

Vol. 37: Rüdiger Schmidt
Advances in Nonlinear Parameter Optimization
VI, 159 pages, 1982

Vol. 38: System Modeling and Optimizati
Proceedings of the 10 th IFIP Conference
New York City, USA, Aug. 31 – Sept. 4, 1
Edited by R.F. Drenick and F. Kozin
XI, 894 pages. 1982

Vol. 39: Feedback Control of Linear and Nonlinear Systems
Proceedings of the Joint Workshop on Feedback and Synthesis of Linear and Nonlinear Systems
Bielefeld/Rom
XIII, 284 pages. 1982

Vol. 40: Y.S. Hung, A.G.J. MacFarlane
Multivariable Feedback: A Quasi-Classical Approach
X, 182 pages. 1982

Vol. 41: M. Gössel
Nonlinear Time-Discrete Systems – A General Approach by Nonlinear Superposition
VIII, 112 pages. 1982

Vol. 42: Advances in Filtering and Optimal Stochastic Control
Proceedings of the IFIP-WG 7/1 Working Conference
Cocoyoc, Mexico, February 1–6, 1982
VIII, 391 pages. 1982